Lactic Acid Fermentation of Fruits and Vegetables

Food Biology Series

Lactic Acid Fermentation of Fruits and Vegetables

Editor
Spiros Paramithiotis

Department of Food Science and Human Nutrition
Agricultural University of Athens, Athens, Greece

CRC Press
Taylor & Francis Group
Boca Raton London New York

CRC Press is an imprint of the
Taylor & Francis Group, an **informa** business

A SCIENCE PUBLISHERS BOOK

CRC Press
Taylor & Francis Group
6000 Broken Sound Parkway NW, Suite 300
Boca Raton, FL 33487-2742

First issued in paperback 2021

© 2017 by Taylor & Francis Group, LLC
CRC Press is an imprint of Taylor & Francis Group, an Informa business

No claim to original U.S. Government works

ISBN-13: 978-0-367-78267-2 (pbk)
ISBN-13: 978-1-4987-2690-0 (hbk)

Visit the Taylor & Francis Web site at
http://www.taylorandfrancis.com

and the CRC Press Web site at
http://www.crcpress.com

Preface to the Series

Food is the essential source of nutrients (such as carbohydrates, proteins, fats, vitamins, and minerals) for all living organisms to sustain life. A large part of daily human efforts is concentrated on food production, processing, packaging and marketing, product development, preservation, storage, and ensuring food safety and quality. It is obvious therefore, our food supply chain can contain microorganisms that interact with the food, thereby interfering in the ecology of food substrates. The microbe-food interaction can be mostly beneficial (as in the case of many fermented foods such as cheese, butter, sausage, etc.) or in some cases, it is detrimental (spoilage of food, mycotoxin, etc.). The *Food Biology* series aims at bringing all these aspects of microbe-food interactions in form of topical volumes, covering food microbiology, food mycology, biochemistry, microbial ecology, food biotechnology and bio-processing, new food product developments with microbial interventions, food nutrification with nutraceuticals, food authenticity, food origin traceability, and food science and technology. Special emphasis is laid on new molecular techniques relevant to food biology research or to monitoring and assessing food safety and quality, multiple hurdle food preservation techniques, as well as new interventions in biotechnological applications in food processing and development.

The series is broadly broken up into food fermentation, food safety and hygiene, food authenticity and traceability, microbial interventions in food bio-processing and food additive development, sensory science, molecular diagnostic methods in detecting food borne pathogens and food policy, etc. Leading international authorities with background in academia, research, industry and government have been drawn into the series either as authors or as editors. The series will be a useful reference resource base in food microbiology, biotechnology, food science and technology for reserchers, teachers, students and food science and technology practitioners.

Ramesh C. Ray
Series Editor

Preface

Lactic acid fermentation has been practised for thousands of years primarily to preserve surplus and perishable foodstuff and also to enhance them organoleptically. Depending on the availability of raw materials and ambient temperatures, the production of a wide range of spontaneously lactic acid fermented fruits and vegetables has been successful; some of them are nowadays recognized as characteristic of certain geographical areas.

Advances in the methodological approach, equipment, as well as the social needs have dictated the direction on which research should be focused. As a result, a wealth of knowledge on technological, physicochemical, microbiological, nutritional and safety aspects has been generated. The aim of this book is to collect, present and discuss all available information regarding lactic acid fermentation of fruits and vegetables. For this purpose, experts from around the globe offer their knowledge and present their topics in a highly informative and comprehensive manner.

The book comprises of thirteen chapters. The first chapter offers an overview of lactic acid fermentation of fruits and vegetables and may be considered as an introduction to the topic. Aspects that apply to all products, i.e. the lactic acid bacteria, nutritional value and safety are integrated into Chapters 2 to 4, setting the basis on which details that only apply to specific products may be distinguished and discussed. Chapters 5 to 8 are dedicated to sauerkraut, kimchi, fermented cucumbers and olives, i.e. products that have been extensively studied and have received commercial significance. In Chapters 9 to 11 information on regional products with great potential from Asia, Europe and Africa is comprehensively presented. Chapter 12 deals with lactic acid fermented juices and smoothies, a very interesting and promising series of products. Finally, in Chapter 13, the fields in which intensive study is expected to take place in the following years are discussed.

I would like to thank Dr. Ramesh C. Ray who honored me with his invitation to edit this book under the book series called 'Food Biology', all the contributing authors for devoting their time and effort as well as the science publishers/CRC Press for publishing this book.

Spiros Paramithiotis

Contents

Chapter 13: The Future of Lactic Acid Fermentation of Fruits and Vegetables

Lactic Acid Fermentation of Fruits and Vegetables: An Overview

Spiros Paramithiotis[1], George Papoutsis[2] and Eleftherios H. Drosinos[1]*

1. Introduction

Lactic acid fermentation has been carried out for thousands of years to preserve surplus and perishable foodstuff as well as to enhance them organoleptically. A wide range of fruits and vegetables have been traditionally used to serve as a substrate for lactic acid fermentation, resulting in a wealth of final products, some of which are characteristic of certain geographical areas.

Dominance of lactic acid bacteria in the developing microecosystem is of principal importance for a desirable outcome of the process. Rapid acidification that results from lactic acid bacteria metabolism suppresses growth of antagonistic microbiota. Salting is a processing step that assists in development of lactic acid bacteria during the beginning of fermentation; proper temperature as well as the quality of raw material also play a decisive role. Failure of acidity to develop quickly, due to the unsuccessful control of the aforementioned parameters, may lead to the dominance of antagonistic biota, mainly *Enterobacteriaceae*, the development of which is perceived as spoilage.

Another effective way to control the fermentation process is to develop starter cultures that have been the epicenter of intensive study, especially regarding the products that hold worldwide commercial significance, i.e. fermented olives, cucumbers, sauerkraut and kimchi. Several bacteriocinogenic strains, that enhance the safety as well as potential probiotic strains that improve the nutritional quality of the final products, have been examined and reported. At the same time, the microecosystem

[1] Department of Food Science and Human Nutrition, Agricultural University of Athens, Athens, Greece.
[2] Directorate of Secondary Education, Ministry of Education, Research and Religious Affairs, Athens, Greece.
* Corresponding author: E-mail: ehd@aua.gr

development of several other spontaneously-fermented fruits and vegetables is helping to improve our understanding of lactic acid bacteria in general and of specific products in particular.

The following paragraphs present an overview of lactic acid fermentation in fruits and vegetables. An historical overview, based on ancient Greek and Roman literature, the main attributes of the production procedure, the safety concerns and the nutritional properties, are also presented.

2. Historical Aspects

A huge variety of fruits and vegetables were cultivated and consumed in the East Mediterranean region during ancient times; references to which may be abundantly retrieved from the *Odyssey* of Homer and from subsequent *Symposium* of Xenophon, *Deipnosophists* of Athenaeus, etc. Additionally, certain types or varieties were highly appreciated, such as the pears from Kea, figs from Attica and Rome, turnips from Magnesia, pomegranates from Voiotia, almonds from Naxos, walnuts from Karystos, etc.

Preservation of fruits and vegetables was mostly practiced by salting, dehydration or acidification. Salt was applied as such or in the form of brine, with the latter being documented more often in ancient Greek literature. Plain brine was used for preservation of fish, meat and olives as well as other plant foods, such as the leaves and roots of oilseed rape (*Brassica napus* L.). Moreover, it was mixed with vinegar as well as other materials such as garlic, thyme, etc. Dehydration was a very common method applied mainly to preservation of fruits such as figs, grapes, dates, etc. Fermentation was often used for production of a series of products such as bread, wine, vinegar, dairy products and beer. More than 50 types of bread are documented in ancient Greek literature, depending on the raw materials, production procedure, and extent and type of fermentation. As for the latter, Atheneaus, in his *Deipnosophists* mentions the sour taste of some breads — a reference that probably implies the use of sourdough. Wine seemed to be a generic term that is referred to alcoholic fermentation in general. However, the term was used as such only when grapes were taken as substrate of fermentation. Use of other raw materials, either exclusively or mixed with grape, was indicated by an adjective added to the generic term. An extended variety of raw materials were used for that purpose, such as *Cydonia oblonga* Mill., *Malus* sp., *Punica granatum* L., *Myrtus communis* L., *Pistacia lentiscus* L., *Phoenix dactylifera* L., *Ficus carica* L., *Laurus nobilis* L., *Pinus* sp., *Origanum dictamnus* L., *Thymus capitatus* L., *Satureja* sp., *Mentha pulegium* L., *Foeniculum vulgare* Mill., *Anethum graveolens* L., *Valeriana celtica* L., *Petroselinon hortense* Hoffm., *Daphne gnidium* L., *Daphne oleides* L., *Apium graveolens* L., giving rise to a tremendous wealth of final products. There are two very special products worth mentioning separately. The first one was made from lotus fruit and the second from dates. The land of the lotus-eaters is mentioned in *Odyssey* of Homer; it is located in the Gabes Gulf in Tunis. Lotus was a name used to describe a plethora of fruits. However, the fruit mentioned by Homer seems to be *Ziziphus lotus* L. According to Herodotus and Polybius, lotus fruits were used for production of wine. However, due to their short shelf life of ten days, only small quantities were made each time. Dates have been used for wine and honey production since the 5th century B.C., according to Herodotus and Xenophon. Ancient Babylon was considered the hub of production and through trade practices, these products reached the Mediterranean basin. However, due to lack of mention in literature, it is believed that these products were either uncommon or not appreciated. Similarly, several vinegar types are documented on the basis of their origin or the raw materials used. The use of barley and wheat for production of alcoholic beverages was also known, but less appreciated in ancient Greece. On the contrary, Egypt, Thrace, Iberia and Britain were mentioned by Aristotle, Ekataios and Pedanius Dioscorides as centers where these products were widely consumed. Moreover, Pedanius Dioscorides, in his work entitled *De Materia Medica*, mentions that these drinks may harm the kidneys and the brain; are diuretic and cause flatulence. This was most probably the reason for their restricted acceptance; wine was much more appreciated as it was considered healthier and without any side-effects.

Regarding lactic acid fermentation of fruits and vegetables as perceived nowadays, it seems to be less widespread mainly due to (a) the availability of fresh fruits and vegetables throughout the year, as mentioned in the texts of the Homeric age, and (b) the preferred use of dehydration and smoking for preservation, as mentioned by Atheneaus in his *Deipnosophists*. However, there was a product, for which lactic acid fermentation may be inferred; it was called αβυρτάκη (abyrtákē) and the land of Medes (Northern Iran) was reported as its origin, from which it was spread throughout the Mediterranean basin. This product was more like sauce and made by mixing and grading a variety of raw materials, such as cardamom and mustard seeds, leek, garlic, raisins, pomegranate and leafy vegetables. Lactic acid fermentation may be inferred because of (a) the very intense and sour taste that characterized this product. Since the addition of vinegar was not mentioned in any Greek text but only in a few subsequent Roman ones, the existence of a variety of similar products may be concluded, and (b) the requirement of qualified personnel that have received training in the place of origin; ordinary cooks were unable to handle such a product.

Thus, in the east Mediterranean region, while Greek and Roman civilizations sequentially flourished until the first century A.D., fermentation was a common process that was adopted for the production of a range of products of everyday use. Lactic acid fermentation, was also applied and was equally appreciated as alcoholic fermentation, through a wide variety of dairy products and types of bread. Judging from the references, it seems that fruits and vegetables were used marginally as a substrate for lactic acid fermentation.

3. Production Procedure

Production of fermented fruits and vegetables essentially involves the following steps:

- *Collection of the raw materials:* Depending on the desired final product, the respective raw materials are collected, e.g. cabbage for sauerkraut.

- *Primary processing:* All essential steps, such as washing, coring, trimming, cutting, peeling, blanching, cooking or other treatments (e.g. lye treatment) are employed prior to fermentation, e.g. lye treatment to hydrolyze oleuropein, a phenolic glucoside that is largely responsible for bitterness in olives and release a proportion of the carbohydrate content of the fruit into the solution. Lye treatment is followed by washing to remove the excess lye solution, so that a suitable final pH value is obtained. In the case of sauerkraut fermentation, chopping of cabbage in an essential step in primary processing.

- *Salting:* During salt application, fermented fruits and vegetables may be divided into three categories, viz. dry-salted, brine-salted and non-salted. In dry-salting, salt is applied on to the already chopped raw materials. Brine is formed upon osmotic extraction of water from the plant tissue. The amount of salt depends upon the type of raw material used but generally it ranges between 2.5–5 percent of vegetable weight. Brine-salting is alternatively used for vegetables from which extraction of sufficient amount of water is not possible. As soon as vegetables are immersed in brine solution (usually 5–20 percent NaCl), carbohydrates and other nutrients are extracted and the substrate for the upcoming fermentation is formed. Finally, in the process called 'pit fermentation', a type of non-salted fermentation occurs, i.e. without any prior salt treatment. This method is usually applied to starchy vegetables and fruits in places like South Pacific, Ethiopia and the Himalayas (Gashe 1987, Tamang et al. 2005).

- *Fermentation:* Provided that salting has been properly performed, the temperature is the most critical parameter that determines the outcome of the fermentation. During fermentation a series of physicochemical changes takes place that along with the dynamics of the microbial populations will be discussed in the following paragraph.

- *Secondary treatment:* A secondary processing step might occur after fermentation involving drying, pasteurization, etc. Once the final product is ready, it is packed and consumed either as it is or after heat treatment, such as boiling or frying.

A series of physicochemical and microbiological changes take place during fermentation, most of which are product specific. Rapid acidity development and drop of the pH value is desired in all cases. This is achieved due to the dominance of lactic acid bacteria. Raw materials of good microbiological quality, proper salting and suitable temperature conditions may ensure the reproducibility of fermentations, even if they are driven spontaneously. *Enterobacteriaceae* may prevail when acidity fails to develop quickly enough and reach populations the metabolic activities of which may be perceived as spoilage.

Table 1: Microorganisms Involved in Some Fruit and Vegetable Fermentations

Product	Microorganisms	References
Fermented cucumber	*Lb. brevis; Lb. plantarum; Pd. pentosaceus*	Fleming et al. (1995), Singh and Ramesh (2008)
Sauerkraut	*Ec. faecalis; Lb. brevis; Lb. confusus; Lb. curvatus; Lb. plantarum; Lb. sakei; Lc. lactis* subsp. *lactis; Ln. fallax; Ln. mesenteroides; Pd. pentosaceus*	Harris (1998), Barrangou et al. (2002)
Kimchi	*B. subtilis; B. mycoides; B. pseudomycoides; Lc. carnosum; Lc. gelidum; Lc. lactis; Lb. sakei; Lb. brevis; Lb. curvatus; Lb. parabrevis; Lb. pentosus; Lb. plantarum; Lb. spicheri; Lc. gelidum; Ln. carnosum; Ln. citreum; Ln. gasicomitatum; Ln. gelidum; Ln. holzapfelii; Ln. inhae; Ln. kimchii; Ln. lactis; Ln. mesenteroides; W. koreensis; W. cibaria; W. confusa; W. soli; W. kandleri; Se. marcescens*	Lee et al. (2005), Cho et al. (2006), Chang et al. (2008), Jung et al. (2013), Jeong et al. (2013a, b), Yeun et al. (2013)
Eggplant	*Lb. brevis; Lb. fermentum; Lb. pantheris; Lb. paracasei; Lb. pentosus; Lb. plantarum; Pd. pentosaceus*	Sesena and Palop (2007), Nguyen et al. (2013)
Mustard	*Lb. alimentarius; Lb. brevis; Lb. coryniformis; Lb. farciminis; Lb. fermentum; Lb. pentosus; Lb. versmoldensis; Ln. citreum; Ln. mesenteroides; Ln. pseudomesenteroides; Pd. pentosaceus; T. halophilus; W. cibaria; W. paramesenteroides*	Tamang et al. (2005), Chen et al. (2006), Chao et al. (2009), Nguyen et al. (2013)
Caper berries	*Ec. casseliflavus; Ec. faecium; Lb. brevis; Lb. fermentum; Lb. paraplantarum; Lb. pentosus; Lb. plantarum; Pd. acidilactici; Pd. pentosaceus*	Pulido et al. (2005)
Leek	*C. maltaromaticum; Lb. brevis; Lb. crustorum; Lb. hammessii/ parabrevis; Lb. nodensis; Lb. parabrevis; Lb. plantarum; Lb. sakei; Lb. sakei/curvatus; Lc. lactis; Lc. raffinolactis; Ln. gasicomitatum; Ln. gelidum; Ln. kimchii; Ln. lactis/ citreum; Ln. lactis/garlicum; Ln. mesenteroides; W. soli*	Wouters et al. (2013a)
Cauliflower	*Ec. faecalis*-group; *Ec. faecium*-group; *Lb. brevis; Lb. plantarum*-group; *Lb. sakei/curvatus; Ln. mesenteroides*-group; *Pd. pentosaceus; W. kimchii; W. viridescens*	Paramithiotis et al. (2010), Wouters et al. (2013b)
Green tomatoes	*Lb. casei; Lb. curvatus; Ln. citreum; Ln. mesenteroides*	Paramithiotis et al. (2014b)
Asparagus	*B. licheniformis; Ec. faecium*-group; *Lb. sakei; W. cibaria; W. viridescens*	Paramithiotis et al. (2014a)
Brovada	*Lb. coryniformis; Lb. higardii; Lb. maltaromicus; Lb. plantarum; Lb. viridescens; Pd. parvulus*	Maifreni et al. (2004)

Lb: Lactobacillus; Lc: Lactococcus; Ln: Leuconostoc; W: Weissella; Se: Serratia; B: Bacillus; Ec: Enterococcus; T: Tetragenococcus; C: Carnobacterium

In Table 1, the microorganisms involved in certain fruit and vegetable fermentations are listed. There is extended diversity at species level that is justified by the diversity of the raw materials used, ambient temperatures, initial pH value as well as NaCl content. *Ln. mesenteroides* is often isolated from the early stages of spontaneous fermentations. This microorganism is characterized by shorter generation time, ability to tolerate a wide range of NaCl concentrations and sensitivity to acidic conditions (Stamer et al. 1971). However, this was not the case during spontaneous fermentation of unripe tomatoes, in which it initiated the fermentation when the initial pH value was between 3.8–4.8 (Paramithiotis et al. 2014b), indicating that there are still many aspects of the physiology of this microorganism that are still unknown. On the other hand, *Lb. plantarum* is mostly associated with the final stages of fermentation since it is more acid tolerant and distinguished by a large metabolic capacity that enables growth on a wide range of carbon sources (Daeschel et al. 1987). The ability to tolerate stressful conditions may justify the presence of several other lactic acid bacteria species such as *Lb. sakei*, *Lb. brevis*, *Pd. pentosaceus* and *Weissella* spp. that can tolerate acidic conditions; *Ln gelidum*, that can grow at temperatures as low as 10°C; *Ln. gasicomitatum*, that has greater acid tolerance compared to *Ln. mesenteroides* and *Ln. citreum*; and *W. koreensis*, that has a psychrophilic character and can grow at even lower temperatures (Daeschel et al. 1987, Cho et al. 2006, Jung et al. 2013, Jeong et al. 2013a). During spontaneous vegetable fermentation, a succession at species level is often reported with the most characteristic one occurring during sauerkraut and cauliflower fermentation in which *Ln. mesenteroides* initiates fermentation and is gradually replaced by *Lb. plantarum*. Moreover, a succession at subspecies level was also reported during spontaneous asparagus and green tomato fermentations (Paramithiotis et al. 2014a, b). This succession was evident for all species involved in the fermentation, suggesting that trophic relationships that drive spontaneous fermentations may be far more complex than expected.

4. Safety Concerns

Generally, food-borne illness is more likely to occur after consumption of fresh produce rather than the respective fermented ones, mainly due to the antagonistic effect of lactic acid bacteria. Indeed, microbial as well as viral gastroenteritis outbreaks have been repeatedly assigned to the consumption of raw vegetables (Basset and McClure 2008). More accurately, analysis of data regarding food-borne illnesses and associated deaths during 1998-2008 held fresh produce accountable for 46 percent and 23 percent cases, respectively (Painter et al. 2013). Vegetables most frequently used as raw materials, i.e. lettuce and cucumbers, are very often implicated as the food vehicle for food-borne disease outbreaks (Table 2). On the contrary, only a weak association between fermented, salted or acidified fruits and vegetables and *Cl. botulinum, L. monocytogenes* and pathogenic *E. coli* has been identified (EFSA 2013).

Table 2: Outbreaks in the U.S. Associated with Consumption of Lettuce and Cucumber (1998-2013) (CDC 2015).

Year	State	Agent	Ill/Hospita-lization/ Deaths
Lettuce			
1999	Nebraska	*E. coli* O157:H7	65/8/0
1999	Minnesota	Norovirus Genogroup I	25/0/0
2000	Connecticut	*C. jejuni*	13/1/0
2001	Minnesota	Norovirus Genogroup I	64/0/0
2006	Rhode Island	Norovirus	99/0/0
2007	multistate	*S.* Typhimurium	76/4/0
2009	New york	Norovirus	24/0/0
2010	Oregon	Norovirus Genogroup II	23/0/0
2010	Illinois	Norovirus Genogroup II	7/1/0
2011	Maryland	Norovirus Genogroup II	93/0/0
2011	Alabama	Norovirus Genogroup I	21/2/0
2012	Minnesota	Norovirus	9/0/0
2013	South Carolina	Norovirus Genogroup I	8/0/0
2013	Arizona	*E. coli* O157:H7	94/22/0
Cucumber			
2011	Georgia	*S.* Saintpaul	14/2/0
2012	multistate	*S.* Javiana	49/14/0
2013	Michigan	*S.* enterica I 4,[5],12:i:-	12/1/0
2013	multistate	*S.* Saintpaul	84/17/0

The efficiency of lactic acid bacteria to inhibit growth of food-borne pathogens in both plant- and meat- derived material has been extensively studied (Drosinos and Paramithiotis 2007). Apart from acidity, which has been recognized as the primary inhibiting agent, antagonism for the available nutrients, as well as production of metabolites, such as carbon dioxide, hydrogen peroxide, ethanol, diacetyl and even bacteriocins may contribute to the suppression of the growth of human pathogens. However, generalizations regarding the microbiological safety of fermented fruits and vegetables can be uncertain due to:

1. The effect of phylloplane niche on plant-microbe interactions: The extended variability in plant surface morphology, which is greatly affected by factors such as plant species, cultivar, surface morphology, tissue composition and metabolic activities (Burnett et al. 2000,

Takeuchi & Frank 2000, 2001, Beuchat 2002, Brandl 2006, Klerks et al. 2007, Mitra et al. 2009, Raybaudi-Massilia et al. 2009) results in a similar diversity of ecological niches in which pathogenic bacteria may interact with epiphytic bacteria, adapt and even penetrate the plant tissue (Brandl and Mandrell 2002, Cooley et al. 2006, Critzer and Doyle 2010, Warriner and Namvar 2010, Kroupitski et al. 2009, 2011).

2. The tendency of the bacterial cells to aggregate between the grooves of the epidermal plant cells and to internalize into the subsurface of the plant material during pre- and post-harvest operations (Seo and Frank 1999, Warriner et al. 2003, Hadjok et al. 2008).

As a result, the fate of major food-borne pathogens during and after fermentation has been reconsidered. Inatsu et al. (2004) studied the survival of *E. coli* O157:H7, *S.* Enteritidis, *St. aureus* and *L. monocytogenes* in commercial and in laboratory-prepared kimchi. The products were inoculated with the pathogens at 5-6 log CFU g^{-1} and incubated at 10°C. It was found that the population of *E. coli* O157:H7 remained at high levels throughout the incubation period in both products; similar results were obtained with *S.* Enteritidis but only for the product prepared in the laboratory. The population decreased to the enumeration limit after 12 days of incubation for *St. aureus*, 16 and 20 days for *L. monocytogenes*, regarding the commercially and laboratory-made products, respectively and after 16 days for *S.* Enteritidis regarding the commercially prepared products. Decrease of *E. coli* O157:H7 and *L. monocytogenes* populations below enumeration limit during the production of whole-head and shredded sauerkraut was reported by Niksic et al. (2005). In this study, the effect of fermentation temperature (18 and 22°C) and salt concentration (1.8, 2.25 and 3 percent) on the survival of the above-mentioned pathogenic microorganisms was assessed. Both pathogens persisted in the fermentation brines for most of the fermentation period but their population decreased below enumeration limit in the end, i.e. after 15 and 27 days for shredded and whole head sauerkrauts, respectively. The fate of *Bacillus cereus*, *L. monocytogenes* and *St. aureus* during kimchi fermentation under two treatments: heat treatment (85°C for 15 min) or neutralization treatment (pH 7) at day 0 or day 3 was evaluated by Kim et al. (2008). Although heat treatment on day 0 had no effect on *B. cereus* population, it resulted in reduction of *L. monocytogenes* and *St. aureus* populations. Significant and complete inactivation was observed for *B. cereus* and *L. monocytogenes*, respectively, but only marginal for *St. aureus* after heat treatment on day 3. Neutralization treatment on day 3 resulted in complete inactivation of *L. monocytogenes*, significant decrease of *B. cereus* counts and only marginal decrease of *St. aureus* population. All these effects were correlated with lactic acid bacteria growth and concomitant pH decrease.

The survival of *E. coli* O157:H7 in cucumber fermentation brines obtained from commercial plants at different stages of fermentation was assessed by Breidt Jr and Caldwell (2011). The time required to obtain the 5-log reduction standard by Food and Drug Administration for *E. coli* O157:H7 was positively correlated with the pH value and the NaCl composition of the brines. *E. coli* O157:H7 was not able to compete with either *Lb. plantarum* or *Ln. mesenteroides* at 30°C, resulting in a decrease of the cell counts of the pathogen to below detection limit within 2-3 days. Cho et al. (2011) studied the survival of *E. coli* O157:H7 and *L. monocytogenes* during cabbage and radish kimchi fermentation supplemented with raw pork meat. The population of both pathogens gradually decreased during fermentation at 4°C and they were no longer detected after 14 and 15 post-fermentation days, for *E. coli* O157:H7 and *L. monocytogenes*, respectively. The fate of *L. monocytogenes* and *S.* Typhimurium during spontaneous cauliflower fermentation was assessed by Paramithiotis et al. (2012), showing that both pathogens survived at low (2.0 log CFU mL^{-1}) and medium (4.0 log CFU mL^{-1}) inoculation levels. *L. monocytogenes* strain was more capable in surviving than *S.* Typhimurium strain, whereas no significant differences were recorded at high (6.0 log CFU mL^{-1}) inoculation level. The survival of *E. coli* O157:H7, *S.* Enteritidis and *L. monocytogenes* during storage of fermented green table olives of cv. Halkidiki in brine was studied by Argyri et al. (2013). In this study, olives were previously fermented either spontaneously or with addition of starter culture, inoculated with a cocktail of each pathogen at ca. 7.0 log CFU mL^{-1}, packaged in polyethylene pouches and stored at 20°C. The population of all pathogens gradually decreased and was detectable until the 19th, 21st and 31st day of storage for *E. coli* O157:H7, *S.* Enteritidis and *L. monocytogenes*, respectively. Survival of *E. coli*, *S.* Enteritidis, *S.* Typhimurium, *L. monocytogenes* and *St. aureus* inoculated at ca. 8 log CFU mL^{-1} into industrial table olive brines during storage at 4°C and room temperature in aerobic or anaerobic conditions was studied by Medina et al. (2013). It was discovered that all pathogens were eliminated at a rate dependent upon the composition of the brines in phenolic compounds, temperature and oxygen availability. More accurately, the absence of alkaline treatment resulted in enhanced concentration of antibacterial compounds, leading to a 5 log reduction time of the pathogens between 5–10 min, regardless of storage temperature or oxygen availability. On the contrary, treatment with NaOH (sodium hydroxide) resulted in lower concentrations of anti-bacterial compounds, leading to a longer 5 log reduction time, i.e. 2.8 days for *L. monocytogenes*, 5.2 days for *E. coli*, 6.2 days for *St. aureus* and 6.7 days for salmonellae during aerobic storage at refrigeration temperature, reaching 17 days for *St. aureus* and 11 days for *E. coli* under anaerobic conditions at refrigerating temperatures.

5. Nutritional Attributes

The multiple effects that lactic acid fermentation has on the nutritional value of fruits and vegetables have been extensively studied. Lactic acid bacteria may modify the level and bioavailability of nutrients through their metabolism, or interact with the gut microbiota, or even the human immune system.

The edible part of several plants may contain a series of compounds that are collectively called antinutrients and may interfere with nutrient assimilation by humans or even confer undesirable physiological effects. As far as the substances that have been characterized by their antinutritional activity are concerned, only lectins are encountered in fruits and vegetables. However, only certain lectins from legumes and cereals have been found to exert toxic effects on humans and animals after oral ingestion whereas others, such as those from tomatoes, lentils, peas, etc. are not toxic (Radberg et al. 2001). Since lectins are heat-sensitive molecules, they may be easily removed if cooked or steamed during production, either before or after fermentation.

Folate is produced by various green leafy vegetables, cereals, legumes and some microorganisms. It is an essential component in the human diet, with vegetables and dairy products being the main source. Its fate during lactic acid fermentation of vegetables was studied to some extent. In a study conducted by Jagerstad et al. (2004) commercial starter cultures designed for milk fermentation were used to ferment beetroot, white cabbage as well as a mixture of other vegetables. It was noted that a starter culture consisting of *Lb. plantarum*, *Lc. lactis/cremoris* and *Leuconostoc* sp. resulted in nearly doubling of folate concentration and mainly as 5-methyl-tetrahydrofolic acid that is a native and bioavailable form. Moreover, *Propionibacterium freudenreichii* was reported to produce vitamin B_{12}. This study highlighted that folate production is strain-dependent, as already concluded by studies on fermented dairy products and sourdoughs (Rao et al. 1984, Lin and Young 2000, Crittenden et al. 2003, Kariluoto et al. 2006).

The raw materials may also contain several other compounds with very interesting properties, such as flavonoids and glucosinolates, for which only scarce information exists regarding their fate during lactic acid fermentation. The former are phenol derivatives widely distributed in plants; a variety of biochemical effects is attributed to them, including action against cardiovascular diseases, cancer, inflammation and allergy (Middleton et al. 2000, Birt et al. 2001, Yao et al. 2004, Halliwell 2007). Glucosinolates are a group of sulphur-containing plant secondary metabolites found in plans of the order *Capparales*, which includes the *Brassicaceae* family, widely used as raw material for vegetable fermentation. Their major hydrolysis products, isothiocyanates, have been found to act protectively against cancer, particularly in the bladder, colon and lung (Mithen et al. 2000, Song and Thornalley, 2007).

The selection of proper starter cultures is of vital importance for the quality of the final product. The current perception of the ideal starter cultures requires that they not only provide with adequate acidification but enhance the nutritional value of the product as well. During growth of lactic acid bacteria, production of lactic acid as a mixture of the optical isomers L-(+) and D-(-) takes place. The latter cannot be metabolized by humans and excessive intake may result to acidosis, i.e. a disturbance of the acid-alkali balance in the blood (Nout and Motarjemi, 1997). Therefore the production of only the former isomer is desired.

The 'probiotic' concept has been the epicenter of intensive study over the last decades; several properties, including the antagonistic activity against potentially pathogenic microorganisms, particularly invasive Gram-negative pathogens, as well as properties that are beneficial to the host such as modulation of immune response, reduction of hypercholesterolemia, prevention of intestinal and urogenital infections, anti-cancer activity as well as protection against traveler's, infantile and antibiotic-induced diarrhoea and allergic diseases have been proposed (Gilliland et al. 1985, Ushiyama et al. 2003, Chen and Walker 2005, Commane et al. 2005, Zarate and Nader-Macias 2006, Koo and DuPond 2006, Guandalini et al. 2000, Kim et al. 2003, Tricon et al. 2006). A wide variety of fruits and vegetables have been examined for their suitability as carriers of probiotic bacteria (Betoret et al. 2003, Yoon et al. 2004, 2005, 2006, Roessle et al. 2010, Fonteles et al. 2011, Alegre et al. 2011). Moreover, the probiotic potential of numerous isolates from fermented products has been reported. However, only limited amount of studies that evaluates the actual potential of these presumptive probiotic bacteria as a functional or an adjunct starter culture currently exists.

Table olives are recognized as a suitable vehicle of microorganisms with probiotic potential. Lavermicocca et al. (2005) studied the survival of *Lb. rhamnosus, Lb. paracasei, Bifidobacterium bifidum* and *Bf. longum* strains on table olives at room temperature. The highest survival was observed for *Lb. paracasei* strain that was further used for successful validation of olives as carriers of bacterial cells through the human gastrointestinal tract. The efficacy of the same strain as a starter culture was evaluated by De Bellis et al. (2010). It successfully colonized the surface of debittered green olives cv. Bella di Cerignola and dominated the indigenous lactic acid bacteria population under different brining conditions as well as fermentation temperatures. The utilization of multifunctional *Lb. pentosus* strains as starter cultures of table olive for Spanish-style fermentation of cv. Manzanilla was studied by Rodríguez-Gómez et al. (2013). Although lactic acid bacteria dominated the fermentation, the *Lb. pentosus* strains were recovered at the end of fermentation and *Enterobacteriaceae* population was suppressed. At strain level a wide diversity was revealed, leading to the conclusion that inoculation with a given strain may not assure its prevalence. Similar conclusions were drawn by Randazzo et al. (2014). In the latter study, *Lb. rhamnosus* strains H25 and GG were used as starter cultures for the fermentation of Giarraffa and Grossa di Spagna cultivars according to the Sicilian traditional method and their ability to survive was demonstrated. Blana et al. (2014) studied the potential of presumptive probiotic *Lb. pentosus* and *Lb. plantarum* strains as starter cultures in

Spanish-style fermentation of cv. Halkidiki at room temperature in 8 percent and 10 percent w/v NaCl brines. *Lb. pentosus* B281 exhibited higher colonization in both salt levels at the end of fermentation, whereas *Lb. plantarum* failed to colonize the surface of the olives at 10 percent NaCl. Moreover, *Lb. pentosus* strain dominated over *Lb. plantarum* strain when they were inoculated in co-culture. Similar results were obtained when the olives were submitted to a heat shock at 80°C for 10 min. in order to reduce the level of indigenous microbiota, thus facilitating the dominance of the inoculated cultures (Argyri et al. 2014). Furthermore, the *Lb. plantarum* strain was able to colonize the surface of the olives at 10 percent NaCl.

The effect of blanching and fermentation by a potential probiotic *Lb. paracasei* strain on the glucosinolate level of cabbage during sauerkraut fermentation was studied by Sarvan et al. (2013). It was reported that 35 percent of the original glucosinolate content remained after 71 hours of fermentation, while at the same time the population of the *Lb. paracasei* strain was enumerated at 8.26 ± 1.2 log CFU g^{-1}. Therefore, the authors rightfully highlighted the possibility to produce an actual functional food out of a typical sauerkraut fermentation.

6. Conclusions and Future Perspectives

Lactic acid fermentation of fruits and vegetables is an ancient procedure that is carried on till now, embodied to the food habits of several cultures. These products hold an excellent safety status, at least from a microbiological point of view and have the potential to contribute in the improvement of the nutritional quality of everyday diet.

The most widely consumed products have been extensively studied. However, there is still an unexplored variety of regional products with unique sensorial characteristics and great potential. The microecosystem of several such products has been described in recent years and several more are extremely likely to be studied in the future. The advent of modern molecular approaches will definitely expand our understanding of the trophic relationships taking place in the microcommunity and the response to biotic and abiotic factors that are encountered. Moreover, the rapidly developing field of nutritional genomics will provide new insights into the nutritional value of these products. Advances in the above-mentioned fields are presented in detail in Chapter 13.

References

(CDC 2015). Food-borne Outbreak Online Database (FOOD) available at http://wwwn.cdc.gov/foodborneoutbreaks/. Accessed 29 July 2015.

Alegre, I., I. Viñas, J. Usall, M. Anguera and M. Abadias. 2011. Microbiological and physicochemical quality of fresh-cut apple enriched with the probiotic strain *Lactobacillus rhamnosus* GG. *Food Microbiology* 28: 59-66.

Argyri A.A., E. Lyra, E.Z. Panagou and C.C. Tassou. 2013. Fate of *Escherichia coli* O157:H7, *Salmonella* Enteritidis and *Listeria monocytogenes* during storage of fermented green table olives in brine. *Food Microbiology* 36: 1-6.

Argyri, A.A., A.A. Nisiotou, A. Malouchos, E.Z. Panagou and C.C. Tassou. 2014. Performance of two potential probiotic *Lactobacillus* strains from the olive microbiota as starters in the fermentation of heat-shocked green olives. *International Journal of Food Microbiology* 171: 68-76.

Barrangou, R., S.S. Yoon, F. Breidt Jr, H.P. Fleming and T.R. Klaenhammer. 2002. Identification and characterization of *Leuconostoc fallax* strains isolated from an industrial sauerkraut fermentation. *Applied and Environmental Microbiology* 68: 2877-2884.

Bassett, J. and P. McClure. 2008. A risk assessment approach for fresh fruits. *Journal of Applied Microbiology* 104: 925-943.

Betoret, N., L. Puente, M.J. Díaz, M.J. Pagán, M.J. García, M.L. Gras, J. Martinez-Monzo and P. Fito. 2003. Development of probiotic enriched dried fruits by vacuum impregnation. *Journal of Food Engineering* 56: 273-277.

Beuchat, L.R. 2002. Ecological factors influencing survival and growth of human pathogens on raw fruits and vegetables. *Microbes and Infection* 4: 413-423.

Birt, D.F., S. Hendrich and W.Q. Wang. 2001. Dietary agents in cancer prevention: Flavonoids and isoflavonoids. *Pharmacology and Therapeutics* 90: 157-177.

Blana, V.A., A. Grounta, C.C. Tassou, G.J. Nychas and E.Z. Panagou. 2014. Inoculated fermentation of green olives with potential probiotic *Lactobacillus pentosus* and *Lactobacillus plantarum* starter cultures isolated from industrially fermented olives. *Food Microbiology* 38: 208-218.

Brandl, M.T. 2006. Fitness of human enteric pathogens on plants and implications for food safety. *Annual Review of Phytopathology* 44: 367-392.

Brandl, M.T. and R.E. Mandrell. 2002. Fitness of *Salmonella enterica* serovar Thompson in the cilantro phyllosphere. *Applied and Environmental Microbiology* 68: 3614-3621.

Breidt Jr, F. and J.M. Caldwell. 2011. Survival of *Escherichia coli* O157:H7 in cucumber fermentation brines. *Journal of Food Science* 76: M198-M203.

Burnett, S.L., J. Chen and L.R. Beuchat. 2000. Attachment of *Escherichia coli* O157:H7 to the surfaces and internal structures of apples as detected by confocal scanning laser microscopy. *Applied and Environmental Microbiology* 66: 4679-4687.

Chao, S.H., R.J. Wu, K. Watanabe and Y.C. Tsai. 2009. Diversity of lactic acid bacteria in suan-tsai and fu-tsai, traditional fermented mustard products of Taiwan. *International Journal of Food Microbiology* 135: 203-210.

Chang, H.W., K.H. Kim, Y.D. Nam, S.W. Roh, M.S. Kim, C.O. Jeon, H.M. Oh and J.W. Bae. 2008. Analysis of yeast and archaeal population dynamics in kimchi using denaturing gradient gel electrophoresis. *International Journal of Food Microbiology* 126: 159-166.

Chen, C.C. and A. Walker. 2005. Probiotics and prebiotics: Role in clinical disease states. *Advances in Pediatrics* 52: 77-113.

Chen, Y.S., F. Yanagida and J.S. Hsu. 2006. Isolation and characterization of lactic acid bacteria from suan-tsai (fermented mustard), a traditional fermented food in Taiwan. *Journal of Applied Microbiology* 101: 125-130.

Cho, G.Y., M.H. Lee and C. Choi. 2011. Survival of *Escherichia coli* O157:H7 and *Listeria monocytogenes* during kimchi fermentation supplemented with raw pork meat. *Food Control* 22: 1253-1260.

Cho, J., D. Lee, C. Yang, J. Jeon, J. Kim and H. Han. 2006. Microbial population dynamics of kimchi, a fermented cabbage product. *FEMS Microbiology Letters* 257: 262-267.

Commane, D., R. Hughes, C. Shortt and I. Rowland. 2005. The potential mechanisms involved in the anti-carcinogenic action of probiotics. *Mutation Research* 591: 276-289.

Cooley, M.B., D. Chao and R.E. Mandrell. 2006. *Escherichia coli* O157:H7 survival and growth on lettuce is altered by the presence of epiphytic bacteria. *Journal of Food Protection* 69: 2329-2335.

Crittenden, R.G., N.R. Martinez and M.J. Playne. 2003. Synthesis and utilisation of folate by yogurt starter cultures and probiotic bacteria. *International Journal of Food Microbiology* 80: 217-222.

Critzer, F.J. and M.P. Doyle. 2010. Microbial ecology of food-borne pathogens associated with produce. *Current Opinion in Biotechnology* 21: 125-130.

Daeschel, M.A., R.E. Andersson and H.P. Fleming. 1987. Microbial ecology of fermenting plant materials. *FEMS Microbiology Reviews* 46: 357-367.

De Bellis, P., F. Valerio, A. Sisto, S.L. Lonigro and P. Lavermicocca. 2010. Probiotic table olives: Microbial populations adhering on the olive surface in fermentation sets inoculated with the probiotic strains *Lactobacillus paracasei* IMPC 2.1 in an industrial plant. *International Journal of Food Microbiology* 140: 6-13.

Drosinos, E.H. and S. Paramithiotis. 2007. Research trends in lactic acid fermentation. Pp. 39-92. *In*: M.V. Palino (Ed.). *Food Microbiology Research Trends*. Nova Science Publishers, NY, USA.

EFSA Panel on Biological Hazards (BIOHAZ). 2013. Scientific opinion on the risk posed by pathogens in food of non-animal origin. Part 1 (outbreak data analysis and risk ranking of food/pathogen combinations). *EFSA Journal* 11(1):3025. p. 138, doi:10.2903/j.efsa.2013.3025. Available online: www.efsa.europa.eu/efsajournal

Fleming, H.P., K.H. Kyung and F. Breidt. 1995. Vegetable fermentations. pp. 629-662. *In*: G. Reed and T.W. Nagodawithana (Eds). *Biotechnology*, second completely revised edition, vol. 9. *Enzymes, Biomass, Food and Feed*. VCH, Weinheim, Germany.

Fonteles, T.V., M.G.M. Costa, A.L.T. Jesus and S. Rodrigues. 2011. Optimization of the fermentation of cataloupe juice by *Lactobacillus casei* NRRL B-442. *Food and Bioprocess Technology* 5: 2819-2826.

Gashe, B.A. 1987. Kocho fermentation. *Journal of Applied Bacteriology* 62: 473-477.

Gilliland, S.E., C.R. Nelson and C. Maxwell. 1985. Assimilation of cholesterol by *Lactobacillus acidophilus*. *Applied and Environmental Microbiology* 49: 377-385.

Guandalini, S., L. Pensabene, M.A. Zikri, J.A. Dias, L.G. Casali, H. Hoekstra, S. Kolacek, K. Massar, D. Micetic-Turk, A. Papadopoulou, J.S. de Sousa, B. Sandhu, H. Szajewska and Z. Weizman. 2000. *Lactobacillus* GG administered in oral rehydration solution to children with acute diarrhea: A multicenter European trial. *Journal of Pediatric Gastroenterology and Nutrition* 30: 54-60.

Hadjok, C., G.S. Mittal and K. Warriner. 2008. Inactivation of human pathogens and spoilage bacteria on the surface and internalized within fresh produce by using a combination of ultraviolet light and hydrogen peroxide. *Journal of Applied Microbiology* 104: 1014-1024.

Halliwell, B. 2007. Dietary polyphenols: Good, bad, or indifferent for your health? *Cardiovascular Research* 73: 341-347.

Harris, L.J. 1998. The microbiology of vegetable fermentations. Pp. 45-72. *In*: B.J.B. Wood (Ed). *Microbiology of Fermented Foods*. Blackie Academic and Professional, London, UK.

Inatsu, Y., M.L. Bari, S. Kawasaki and K. Isshiki. 2004. Survival of *Escherichia coli* O157:H7, *Salmonella* Enteritidis, *Staphylococcus aureus*, and *Listeria monocytogenes* in kimchi. *Journal of Food Protection* 67: 1497-1500.

Jagerstad, M., J. Jastrebova and U. Svensson. 2004. Folates in fermented vegetables—A pilot study. *Lebensmittel-Wissenschaft und-Technologie* 37: 603-611.

Jeong, S.H., J.Y. Jung, S.H. Lee, H.M. Jin and C.O. Jeon. 2013a. Microbial succession and metabolite changes during fermentation of dongchimi, traditional Korean watery kimchi. *International Journal of Food Microbiology* 164: 46-53.

Jeong, S.H., S.H. Lee, J.Y. Jung, E.J. Choi and C.O. Jeon. 2013b. Microbial succession and metabolite changes during long-term storage of kimchi. *Journal of Food Science* 78: M763-M769.

Jung, J.Y., S.H. Lee, H.M. Jin, Y. Hahn, E.L. Madsen and C.O. Jeon, 2013. Metatranscriptomic analysis of lactic acid bacterial gene expression during kimchi fermentation. *International Journal of Food Microbiology* 163: 171-179.

Kariluoto, S., M. Aittamaa, M. Korhola, H. Salovaara, L. Vahteristoand V. Piironen. 2006. Effects of yeasts and bacteria on the levels of folates in rye sourdoughs. *International Journal of Food Microbiology* 106: 137-143.

Kim, H.J., M. Camilleri, S. McKinzie, M.B. Lempke, D.D. Burton, G.M. Thomforde and A.R. Zinsmeister. 2003. A randomized controlled trial of a probiotic, VSL#3, on gut transit and symptoms in diarrhoea-predominant irritable bowel syndrome. *Alimentary Pharmacology and Therapeutics* 17: 895-904.

Kim, Y.S., Z.B. Zheng and D.H. Shin. 2008. Growth inhibitory effects of kimchi (Korean traditional fermented vegetable product) against *Bacillus cereus*, *Listeria monocytogenes* and *Staphylococcus aureus*. *Journal of Food Protection* 71: 325-332.

Klerks, M.M., E. Franz, M. van Gent-Pelzer, C. Zijlstra and A.H.C. van Bruggen. 2007. Differential interaction of *Salmonella enterica* serovars with lettuce cultivars and plant-microbe factors influencing the colonization efficiency. *The ISME Journal* 1: 620-631.

Koo, H.L. and H.L. DuPond. 2006. Current and future developments in travelers' diarrhea therapy. *Expert Review of Anti-Infective Therapy* 4: 417-427.

Kroupitski, Y., D. Golberg, E. Belausov, R. Pinto, D. Swartzberg, D. Granot and S. Sela. 2009. Internalization of *Salmonella enterica* in leaves is induced by light and involves chemotaxis and penetration through open stomata. *Applied and Environmental Microbiology* 75: 6076-6086.

Kroupitski, Y., R. Pinto, E. Belausov and S. Sela. 2011. Distribution of *Salmonella typhimurium* in romaine lettuce leaves. *Food Microbiology* 28: 990-997.

Lavermicocca, P., F. Valerio, S.L. Lonigro, M. De Angelis, L. Morelli, M.L. Callegari, C.G. Rizzello and A. Visconti. 2005. Study of adhesion and survival of lactobacilli and bifidobacteria on table olives with the aim of formulating a new probiotic food. *Applied and Environmental Microbiology* 71: 4233-4240.

Lee, J.S., G.Y. Heo, J.W. Lee, Y.J. Oh, J.A. Park, Y.H. Park, Y.R. Pyun and J.S. Ahn. 2005. Analysis of kimchi microflora using denaturing gradient gel electrophoresis. *International Journal of Food Microbiology* 102: 143-150.

Lin, M.Y. and C.M. Young. 2000. Folate levels in cultures of lactic acid bacteria. *International Dairy Journal* 10: 409-413.

Maifreni, M., M. Marino and L. Conte. 2004. Lactic acid fermentation of *Brassica rapa*: Chemical and microbial evaluation of a typical Italian product (brovada). *European Food Research and Technology* 218: 469-473.

Medina, E., M. Brenes, C. Romero, E. Ramírez and A. de Castro. 2013. Survival of food-borne pathogenic bacteria in table olive brines. *Food Control* 34: 719-724.

Middleton, E., C. Kandaswami and T.C. Theoharides. 2000. The effects of plant flavonoids on mammalian cells: Implications for inflammation, heart disease and cancer. *Pharmacological Reviews* 52: 673-751.

Mithen, R.F., M. Dekker, R. Verkerk, S. Rabot and I.T. Johnson. 2000. The nutritional significance, biosynthesis and bioavailability of glucosinolates in human foods. *Journal of the Science of Food and Agriculture* 80: 967-984.

Mitra, R., E. Cuesta-Alonso, A. Wayadande, J. Talley, S. Gilliland and J. Fletcher. 2009. Effect of route of introduction and host cultivar on the colonization, internalization, and movement of the human pathogen *Escherichia coli* O157: H7 in spinach. *Journal of Food Protection* 72: 1521-1530.

Nguyen, D.T.L., K. Van Hoorde, M. Cnockaert, E. De Brandt, M. Aerts, L. Thanh and P. Vandamme. 2013. A description of the lactic acid bacteria microbiota associated with the production of traditional fermented vegetables in Vietnam. *International Journal of Food Microbiology* 163: 19-27.

Niksic, M., S.E. Niebuhr, J.S. Dickson, A.F. Mendonca, J.J. Koziczkowski and J.L.E. Ellingson. 2005. Survival of *Listeria monocytogenes* and *Escherichia coli* O157:H7 during sauerkraut fermentation. *Journal of Food Protection* 68: 1367-1374.

Nout, M.J.R. and Y. Motarjemi. 1997. Assessment of fermentation as a household technology for improving food safety: A joint FAO/VVHO workshop. *Food Control* 8: 221-226.

Painter, J.A., R.M. Hoekstra, T. Ayers, R.V. Tauxe, C.R. Braden, F.J. Angulo and P.M. Griffin. 2013. Attribution of food-borne illnesses, hospitalizations and deaths to food commodities by using outbreak data, United States, 1998-2008. *Emerging Infectious Diseases* 19: 407-415.

Paramithiotis, S., A.I. Doulgeraki, I. Tsilikidis, G.J.E. Nychas and E.H. Drosinos. 2012. Fate of *Listeria monocytogenes* and *Salmonella* sp. during spontaneous cauliflower fermentation. *Food Control* 27: 178-183.

Paramithiotis S., K. Kouretas and E.H. Drosinos. 2014b. Effect of ripening stage on the development of the microbial community during spontaneous fermentation of green tomatoes. *Journal of the Science of Food and Agriculture* 94: 1600-1606.

Paramithiotis, S., A.I. Doulgeraki, A. Karahasani and E.H. Drosinos. 2014a. Microbial population dynamics during spontaneous fermentation of *Asparagus officinalis* L. young sprouts. *European Food Research and Technology* 239: 297-304.

Paramithiotis, S., O.L. Hondrodimou and E.H. Drosinos. 2010. Development of the microbial community during spontaneous cauliflower fermentation. *Food Research International* 43: 1098-1103.

Pulido, R.P., N. Ben Omar, H. Abriouel, R.L. Lopez, M. Martınez Canamero and A. Galvez. 2005. Microbiological study of lactic acid fermentation of caper berries by molecular and culture-dependent methods. *Applied and Environmental Microbiology* 71: 7872-7879.

Radberg, K., M. Biernat, A. Linderoth, R. Zabielski, S.G. Pierzynowski and B.R. Westrom. 2001. Enteral exposure to crude red kidney bean lectin induces maturation of the gut in suckling pigs. *Journal of Animal Science* 79: 2669-2678.

Randazzo C.L., A. Todaro, A. Pino, I. Pitino, O. Corona, A. Mazzaglia and C. Caggia. 2014. Giarraffa and Grossa di Spagna naturally fermented table olives: Effect of starter and probiotic cultures on chemical, microbiological and sensory traits. *Food Research International* 82: 1154-1164.

Rao, D.R., A.V. Reddy, S.R. Pulusani and P.E. Cornwell. 1984. Biosynthesis and utilization of folic acid and vitamin B_{12} by lactic acid cultures in skim milk. *Journal of Dairy Science* 67: 1169-1174.

Raybaudi-Massilia, R.M., J. Mosqueda-Melgar, R. Soliva-Fortuny and O. Martin-Belloso. 2009. Control of pathogenic and spoilage microorganisms in fresh-cut fruits and fruit juices by traditional and alternative natural antimicrobials. *Comprehensive Reviews in Food Science and Food Safety* 8: 157-180.

Rodríguez-Gómez, F., J. Bautista-Gallego, F.N. Arroyo-López, V. Romero-Gil, R. Jiménez-Díaz, A. Garrido-Fernández and P. Garcia-Garcia. 2013. Table olive fermentation with multifunctional *Lactobacillus pentosus* strains. *Food Control* 34: 96-105.

Roessle, C., M.A.E. Auty, N. Brunton, R.T. Gormley and F. Butler. 2010. Evaluation of fresh-cut apple slices enriched with probiotic bacteria. *Innovative Food Science and Emerging Technologies* 11: 203-209.

Sarvan, I., F. Valerio, S.L. Lonigro, S. de Candia, R. Verkerk, M. Dekker and P. Lavermicocca. 2013. Glucosinolate content of blanched cabbage (*Brassica oleracea* var. *capitata*) fermented by the probiotic strain *Lactobacillus paracasei* LMG-P22043. *Food Research International* 54: 706-710.

Seo, K.H. and J.F. Frank. 1999. Attachment of *Escherichia coli* O157:H7 to lettuce leaf surface and bacterial viability in response to chlorine treatment as demonstrated by using confocal scanning laser microscopy. *Journal of Food Protection* 62: 3-9.

Sesena, S. and M.Ll. Palop. 2007. An ecological study of lactic acid bacteria from Almagro eggplant fermentation brines. *Journal of Applied Microbiology* 103: 1553-1561.

Singh, A.K. and A. Ramesh. 2008. Succession of dominant and antagonistic lactic acid bacteria in fermented cucumber: Insights from a PCR-based approach. *Food Microbiology* 25: 278-287.

Song, L. and P.J. Thornalley. 2007. Effect of storage, processing and cooking on glucosinolate content of *Brassica* vegetables. *Food and Chemical Toxicology* 45: 216-224.

Stamer, J.R., B.O. Stoyla and B.A. Dunckel. 1971. Growth rates and fermentation patterns of lactic acid bacteria associated with the sauerkraut fermentation. *Journal of Milk and Food Technology* 34: 521-525.

Takeuchi, K. and J.F. Frank. 2000. Penetration of *Escherichia coli* O157:H7 into lettuce tissues as affected by inoculum's size and temperature and the effect of chlorine treatment on cell viability. *Journal of Food Protection* 63: 434-440.

Takeuchi, K. and J.F. Frank. 2001. Quantitative determination of the role of lettuce leaf structures in protecting *Escherichia coli* O157:H7 from chlorine disinfection. *Journal of Food Protection* 64: 147-151.

Tamang, J.P., B. Tamang, U. Schillinger, C.M.A.P. Franz, M. Gores and W.H. Holzapfel. 2005. Identification of predominant lactic acid bacteria isolated from traditionally fermented vegetable products of the Eastern Himalayas. *International Journal of Food Microbiology* 105: 347-356.

Tricon, S., S. Willersw, H.A. Smitw, P.G. Burneyz, G. Devereux, A.J. Frew, S. Halkenz, A. Hostz, M. Nelsonz, S. Shaheenz, J.O. Warner and P.C. Calder. 2006. Nutrition and allergic diseases. *Clinical and Experimental Allergy Reviews* 6: 117-188.

Ushiyama, A., K. Tanaka, Y. Aiba, T. Shiba, A. Takagi, T. Mine and Y. Koga. 2003. *Lactobacillus gasseri* OLL2716 as a probiotic in clarithromycin-resistant *Helicobacter pylori* infection. *Journal of Gastroenterology and Hepatology* 18: 986-991.

Warriner, K. and A. Namvar. 2010. The tricks learnt by human enteric pathogens from phytopathogens to persist within the plant environment. *Current Opinion in Biotechnology* 21: 131-136.

Warriner, K., F. Ibrahim, M. Dickinson, C. Wright and W.M. Waites. 2003. Internalization of human pathogens within growing salad vegetables. *Biotechnology and Genetic Engineering Reviews* 20: 117-134.

Wouters, D., N. Bernaert, W. Conjaerts, B. Van Droogenbroeck, M. De Loose and L. De Vuyst. 2013a. Species diversity, community dynamics and metabolite kinetics of spontaneous leek fermentations. *Food Microbiology* 33: 185-196.

Wouters, D., S. Grosu-Tudor, M. Zamfir and L. De Vuyst. 2013b. Bacterial community dynamics, lactic acid bacteria species diversity and metabolite kinetics of traditional Romanian vegetable fermentations. *Journal of the Science of Food and Agriculture* 93: 749-760.

Yao, L.H., Y.M. Jiang, J. Shi, F.A. Tomas-Barberan, N. Datta, R. Singanusong and S.S. Chen. 2004. Flavonoids in food and their health benefits. *Plant Foods for Human Nutrition* 59: 113-122.

Yeun, H., H.S. Yang, H.C. Chang and H.Y. Kim. 2013. Comparison of bacterial community changes in fermenting kimchi at two different temperatures using a denaturing gradient gel electrophoresis analysis. *Journal of Microbiology and Biotechnology* 23: 76-84.

Yoon, K.Y., E.E. Woodams and Y.D. Hang. 2004. Probiotication of tomato juice by lactic acid bacteria. *Journal of Microbiology* 42: 315-318.

Yoon, K.Y., E.E. Woodams and Y.D. Hang. 2006. Production of probiotic cabbage juice by lactic acid bacteria. *Bioresource Technology* 97: 1427-1430.

Yoon, K.Y., E.E. Woodams and Y.D. Hang. 2005. Fermentation of beet juice by beneficial lactic acid bacteria. *LWT-Food Science and Technology* 38: 73-75.

Zarate, G. and M.E. Nader-Macias. 2006. Influence of probiotic vaginal lactobacilli on *in vitro* adhesion of urogenital pathogens to vaginal epithelial cells. *Letters in Applied Microbiology* 43: 174-180.

2

Lactic Acid Bacteria of Fermented Fruits and Vegetables

Pasquale Russo[1], Graziano Caggianiello[2], Mattia Pia Arena[1], Daniela Fiocco[2], Vittorio Capozzi[1] and Giuseppe Spano[1]*

1. Introduction

Fermented foods are common throughout the world and for centuries the fermentation processes have been adopted for preserving perishable foods, such as those of plant origin. Over the years, the wide variety of fermented foods has been influenced by socio-cultural aspects of different ethnic groups as well as by the availability of variety of plant resources depending on the geographic areas (Tamang and Samuel 2010). Common vegetables that can be fermented include cabbage, cauliflower, leafy mustard, radish, carrot, cucumber, green onion, pumpkin, tomato, spinach, asparagus, lettuce, broccoli, ginger, bamboo shoot, in addition to several juices extracted from vegetables and fruits (Di Cagno et al. 2013, Marsh et al. 2014).

Vegetable fermented foods and beverages are obtained through the metabolism of yeasts and bacteria, mainly represented by the lactic acid bacteria (LAB) group. These microorganisms ferment the chemical components of raw materials, thus improving some aspects such as organoleptic, sensory properties and nutritional value of foods, while increasing digestibility and biodegradation of anti-nutritive factors or other undesirable compounds, and causing the release of release of antimicrobial substances and antioxidants. Besides, probiotic LAB can confer functional properties on the final product. (Drosinos and Paramithiotis 2012, Di Cagno et al. 2013, Filannino et al. 2015).

The interest for LAB has evolved over time. Besides the pro-technological aspect related to their role as starter cultures in fermentation processes, their probiotic activities are also gaining considerable attention

[1] Department of Agriculture, Food and Environmental Sciences, via Napoli 25, 71122, University of Foggia, Foggia, Italy.
[2] Department of Clinical and Experimental Medicine, University of Foggia, Foggia, Italy.
* Corresponding author: giuseppe.spano@unifg.it

(Gupta and Abu-Ghannam 2012, Marsh et al. 2014). Accordingly, a high number of LAB occurring in fermented foods have been investigated for their probiotic attributes.

In this chapter, we shall provide an overview of the key role played by lactic acid bacteria in vegetable fermentation, focusing on both traditional and innovative applications in the food industry.

2. Vegetable Fermentation: Spontaneous or Controlled?

Lactobacillus, *Leuconostoc*, *Weissella*, *Enterococcus* and *Pediococcus* are the genera that mainly occur in vegetables fermentation. When favorable conditions of anaerobiosis, temperature, water activity and salt concentration are established, epiphytic LAB begin a spontaneous fermentation. Generally, this process is characterized by a progressive participation of both heterofermentative and homofermentative LAB species, besides the contribution of some yeasts (Di Cagno et al. 2013). During spontaneous fermentation, almost all carbohydrates are converted into lactic acid, ethanol, mannitol and acetic acid. The resulting progressive drop in pH inhibits the growth of gram-negative bacteria and establishes favorable conditions for development of LAB microflora.

In contrast to other fermented foods, such as alcoholic beverages (wine and beer), bakery and dairy-based foods, for the production of which the employment of starter cultures is a well-established technology, the fermentation of vegetables usually relies on indigenous microbes. This is mainly due to a great variability in the composition of the substrates, to the absence of available technology in providing microorganisms for industrial applications and to the household level of traditional productions that do not require the scalability of the process. Moreover, community dynamics during spontaneous vegetable fermentation makes it difficult to reproduce the microbial ecology for these processes.

In the last few years, several works investigated the microbial composition of spontaneous fermented vegetables with the aim to increase knowledge on the process and to select starter cultures for controlling food fermentation and quality (Yang et al. 2014). It was reported that some fermented vegetables are characterized by occurrence of dominant microbial species or genera. It is the case of yacon fermentation that was related mainly to the metabolic activity of *Leuconostoc* spp. (Reina et al. 2015). In contrast, changes in LAB microflora were reported throughout the spontaneous fermentation of Chinese sauerkraut (Xiong et al. 2012). Similarly, *Enterobacteriaceae* and a wide LAB species diversity encompassing *Weissella* spp., dominate the first phase of fermentation of cauliflower and mixed vegetables, followed by a second phase wherein *Ln. citreum* and *Lb. brevis* occur, and a final phase characterized by the prevalence of *Lb. brevis* and *Lb. plantarum* (Wouters et al. 2013a). Spontaneous kimchi fermentation in contaminated raw materials leads to the growth of several

LAB species, which result in variations in the taste and sensory qualities of kimchi products. Difficulties in standardization of the process aimed at the industrial production of kimchi with the same properties have been discovered (Jung et al. 2014).

Although the use of starter cultures was expected to play a key role in the industrial production of standardized fermented vegetables, several factors should be carefully evaluated for the successful application of starter cultures including the contribution on the organoleptic characteristics, safety, health benefits, and shelf life (Bourdichon et al. 2012). In general, the microbial population prevailing over endogenous LAB communities at the end of fermentation appears to comprise good candidate starter cultures for controlled vegetable fermentation processes (Wouters et al. 2013b). Nowadays, green table olives, sauerkraut, pickles and kimchi are fermented vegetable foods produced on industrial scale for which the use of starter cultures is established (Hurtado et al. 2012, Lee et al. 2015). Some vegetable fermented foods throughout the world and the main LAB species involved in their fermentation are summarized in Table 1.

3. Fermented Vegetables Throughout the World

Olives, sauerkraut, pickles cucumber, kimchi and some fermented vegetable juices are among the most widespread fermented vegetables to be used in the world. However, there are several fermented vegetables which are lesser known in Western countries, but more popular in Asian and African regions (Demarigny 2012, Satish et al. 2103, Franz et al. 2014).

Table olives are obtained from the fruit of the olive tree typical of the Mediterranean countries such as Italy, Spain, Greece, Turkey and Morocco, with variations in their preparation depending on the geographical area. The olive-processing methods are mainly three: i) green Spanish olives are treated with NaOH (sodium hydroxide) and then fermented; ii) Greek olives, which are naturally black, are fermented without NaOH treatment; iii) California olives (ripe black or green) are not fermented but simply treated with sodium hydroxide and may be subjected to aeration in order to promote pigment oxidation, thus converting the green to black color. In both Greek and Spanish methods, the fermentation is led by native microflora (Hutkins 2008) where the brine solution increases the shelf-life and sensory properties (Brenes 2004). The necessity to transform olives arises due to the presence of a polyphenol, namely oleuropein, which confers a bitter taste to the fruit, rendering them inedible before processing (Tamang and Kailasapathy 2010). The bacterial flora constitutes a crucial component in the successful fermentation and final flavour of table olives (Sabatini et al. 2008), with *Lb. plantarum*, *Lb. casei* and *Lb. pentosus* being the main LAB of this process (Ercolini et al. 2006). The use of starter cultures during fermentation is a practice mainly used in industrial production of

Table 1. Examples of Fermented Vegetables and the Main LAB Species Involved in the Fermentation Process

Fermented food	Raw food	Dominant LAB species	Country
Fu-tsai and suan-tsai	Mustard	*Pd. pentosaceus, Tetragenococcus halophilus, Lb. farciminis, Ln. mesenteroides, Ln. pseudomesenteroides, W. cibaria, W. paramesenteroides*	Taiwan
Goyang	Magane-saag	*Lb. plantarum, Lb. brevis, Lc. lactis, Ec. faecium, Pd. pentosaceus*	Nepal
Gundruk	Rayo-sag mustard cauliflower	*Lb. fermentum, Lb. plantarum, Lb. casei, Lb. casei* subsp. *pseudoplantarum, Pd. pentosaceus*	Himalayas
Inziangsang	Mustard leaves	*Lb. plantarum, Lb. brevis, Pd. acidilactici*	India
Jeruk	Gherkin and cucumber, ginger, onion, chilli, bamboo shoot, mustard leaves	*Leuconostoc, Lactobacillus, Pediococcus* and *Enterococcus* spp.	Malaysia
Khalpi	Cucumber	*Ln. fallax, Pd. pentosaceus, Lb. brevis, Lb. plantarum*	Himalayas
Kimchi	Pickle	*Ln. mesenteroides, Ln. pseudomesenteroides, Ln. lactis, Lb. brevis, Lb. plantarum*	Korea
Mesu	Bamboo shoot	*Lb. plantarum, Lb. brevis, Lb. curvatus, Ln. citreum, Pd. pentosaceus*	Himalayas
Olives	Olive	*Lb. plantarum, Lb. pentosus, Lb. casei, Lb. brevis*	Mediterranean
Pak-gard-dong	Leaf of mustard	*Pd. pentosaceus, Lb. brevis, Lb. plantarum*	Thailand
Pak-sian-dong	Leaves of pak-sian	*Ln. mesenteroides, Lb. fermentum, Lb.buchneri, Lb. plantarum, Lb. brevis, Pd. pentosaceus*	Thailand
Pickled cucumbers	Cucumber	*Lb. plantarum, Lb. pentosus, Leuconostoc, Pediococcus* spp.	Europe, USA
Sauerkraut	Cabbage	*Ln. mesenteroides, Lb. brevis, Lb. plantarum, Pd. pentosaceus*	Europe, USA
Sayurasin	Mustard cabbage leaf	*Ln. mesenteroides, Lb. confusus, Lb. plantarum, Pd. pentosaceus*	Indonesia
Sinki	Radish tap root	*Lb. plantarum, Lb. brevis, Lb. casei, Ln. fallax*	Himalayas
Sunki	Leaves and stems of red turnip	*Lb. plantarum, Lb. brevis, Lb. buchneri, Ec. faecalis, Pd. pentosaceus*	Japan

table olives, and their criteria of selection include acid production, salt and temperature tolerance (Heperkan 2013). Starter cultures reduce the risk of deterioration and induce a faster acidification of brine with lesser metabolic energy consumption by the bacterial cells, greater predictability of the fermentation process and better organoleptic characteristics (Tamang and Kailasapathy 2010, Hurtado et al. 2012).

Caper is a wild shrub growing in semi-arid regions and very commonly found in Mediterranean countries. It is obtained from the buds of the plant and to a lesser extent from the berries. The production is based mainly on wild rather than cultivated plants. Therefore, the production costs are quite high as its production is dependent on the season. The fermentation of capers in brine is a spontaneous process by LAB (Pulido et al. 2012). *Lb. plantarum, Lb. paraplantarum, Lb. pentosus, Lb. brevis, Lb. fermentum, Pd. pentosaceus* and *Pd. acidilactici* are the representative LAB, with *Lb. plantarum* being the most abundant species (Pulido et al. 2005, Hui and Evranuz 2012).

Sauerkraut, originating in Northern Europe, derives its name from the German word, *kraut*, although it seems to have been imported from China. Indeed, the lactic fermentation of vegetables is a very common practice of preservation in the Asian countries. Sauerkraut is obtained from dried and salted cabbage cut into thin strips and fermented by wild flora of LAB (Halász et al., 1999). The fermentation process includes a hetero-fermentative phase, in which the less acid-tolerant LAB occur, followed by a homo-fermentative step, which is dominated by acid-tolerant *Ln. mesenteroides, Lb. brevis, Lb. plantarum* and *Pd. pentosaceus*. Glucose and fructose are the most abundant fermentable sugars found in cabbage, followed by sucrose, which represents less than 0.2 per cent (Plengvidhya et al. 2004). Different spices are added during the preparation of sauerkraut. The high salt concentration and high fermentation temperatures lead to increased acidity of the product (Guizani 2011). The addition of salt promotes the extraction of plant fluids, which are necessary for the development of LAB (Tamang and Kailasapathy 2010). The consequent production of lactic acid and acetic acid lowers the pH and the production of carbon dioxide replaces oxygen, with acidity and anaerobic conditions preventing the development of spoilage microorganisms (Halász et al. 1999).

Pickling cucumbers are made from unripe *Cucumis sativus* and are obtained through fermentation and treatment with dill, various spices and immersion in brine (5–8 per cent), often acidified with acetic acid to pH 4.5 or supplemented with potassium sorbate to prevent the development of spoilage fungi (Lahtinen et al. 2011). The fermentation, which lasts up to four weeks, depending on the temperature, positively affects the organoleptic characteristics and stabilization of the final product, by lowering the pH and concentration of fermentable sugars and thereby reducing the possibility of growth of spoilage microorganisms. At the beginning of fermentation, an aerobic microbial population prevails, with a low concentration of LAB, which act later in the process (Tamang and Kailasapathy 2010). *Lb. plantarum* is a dominant species at all stages of fermentation and during storage. *Lb. pentosus* and *Leuconostoc* spp. were isolated throughout fermentation, while *Pediococcus* spp. occured in the late phases (Tamminen et al. 2004, Singh and Ramesh 2008). LAB play a protective role against undesirable microorganisms through the release of different bacteriocins and anti-microbial peptides (Di Cagno et al. 2013). In order to plan production and predict the quality, the use of starter cultures in the manufacture of fermented cucumbers is preferred with respect to the spontaneous fermentation performed by autochthonous bacterial flora (Desai and Sheth 1997).

Some other vegetable fermented foods, such as kimchi, gundruk, sinki, and khalpi (reviewed by Swain et al. 2014) are mainly produced and consumed in the Eastern world.

Kimchi is a traditional, pickled, fermented Korean food and obtained by vegetables such as salted cabbage, radish and cucumber through addition of spices like chili pepper, garlic, ginger and other ingredients. The use of chili in the processing of kimchi has been carried on for centuries (Oum 2005). Kimchi is considered a very healthy food due to its high content of nutrients (vitamins, minerals, fibers and phytochemicals) produced by LAB during fermentation (Lee et al. 2011).

Kimchi fermentation takes place at low temperatures and is carried out by LAB, which produce organic acids and other substances such as CO_2 (carbon dioxide), ethanol, mannitol, bacteriocins, and fatty acids, thus improving the stability and health effects of the product. Generally, in the initial stages of kimchi fermentation, the population of LAB cells is about 4–5 log CFU/g. Aerobic microorganisms are present in the early stages of fermentation. Then anaerobic bacteria take over and *Leuconostoc* spp., *Lactobacillus* spp. and *Weissella* spp. occur. *Ln. mesenteroides* is seen as the most predominant microorganism during the maturation phase at low temperatures, while *Lb. plantarum* prevails in the final stage of fermentation. *Lb. plantarum*, together with *Lb. sakei* causes over-ripening of kimchi with a consequent reduction in the quality of the product. This can also be affected by yeasts that appear especially in the final stage of fermentation of kimchi. In order to avoid a decrease in the quality of kimchi and to prolong its shelf-life, the use of starter cultures and their selection criteria have been proposed (Lee et al. 2015).

Gundruk is a typical fermented product of the Himalayas made from fresh leaves of a plant called rayo-sag (*Brassica rapa*, subspecies *campestris*, variety *cuneifolia*), mustard (*Brassica juncea*) and cauliflower (*Brassica oleracea* variety *botrytis*). The leaves are withered, crushed and lightly pressed, then spontaneously fermented for about ten days. Finally, the product is dried for four days and stored for about two years. The native LAB include mainly *Lb. fermentum*, *Lb. plantarum*, *Lb. casei*, *Lb. casei* subsp.

pseudoplantarum and *Pd. pentosaceus*, while some yeasts are present in the early stages of the fermentation. Since uncontrolled fermentation can affect the product quality, the use of starter cultures as an alternative to spontaneous fermentation has been proposed (Tamang and Kailasapathy 2010, Tamang et al. 2012).

Sinki is another ethnic fermented product from the Himalayans, and obtained from radish (*Raphanus sativus*) taproot. Usually, it is prepared during the winter when the weather is less humid and there is greater availability of the raw material (Das and Deka 2012). Sinki fermentation takes place in a pit with a diameter of 2–3 ft, covered with mud and subsequently burned. Ash is removed from the pit which is covered with sheaths of bamboo and rice straw. The radish is wilted, pressed and immersed in warm water, the pit covered with dry leaves and weighed down by heavy planks or stones before being closed with mud. The fermentation lasts about 30 days, after which the sinki is cut and dried in the sun. For this reason, it can be stored at room temperature (Swain et al. 2014). Dry sinki can be stored for up to two years in hermetic conditions. *Lb. plantarum*, *Lb. brevis*, *Lb. casei* and *Ln. fallax* were isolated from sinki (Tamang and Kailasapathy 2010).

Khalpi is an ethnic, fermented cucumber product of the Himalayas. After harvesting, cucumbers are cut and dried in the sun for a couple of days. They are then put in a bamboo vessel hermetically sealed by using dried leaves and left to ferment for five days. Prolonged fermentation determines an unwanted and higher acidity of the finished product. In the early stages of fermentation, hetero-fermentative LAB such as *Ln. fallax*, *Pd. pentosaceus*, *Lb. brevis* are formed while *Lb. plantarum* predominates if the fermentation is prolonged (Tamang and Kailasapathy 2010).

Several fermented vegetable juices are obtained from either spontaneous fermentation or after starter-culture inoculation of juices previously pasteurized. Usually during fermentation, the temperature reaches about 25°C, although lower temperatures have a positive impact on the beverage quality. A rapid pH decrease minimizes the risk of disadvantageous microbial contamination; generally the final pH is between 3.8 and 4.5 (Lahtinen et al. 2011).

4. Lactic Acid Bacteria for Innovative Foods: Challenges and Perspectives

Lactic acid bacteria create a harmonious fusion between innovative and traditional applications into the food field since in the last few years the employment of LAB for elaboration of functional foods has overlapped their ancestral key role in food fermentation.

Probiotic fortification is the best approach to produce functional foods. In this framework, probiotic fermented foods of vegetable origin are increasing in the market as they are considered a promising alternative to probiotic dairy products (Gupta and Abu-Ghannam 2012). Moreover, probiotic fermented vegetables and fruits can satisfy, better than their dairy counterparts, the tastes of special consumers such as vegetarians and vegans, lactose intolerants, individuals with low-cholesterol intake needs, or allergic to animal proteins (Molin 2001). Some probiotic LAB strains proposed for the production of vegetable fermented foods and beverages are summarized in Table 2.

5. Probiotic Fermented Vegetables

To select starter with probiotic features, LAB strains can be isolated from spontaneously fermented foods and subsequently screened for their probiotic aptitude, including *in vitro, in vivo*, genetic and 'omic' analyses (Papadimitriou et al. 2015). In this way, some quality attributes of the fermented product as well as its technological, nutritional and functional parameters can be improved. As an alternative or as a complementary approach to the selection of autochthonous microorganisms, one can resort to technological characterization of microbial strains previously isolated from different non-food related sources and selected for their probiotic potential. Based on such methodology, strains of human origin would be excellent probiotic candidates, due to their better adaptation to the gut environmental conditions. Accordingly, a clinical isolate of *Lb. paracasei* was investigated as a probiotic culture in the fermentation of green and ripe olives and it showed promising results at both the laboratory and industrial plant levels (De Bellis et al. 2010, Sisto and Lavermicocca 2010).

Among fermented vegetables, the production of green table olives as carriers of probiotic microorganisms is more widespread. From the technological point of view, the main parameters in selecting multifunctional starters include fast acidification in the fermentation brine, predominance over the native microbial populations and ability to colonize olive epidermis (Rodríguez-Gómez et al. 2014a). Thus, LAB strains selected for their technological performance are characterized for their potential as probiotic starter and suggested for the production of functional Spanish-style green table olives on a pilot scale (Abriouel et al. 2012, Argyri et al. 2013, Rodríguez-Gómez et al. 2014a). Similarly, the employment of autochthonous *Lb. plantarum* was proposed as a functional starter for the Italian *Bella di Cerignola* table olives (Bevilacqua et al. 2010). The effects of select starter cultures and probiotic strains on the trend of volatile compounds and on the final sensorial traits of green table olives suggested that the starter selection should take into account the olive cultivar (Randazzo et al. 2014). Since the importance as a probiotic carrier depends on the microbial concentration at consumption time, several studies revealed the influence of the main technological factors on their

Table 2. Some Probiotic LAB Strains Proposed for the Production of Vegetable and Fruit Fermented Foods and Beverages

Probiotic strain	Source of isolation	Fermented food	Techno-logical conditions	Viability (cfu/mL)	Reference
Lb. plantarum *Lb. acidophilus*	Human milk	Orange, apple, grapes and tomato juices			Nagpal et al. (2012)
Lb. plantarum PCS26	Slovenian cheese	Apple juice	Ca-alginate-encapsulation		Dimitrovski et al. (2015)
Lb. acidophilus *Lb. casei* *Lb. delbrueckii* *Lb. plantarum*	Non auto-chthonous	Beet juice	4°C - 4 weeks	$-$ 10^6–10^8 10^6–10^8 10^6–10^8	Yoon et al. (2005)
Lb. plantarum C3 *Lb. casei* A4 *Lb. delbrueckii* D7	Non auto-chthonous	Cabbage juice	4°C - 4 weeks	$4.1 \cdot 10^7$ $-$ $4.5 \cdot 10^5$	Yoon et al. (2006)
Lb. casei *Lb. plantarum* *Bf. longum*	Non auto-chthonous	None juice	4°C - 4 weeks	$-$	Wang et al. (2009)
Lb. casei NRRL B-442	Non auto-chthonous	Cashew apple juice	4°C - 42 days	10^8	Pereira et al. (2011)
Lb. reuteri	Non auto-chthonous	Pumpkin juice	4°C - 4 weeks	10^9	Semjonovs et al. (2013)
Lb. paraplantarum SF9 *Lb. brevis* SF15	Auto-chthonous	Sauerkraut			Beganović et al. (2014)
Lb. plantarum L4 *Ln. mesenteroides* LMG 7954	Non auto-chthonous	Sauerkraut	14 days	10^6	Beganović et al. (2011)
Lb. pentosus TOMC-LAB2	Auto-chthonous	Green table olives	200 days, spontaneously fermented, thermally treated, inoculum into the packing brines	100.0%	Rodríguez-Gómez et al. (2014)
Lb. pentosus B281 *Lb. plantarum* B282	Industrial fermented olives	Green table olives	4°C - 12 months modified atmospheres	35.0% 13.3%	Argyri et al.(2015)
Lb. paracasei IMPC2	Human isolate	Green table olives	4°C and 20°C 4% and 8% NaCl		De Bellis et al. (2010)

viability. Thus, a shelf-life of six months at refrigerating temperature has been proposed as the best condition to deliver probiotic green table olives to the consumer (Argyri et al. 2015), while 10 per cent NaCl (sodium chloride) brines negatively affected the ability of *Lb. plantarum* B282 to colonize the surface of olives (Blana et al. 2014). Interestingly, a promising strategy was the fortification of previously fermented green Spanish olives with autochthonous putative probiotic bacteria, *Lb. pentosus* TOMC-LAB2, by inoculating the bacteria into the packing brines. This approach allowed a cent per unit recovery of viable bacteria under packing conditions, after six months of storage (Rodríguez-Gómez et al. 2014b).

In addition to green table olives, other fermented vegetables were also proposed as new functional foods. *Lb. plantarum* S4-1 isolated and identified from naturally fermented Chinese sauerkraut was examined *in vitro* for its potential probiotic properties and *in vivo* for its cholesterol-lowering effect. The results showed its ability to effectively reduce serum cholesterol level in mice (Yu et al. 2013). In a technological assay, two potential probiotic starter cultures were selected from autochthonous LAB sauerkraut population in order to add probiotic attributes, including health-promoting, nutritional, technological and economic advantages in large-scale industrial sauerkraut production (Beganović et al. 2014). Similarly, the application of probiotic *Lb. plantarum* L4 and *Ln. mesenteroides* LMG 7954 improved the technological features during controlled fermentation of cabbage heads, by lowering NaCl concentrations and considerably accelerating the fermentation process, thus resulting in an enhanced quality of sauerkraut (Beganović et al. 2011).

In the last few years, increase in interest is seen in traditional fermented vegetables originating in India and other Asian countries, such as kimchi, gundruk, khalpi, sinki, goyang and kanjika among others. These foods are suggested as a promising source of probiotics as they harbour several poorly-investigated communities of LAB (reviewed by Satish et al. 2013, Swain et al. 2014). Similarly, many typical African foods predominantly fermented by LAB can make a profound impact on the health and well-being of adults and children (Franz et al. 2014).

6. Probiotic Fermented Juices

Despite their great potential as probiotic carriers, less attention has been paid to fermentation of vegetable and fruit juices. Most studies are based on the microbial addition of probiotic strains to products of vegetables origin to obtain non-fermented functional foods (Martins et al. 2013). Interestingly, fresh-cut apples and pineapples were found to be new vehicles for probiotic microorganisms (Rössle et al. 2010, Alegre et al. 2011, Russo et al. 2014). However, there is popular demand for fermented juices by vegetarians and lactose-allergic consumers (Prado et al. 2008, Di Cagno et al. 2013, Swain et al. 2014). In most of these studies, previously characterized for their probiotic features, LAB are exogenously added as starter cultures to achieve a safe fermentation of the raw material (Sheehan et al. 2007). At the same time, the probiotic microbial cultures should survive at cold storage conditions throughout the shelf-life of the delivery product without negatively affecting the technological parameters. In any case, sensorial evaluation is mandatory before marketing, since the occurrence of foreign molecules metabolized by the probiotic bacteria might influence the final taste of the product (Marsh et al. 2014).

Due to their ability to quickly lower pH and survive the high acidity conditions during four weeks at refrigerated temperatures, *Lb. plantarum* and *Lb. delbrueckii* were suggested as probiotic cultures for the production of fermented cabbage juice (Yoon et al. 2006). With a similar approach, Wang and co-authors (2009) identified *Bifidobacterium longum* and *Lb. plantarum* as optimal probiotic candidates for fermentation of noni juice, a typical fruit of Southeast Asia. In the same way, red beets were evaluated as a potential substrate for the production of probiotic beet juice through four species of LAB, including *Lb. acidophilus*, *Lb. casei*, *Lb. delbrueckii*, and *Lb. plantarum*. Though *Lb. acidophilus* was found interesting from a technological point of view due to its greater production of lactic acid, its viability was low during cold storage at 4°C, thus rejecting its use as probiotic (Yoon et al. 2005). Recently, *Lb. reuteri* strains were selected for the development of a fermented, potentially probiotic, pumpkin juice-based beverage. Salt tolerance and influence on the sensory properties of the juice were the main technological parameters analyzed (Semjonovs et al. 2013). In the past, functional beverages based on tomato juice fermented with a mix of LAB showed a high viability of the probiotic strains during four weeks and 56 days under cold storage (Yoonet al. 2004). However, recently, strains isolated from different juices were exploited for their probiotic features and their employment as starter cultures suggested for production of ethnic fermented beverage, gilaburu juice as well as for the elaboration of different fermented foods (Das and Goyal 2014, Sagdic et al. 2014, Shukla et al. 2014).

7. Fermented Vegetables as Synbiotic Foods

Fruits and vegetables are ideal substrates to be examined for their ability to support both probiotic and prebiotic delivery in humans. Prebiotics are non-digestible food ingredients that benefit the host by selectively stimulating the growth or activity of a limited number of beneficial microbes in the colon (Al-Sheraji et al. 2013). Therefore, it is reasonable to speculate that if prebiotic and probiotic are provided simultaneously, they could exert additive or even synergistic effects (de Vrese and Schrezenmeir 2008). Many vegetable, root and tuber crops as well as some fruit crops are sources of prebiotic carbohydrates and their combination with probiotic microorganisms, added as starters, could be a successful strategy to obtain synbiotic foods (Rodríguez-Gómez et al. 2014a). For example, inulin is the main carbohydrate source in the tubers of Jerusalem artichoke, suggesting that lactobacilli can be used for fermentation of this product (Zalán et al. 2011).

Vegetable milks are new functional formulations that can contain prebiotics which provide technological benefits to the fermentation process, and might have a synergic effect on probiotic survival during processing and storage (Bernat et al. 2014).

Nazzaro et al. (2008) investigated the formulation of a health-improving carrot juice manufactured with the probiotic *Lb. delbrueckii* subsp. *bulgaricus* and *Lb. rhamnosus*, and to which inulin and fructooligosaccharides were added as prebiotic components. This formulation was a good source of probiotic *Lactobacillus* spp. and nutritional components after four weeks of storage at 4°C, without altering the biochemical properties of the fermented juice (Nazzaro et al. 2008). Similarly, the viability of probiotic *B. lactis* BB12 increased in fructan-supplemented fermented cabbage juice (Semjonovs et al. 2014). An ingenious way to provide prebiotic substrates was the encapsulation of probiotic bacteria in an inulin-containing matrix. This formulation not only promoted survival of *Lb. acidophilus* in fermented carrot juice, but also enhanced its viability after exposure to artificial gastrointestinal conditions (Nazzaro et al. 2009).

Recently, it was suggested that oligosaccharides from microbial origin could improve some probiotic attributes of selected LAB (Russo et al. 2012). We know that strains of *Ln. mesenteroides* are employed for the industrial production of dextran and dextransucrase. Therefore, these strains were used as starter to ferment clarified cashew apple juice and at the same time to enrich the substrate with prebiotic oligosaccharides. Interestingly, the growth of a probiotic *Lb. johnsonii* was three-fold higher in cashew apple juice previously fermented by *Ln. mesenteroides* (Vergara et al. 2010).

8. Technological Approaches for Functional Fermented Vegetables and Fruits

Spray-drying of fruit juices containing probiotic bacteria is emerging as innovative strategy in the field of functional vegetables and fruits. The fermented juice powder is highly desirable because the dried product potentially has both a longer shelf-life and lower transportation cost. Moreover, probiotic fruit and vegetable juice powders are suggested for use in probiotic drinks, ice creams, syrups and prepared soups (Pereira et al. 2014). So, a promising attempt was made by using a spray-dried watermelon and carrot juice mixture fermented with a probiotic *Lb. acidophilus* to produce an innovative non-dairy-based food formulation with good flavor and high nutritional and functional value (Mestry et al. 2011).

Similarly, after optimizing the fermentation conditions of cashew apple with probiotic *Lb. casei* NRRL B-442 and evaluating its storage stability at 4°C for 42 days (Pereira et al. 2011, 2013), the potential of spray-dried cashew apple juice as a carrier of the probiotic starter was investigated (Pereira et al. 2014). An improvement in the shelf-life was observed for the powder obtained using maltodextrin (20%) and stored at 25°C, suggesting that different conditions of storage temperature and the selection of drying carriers can noticeably influence the survival rate of probiotics (Pereira et al. 2014).

In a different approach, the use of sonicated pineapple juice was evaluated as fermentable substrate for producing a probiotic beverage by *Lb. casei* NRRL B-442. It showed that sonication can be applied as an inexpensive pre-treatment to increase the shelf-life and to preserve nutritional and organoleptic qualities of juice products (Costa et al. 2013).

Physical protection of probiotics by microencapsulation has aroused interest as it protects the bacteria during its incorporation into the food matrix, storage and after consumption (Martín et al. 2015). Accordingly, it was demonstrated that encapsulation preserved viability and activity of the probiotic *Lb. acidophilus* used to ferment carrot juice after eight weeks of cold storage (Nazzaro et al. 2009). Moreover, *Lb. acidophilus* encapsulated in a mixture of alginate-inulin-xanthan gum showed higher tolerance than free cells when exposed to artificial gastrointestinal conditions (Nazzaro et al. 2009). In an attempt to optimize the encapsulation conditions, apple juice was fermented with free and Ca-alginate-embedded probiotic *Lb. plantarum* PCS 26 at different initial pH values, as well as with whey supplement as a growth enhancer. In contrast to free bacteria, entrapment of cells into a Ca-alginate matrix caused considerably slower growth at the end of fermentation. Stability of the fermented product improved during storage at 4–7°C and the survival of immobilized bacteria significantly increased compared with free-cell fermentation (Dimitrovski et al. 2015).

9. Concluding Remarks and Future Perspectives

Fermented vegetable foods and beverages are traditional components of the dietary culture of every community in the world. Fermentation is an ancestral technology to store perishable foods by exploiting the microbial transformation of chemical constituents of raw materials performed by LAB. However, compared with other fermented foodstuffs, most of the vegetable fermented foods and beverages are produced by natural fermentation, whereby the microbial component is neglected. Critical points that hinder the employment of fermentation under microbe-controlled conditions include the household level of the production, the wide range and diversity of fermentable substrates, the lack of commercialized starter cultures and of methods to reduce the initial microbial contamination in raw materials. Undoubtedly, a fermentation that relies on spontaneously occurring LAB is a valuable approach to preserve the characteristics of the foodstuff. However, in order to meet the increasing global demand for fermented vegetables, it is necessary to develop reliable methods that yield a stable and high-quality product.

In the food industry, the use of starter cultures is recommended to obtain standard products and to reduce the safety risk. Moreover, in recent years, the traditional role of LAB in food fermentation has reinforced their potential as probiotic in the elaboration of new functional foods. Since fermented vegetables are accepted worldwide, these innovative foods can be considered as optimal carriers to deliver beneficial microbes. Moreover, if such products are designed especially for household fermented food preparations, this could have a profound impact on the health and well-being of the people of emerging countries. Hence, more efforts should be devoted to further characterize the microbial ecology of fermented vegetable foods and beverages. An improved knowledge on the fermentation relationships is the first step in the selection of new LAB strains for advances in the food industry.

References

Abriouel, H., N. Benomar, A. Cobo, N. Caballero, M.A. Fernández Fuentes, R. Pérez-Pulido and A. Gálvez. 2012. Characterization of lactic acid bacteria from naturally-fermented Manzanilla Aloreña green table olives. *Food Microbiology* 32: 308-316.

Alegre, I., I. Viñas, J. Usall, M. Anguera and M. Abadias. 2011. Microbiological and physicochemical quality of fresh-cut apple enriched with the probiotic strain *Lactobacillus rhamnosus* GG. *Food Microbiology* 28: 59-66.

Al-Sheraji, S.H., A. Ismail, M.Y. Manap, S. Mustafa, R.M. Yusof and F.A. Hassan. 2013. Prebiotics as functional foods: A review. *Journal of Functional Foods* 5: 1542-1553.

Argyri, A.A., A.A. Nisiotou, P. Pramateftaki, A.I. Doulgeraki, E.Z. Panagou and C.C. Tassou. 2015. Preservation of green table olives fermented with lactic acid bacteria with probiotic potential under modified atmosphere packaging. *LWT-Food Science and Technology* 62: 783-790.

Argyri, A.A., G. Zoumpopoulou, K.A.G. Karatzas, E. Tsakalidou, G.J.E. Nychas, E.Z. Panagou and C.C. Tassou. 2013. Selection of potential probiotic lactic acid bacteria from fermented olives by *in vitro* tests. *Food Microbiology* 33: 282-291.

Beganović, J., A.L. Pavunc, K. Gjuračić, M. Spoljarec, J. Šušković and B. Kos. 2011. Improved sauerkraut production with probiotic strain *Lactobacillus plantarum* L4 and *Leuconostoc mesenteroides* LMG 7954. *Journal of Food Science* 76: 124-129.

Beganović, J., B. Kos, A. LebošPavunc, K. Uroić, M. Jokić and J. Šušković. 2014. Traditionally produced sauerkraut as source of autochthonous functional starter cultures. *Microbiological Research* 169: 623-632.

Bernat, N., M. Cháfer, A. Chiralt and C. González-Martínez. 2014. Hazelnut milk fermentation using probiotic *Lactobacillus rhamnosus* GG and inulin. *International Journal of Food Science and Technology* 49: 2553-2562.

Bevilacqua, A., C. Altieri, M.R. Corbo, M. Sinigaglia and L.I. Ouoba. 2010. Characterization of lactic acid bacteria isolated from Italian Bella di Cerignola table olives: Selection of potential multifunctional starter cultures. *Journal of Food Science* 75: 536-544.

Blana, V.A., A. Grounta, C.C. Tassou, G.J. Nychas and E.Z. Panagou. 2014. Inoculated fermentation of green olives with potential probiotic *Lactobacillus pentosus* and

Lactobacillus plantarum starter cultures isolated from industrially-fermented olives. *Food Microbiology* 38: 208-218.

Bourdichon, F., S. Casaregola, C. Farrokh, J.C. Frisvad, M.L. Gerds, W.P. Hammes, J. Harnett, G. Huys, S. Laulund, A. Ouwehand, I.B. Powell, J.B. Prajapati, Y. Seto, E. Ter Schure, A. Van Boven, V. Vankerckhoven, A. Zgoda, S. Tuijtelaars and E.B. Hansen. 2012. Food fermentations: microorganisms with technological beneficial use. *International Journal of Food Microbiology* 154: 87-97.

Brenes, M. 2004. Olive fermentation and processing: Scientific and technological challenges. *Journal of Food Science* 69: 33-34.

Costa, M.G., T.V. Fonteles, A.L. de Jesus and S. Rodrigues. 2013. Sonicated pineapple juice as substrate for *L. casei* cultivation for probiotic beverage development: process optimisation and product stability. *Food Chemistry* 139: 261-266.

Das, A. and S. Deka. 2012. Fermented foods and beverages of northeast India. *International Food Research Journal* 19: 377-392.

Das, D. and A. Goyal. 2014. Potential probiotic attributes and antagonistic activity of an indigenous isolate *Lactobacillus plantarum* DM5 from an ethnic fermented beverage 'Marcha' of north-eastern Himalayas. *International Journal of Food Sciences and Nutrition* 65: 335-344.

De Bellis, P., F. Valerio, A. Sisto, S.L. Lonigro and P. Lavermicocca. 2010. Probiotic table olives: Microbial populations adhering on olive surface in fermentation sets inoculated with the probiotic strain *Lactobacillus paracasei* IMPC2.1 in an industrial plant. *International Journal of Food Microbiology* 140: 6-13.

de Vrese, M. and J. Schrezenmeir. 2008. Probiotics, prebiotics, and synbiotics. *Advances in Biochemical Engineering/Biotechnology* 111: 1-66.

Demarigny, Y. 2012. Fermented food products made with vegetable materials from tropical and warm countries: Microbial and technological considerations. *International Journal of Food Science and Technology* 47: 2469-2476.

Desai, P. and T. Sheth. 1997. Controlled fermentation of vegetables using mixed inoculum of lactic cultures. *Journal of Food Science and Technology* 34: 155-158.

Di Cagno, R., R. Coda, M. De Angelis and M. Gobbetti. 2013. Exploitation of vegetables and fruits through lactic acid fermentation. *Food Microbiology* 33: 1-10.

Dimitrovski, D., E. Velickova, T. Langerholc and E. Winkelhausen. 2015. Apple juice as a medium for fermentation by the probiotic *Lactobacillus plantarum* PCS 26 strain. *Annals of Microbiology* DOI:10.1007/s13213-015-1056-7.

Drosinos, E.H. and S. Paramithiotis. 2012. Nutritional attributes of lactic acid fermented fruits and vegetables. *Agro Food Industry Hi Tech* 23: 46-48.

Ercolini, D., F. Villani, M. Aponte and G. Mauriello. 2006. Fluorescence *in situ* hybridization detection of *Lactobacillus plantarum* group on olives to be used in natural fermentations. *International Journal of Food Microbiology* 112: 291-296.

Filannino, P., Y. Bai, R. Di Cagno, M. Gobbetti and M.G. Gänzle. 2015. Metabolism of phenolic compounds by *Lactobacillus* spp. during fermentation of cherry juice and broccoli puree. *Food Microbiology* 46: 272-279.

Franz, C.M., M. Huch, J.M. Mathara, H. Abriouel, N. Benomar, G. Reid, A. Galvez and W.H. Holzapfel. 2014. African fermented foods and probiotics. *International Journal of Food Microbiology* 190: 84-96.

Guizani, N. 2011.Vegetable Fermentation and Pickling. Pp. 351-367. In: N.K. Sinha, Y.H. Hui, E.O. Evranuz, M. Siddiq, J. Ahmed (Eds.). *Handbook of Vegetables and Vegetable Processing*. Blackwell Publishing Ltd., Oxford, UK.

Gupta, S. and N. Abu-Ghannam. 2012. Probiotic fermentation of plant based products: Possibilities and opportunities. *Critical Reviews in Food Science and Nutrition* 52: 183-199.

Halász, A., A. Baráth and W.H. Holzapfel. 1999. The influence of starter culture selection on sauerkraut fermentation. *European Food Research and Technology* 208: 434-438.

Heperkan, D. 2013. Microbiota of table olive fermentations and criteria of selection for their use as starters. *Frontiers in Microbiology* 4: 143.

Hui, Y.H. and E.Ö. Evranuz. 2012. *Handbook of Plant-based Fermented Food and Beverage Technology*. CRC Press. Boca Raton, London, New York.

Hurtado, A., C. Reguant, A. Bordons and N. Rozès. 2012. Lactic acid bacteria from fermented table olives. *Food Microbiology* 31: 1-8.

Hutkins, R.W. 2008. *Microbiology and Technology of Fermented Foods*. Blackwell Publishing Ltd., Oxford, UK.

Jung, J.Y., S.H. Lee and C.O. Jeon. 2014. Kimchimicroflora: history, current status and perspectives for industrial kimchi production. *Applied Microbiology and Biotechnology* 98: 2385-2393.

Lahtinen, S., A.C. Ouwehand, S. Salminen and A. von Wright. 2011. *Lactic Acid Bacteria: Microbiological and Functional Aspects*. CRC Press. Boca Raton, London, New York.

Lee, H., H. Yoon, Y. Ji, H. Kim, H. Park, J. Lee, H. Shin and W. Holzapfel. 2011. Functional properties of *Lactobacillus* strains isolated from kimchi. *International Journal of Food Microbiology* 145: 155-161.

Lee, M., J. Jang, J. Lee, H. Park, H. Choi and T. Kim. 2015. Starter cultures for kimchi fermentation. *Journal of Microbiology and Biotechnology* 25: 559-568.

Marsh, A.J., C. Hill, R.P. Ross and P.D. Cotter. 2014. Fermented beverages with health-promoting potential: Past and future perspectives. *Trends in Food Science and Technology* 38: 113-124.

Martín, M.J., F. Lara-Villoslada, M.A. Ruiz and M.E. Morales. 2015. Microencapsulation of bacteria: A review of different technologies and their impact on the probiotic effects. *Innovative Food Science and Emerging Technology* 27: 15-25.

Martins, E.M.F., A.M. Ramos, E.S.L. Vanzela, P.C. Stringheta, C.L. de Oliveira Pinto and J.M. Martins. 2013. Products of vegetable origin: A new alternative for the consumption of probiotic bacteria. *Food Research International* 51: 764-770.

Mestry, A.P., A.S. Mujumdar, and B.N. Thorat. 2011. Optimization of spray-drying of an innovative functional food: fermented mixed juice of carrot and watermelon. *Drying Technology* 29: 1121-1131.

Molin, G. 2001. Probiotics in foods not containing milk or milk constituents, with special reference to *Lactobacillus plantarum* 299v. *American Journal of Clinical Nutrition* 73: 380S-385S.

Nagpal, R., A. Kumar and M. Kumar. 2012. Fortification and fermentation of fruit juices with probiotic *lactobacilli*. *Annals of Microbiology* 62: 1573-1578.

Nazzaro, F., F. Fratianni and R. Coppola. 2009. Fermentative ability of alginate-prebiotic encapsulated *Lactobacillus acidophilus* and survival under simulated gastrointestinal conditions. *Journal of Functional Foods* 1: 319-323.

Nazzaro, F., F. Fratianni, A. Sada and P. Orlando. 2008. Synbiotic potential of carrot juice supplemented with *Lactobacillus* spp. and inulin or fructooligosaccharides. *Journal of the Science of Food and Agriculture* 88: 2271-2276.

Oum, Y.R. 2005. Authenticity and representation: cuisines and identities in Korean-American diaspora. *Postcolonial Studies* 8: 109-125.

Papadimitriou, K., G. Zoumpopoulou, B. Foligné, V. Alexandraki, M. Kazou, B. Pot and E. Tsakalidou. 2015. Discovering probiotic microorganisms: *In vitro, in vivo*, genetic and omics approaches. *Frontiers in Microbiology* 6: 58.

Pereira, A.L.F., T.C. Maciel and S. Rodrigues. 2011. Probiotic beverage from cashew apple juice fermented with *Lactobacillus casei*. *Food Research International* 44: 1276-1283.

Pereira, A.L.F., F.D.L. Almeida, A.L.T. Jesus, and S. Rodrigues. 2013. Storage stability of probiotic beverage from cashew apple juice. *Food and Bioprocess Technology* 6: 3155-3165.

Pereira, A.L.F., F.D.L. Almeida, M.A. Lima, J.M.C. da Costa and S. Rodrigues. 2014. Spray-drying of probiotic cashew apple juice. *Food and Bioprocess Technology* 7: 2492-2499.

Plengvidhya, V., F. Breidt, H, Fleming. 2004. Use of RAPD-PCR as a method to follow the progress of starter cultures in sauerkraut fermentation. *International Journal of Food Microbiology* 93: 287-296.

Prado, F.C., J.L. Parada, A. Pandey and C.R. Soccol. 2008. Trends in non-dairy probiotic beverages. *Food Research International* 41: 111-123.

Pulido, R.P., N. Benomar, M.M. Cañamero, H. Abriouel and A. Gálvez. 2012. Fermentation of caper products. Pp. 201-208. *In*: Hui, Y.H. (Eds.). *Handbook of Plant-based Fermented Food and Beverage Technology*, second ed. CRC Press. Boca Raton, London, New York.

Pulido, R.P., N.B. Omar, H. Abriouel, R.L. López, M.M. Cañamero and A. Gálvez. 2005. Microbiological study of lactic acid fermentation of caper berries by molecular and culture-dependent methods. *Applied and Environmental Microbiology* 71: 7872-7879.

Randazzo, C.L., A. Todaro, A. Pino, I. Pitino, O. Corona, A. Mazzaglia and C. Caggia. 2014. Giarraffa and Grossa di Spagna naturally fermented table olives: Effect of starter and probiotic cultures on chemical, microbiological and sensory traits. *Food Research International* 62: 1154-1164.

Reina, L.D., I.M. Pérez-Díaz, F. Breidt, M.A. Azcarate-Peril, E. Medina and N. Butz. 2015. Characterization of the microbial diversity in yacon spontaneous fermentation at 20°C. *International Journal of Food Microbiology* 203: 35-40.

Rodríguez-Gómez, F., V. Romero-Gil, J. Bautista-Gallego, P. García-García, A. Garrido-Fernández and F.N. Arroyo-López. 2014a. Production of potential probiotic Spanish-style green table olives at pilot plant scale using multifunctional starters. *Food Microbiology* 44: 278-287.

Rodríguez-Gómez, F., V. Romero-Gil, P. García-García, A. Garrido-Fernández and F.N. Arroyo-López. 2014b. Fortification of table olive packing with the potential probiotic bacteria *Lactobacillus pentosus*TOMC-LAB2. *Frontiers in Microbiology* 5: 467.

Rössle, C., M.A.E. Auty, N. Brunton, R.T. Gormley and F. Butler. 2010. Evaluation of fresh-cut apple slices enriched with probiotic bacteria. *Innovative Food Science and Emerging Technologies* 11: 203-209.

Russo, P., M.L.V. de Chiara, A. Vernile, M.L. Amodio, M.P. Arena, V. Capozzi, S. Massa and G. Spano. 2014. Fresh-cut pineapple as a new carrier of probiotic lactic acid bacteria. *BioMed Research International* 2014: 309183.

Russo, P., P. López, V. Capozzi, P. Fernández de Palencia, M.T. Dueñas, G. Spano and D. Fiocco. 2012. Beta-glucans improve growth, viability and colonization of probiotic microorganisms. *International Journal of Molecular Sciences* 13: 6026-6039.

Sabatini, N., M.R. Mucciarella and V. Marsilio. 2008. Volatile compounds in uninoculated and inoculated table olives with *Lactobacillus plantarum* (*Oleaeuropaea* L., cv. Moresca and Kalamata). *LWT-Food Science and Technology* 41: 2017-2022.

Sagdic, O., I. Ozturk, N. Yapar and H. Yetim. 2014. Diversity and probiotic potentials of lactic acid bacteria isolated from gilaburu, a traditional Turkish fermented European cranberrybush (*Viburnum opulus* L.) fruit drink. *Food Research International* 64: 537-545.

Satish, K.R., P. Kanmani, N. Yuvaraj, K.A. Paari, V. Pattukumar and V. Arul. 2013. Traditional Indian fermented foods: A rich source of lactic acid bacteria. *International Journal of Food Science and Nutrition* 64: 415-28.

Semjonovs, P., I. Denina, A. Fomina, L. Sakirova, L. Auzina, A. Patetko and D. Upite. 2013. Evaluation of *Lactobacillus reuteri* strains for pumpkin (*Cucurbita pepo* L.) juice fermentation. *Biotechnology* 12: 202-208.

Semjonovs, P., L. Shakizova, I. Denina, E. Kozlinskis and D. Unite. 2014. Development of a fructan-supplemented synbiotic cabbage juice beverage fermented by *Bifidobacterium lactis* BB$_{12}$. *Research Journal of Microbiology* 9: 129-141.

Sheehan, V.M., P. Ross and G.F. Fitzgerald. 2007. Assessing the acid tolerance and the technological robustness of probiotic cultures for fortification in fruit juices. *Innovative Food Science and Emerging Technologies* 8: 279-284.

Shukla, R. and A. Goyal. 2014. Probiotic Potential of *Pediococcuspentosaceus* CRAG$_3$: A new isolate from fermented cucumber. *Probiotics and Antimicrobial Proteins* 6: 11-21.

Singh, A.K. and A. Ramesh. 2008. Succession of dominant and antagonistic lactic acid bacteria in fermented cucumber: insights from a PCR-based approach. *Food Microbiology* 25: 278-287.

Sisto, A. and P. Lavermicocca. 2012. Suitability of a probiotic *Lactobacillus paracasei* strain as a starter culture in olive fermentation and development of the innovative patented product 'probiotic table olives'. *Frontiers in Microbiology* 3: 174.

Swain, M.R., M. Anandharaj, R.C. Ray and R. Praveen Rani. 2014. Fermented Fruits and Vegetables of Asia: A Potential Source of Probiotics. *Biotechnology Research International* 2014: 250-424.

Tamang, J.P. and D. Samuel. 2010. Dietary Cultures and Antiquity of Fermented Foods and Beverages. Pp. 1-41. *In*: J.P. Tamangand K Kailasapathy (Eds.). *Fermented Foods and Beverages of the World*. CRC Press. Boca Raton, London, New York.

Tamang, J.P. and K. Kailasapathy. 2010. *Fermented Foods and Beverages of the World*. CRC Press. Boca Raton, London, New York.

Tamang, J.P., N. Tamang, S. Thapa, S. Dewan, B. Tamang, H. Yonzan, A.K. Rai, R. Chettri, J. Chakrabarty and N. Kharel. 2012. Microorganisms and nutritional value of ethnic fermented foods and alcoholic beverages of northeast India. *Indian Journal of Traditional Knowledge* 11: 7-25.

Tamminen, M., T. Joutsjoki, M. Sjöblom, M. Joutsen, A. Palva, E. Ryhänen and V. Joutsjoki. 2004. Screening of lactic acid bacteria from fermented vegetables by carbohydrate profiling and PCR–ELISA. *Letters in Applied Microbiology* 39: 439-444.

Vergara, C.M.d.A.C., T.L. Honorato, G.A. Maia and S. Rodrigues. 2010. Prebiotic effect of fermented cashew apple (*Anacardium occidentale* L) juice. *LWT-Food Science and Technology* 43: 141-145.

Wang, C.Y., C.C. Ng, H. Su, W.S. Tzeng and Y.T. Shyu. 2009. Probiotic potential of noni juice fermented with lactic acid bacteria and bifidobacteria. *International Journal of Food Sciences and Nutrition* 60: 98-106.

Wouters, D., S. Grosu-Tudor, M. Zamfir and L. De Vuyst. 2013a. Bacterial community dynamics, lactic acid bacteria species diversity and metabolite kinetics of traditional Romanian vegetable fermentations. *Journal of the Science of Food and Agriculture* 93: 749-760.

Wouters, D., S. Grosu-Tudor, M. Zamfir and L. De Vuyst. 2013b. Applicability of *Lactobacillus plantarum* IMDO 788 as a starter culture to control vegetable fermentations. *Journal of the Science of Food and Agriculture* 93: 3352-3361.

Xiong, T., Q. Guan, S. Song, M. Hao and M. Xie. 2012. Dynamic changes of lactic acid bacteria flora during Chinese sauerkraut fermentation. *Food Control* 26: 178-181.

Yang, H., H. Zou, C. Qu, L. Zhang, T. Liu, H. Wu and Y. Li. 2014. Dominant microorganisms during the spontaneous fermentation of suan cai, a Chinese fermented vegetable. *Food Science and Technology Research* 20: 915-926.

Yoon, K.Y., E.E. Woodams and Y.D. Hang. 2004. Probiotication of tomato juice by lactic acid bacteria. *Journal of Microbiology* 42: 315-318.

Yoon, K.Y., E.E. Woodams and Y.D. Hang. 2005. Fermentation of beet juice by beneficial lactic acid bacteria. *LWT-Food Science and Technology* 38: 73-75.

Yoon, K.Y., E.E. Woodams and Y.D. Hang. 2006. Production of probiotic cabbage juice by lactic acid bacteria. *Bioresource Technology* 97: 1427-1430.

Yu, Z., X. Zhang, S. Li, C. Li, D. Li and Z. Yang. 2013. Evaluation of probiotic properties of *Lactobacillus plantarum* strains isolated from Chinese sauerkraut. *World Journal of Microbiology and Biotechnology* 29: 489-498.

Zalán, Z., J. Hudáček, M. Tóth-Markus, E. Husová, K. Solichová, F. Hegyi, M. Plocková, J. Chumchalová and A. Halász. 2011. Sensorically and anti-microbially active metabolite production of *Lactobacillus* strains on Jerusalem artichoke juice. *Journal of the Science of Food and Agriculture* 91: 672-679.

3

Nutritional Values and Bioactive Compounds in Lactic Acid Fermented Vegetables and Fruits

Manas R. Swain[1]* and Ramesh C. Ray[2]

1. Introduction

Worldwide, numerous varieties of vegetables and fruits for consumed by human but only a very portion is preserved through lactic acid fermentation to avoid perishability, improving palatability and enhancing nutritional quality. Fermented vegetables and fruits are part of the human diet throughout the world. In some places they constitute a minor 5 per cent of the dietary intake (Swain et al. 2014), and are often used during times of hardship, such as famine and drought. This ancient preparation and preservation technique has several advantages like increased digestibility, immunity, lowering of serum cholesterol, inhibiting carcinogenic compounds in the gastro-intestinal tract by reducing the fecal bacterial enzyme activity, etc. (Montet et al. 2014). During the process of lactic acid fermentation, carbohydrates are broken down by single or mixed forms of lactic acid bacteria (LAB) into lactic acid. The fermented products can be produced by using native knowledge, skill and traditional cultures, or by modern techniques employing defined starter culture (Bevilacqua et al., 2010). Fermented fruits and vegetables are vital foods of choice due to their excellent functional and nutritional properties aided by fermenting microorganisms, and novel bioactive compounds from these foods are an emerging area of research showing great promise.

Bioactive compounds may be described as 'food-derived components that in addition to their nutritional value exert a physiological effect on the body (Vermeirssen et al. 2014). Most lactic acid fermenting microorganisms produce organic acids, such as lactate, acetate, propionate and butyrate,

[1] Department of Biotechnology, Indian Institute of Technology- Madras, Chennai-600036, Tamil Nadu, India.

[2] ICAR-Central Tuber Crops Research Institute (Regional Center), Bhubaneswar 751019, Odisha, India.

* Corresponding author: manas.swain@gmail.com

exerting anti-microbial compounds. Moreover, LAB are the predominant microbes in the fermented food products that produce various low-molecular-carbon mass compounds, such as hydrogen dioxide, carbon dioxide, diacetyl (2, 3-butanedione), acetaldehyde, etc. and high-molecular-mass compounds such as anti-microbial peptides and bacteriocins (Ray and Joshi 2014). Yeasts and moulds are associated with biosynthesis of enzymes, antioxidants and melatonin (Mas et al. 2014). An overview of bioactive compounds synthesized by microorganisms in fermented vegetables and fruits is depicted in Fig. 1.

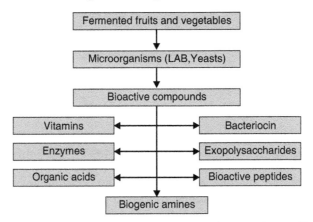

Fig. 1: Bioactive Compounds from the Fermented Fruits and Vegetables

In this chapter, the nutritional values of fermented vegetables and important bioactive compounds produced by fermenting microorganisms are discussed.

2. Lactic Acid Fermentation of Vegetables and Fruits

Traditionally, fermented vegetables and fruits are produced by considering natural, resident microflora present on the raw material. This practice is found to be effective when the dominant microflora (i.e., LAB) is used as the starter during the process of fermentation. Starter culture can be defined as microbial preparation of a large number of cells of single or a mixture of microorganisms. The starter culture is added to the raw material prior to fermentation for accelerating the process and for the dominance of a particular microflora on the raw material to avoid further pathogen contamination (Holzapfel 2002). In general, LAB found in the raw materials act as starter culture during vegetable fermentation. Vegetables are found to be an excellent source of carbohydrates that are converted into lactic acid. During vegetable fermentation, the 2–5 per cent brine solutions are added to provide hypertonic conditions that inhibit the formation of spoilage-causing organisms during and after the fermentation process. As per the climatic condition, raw material composition and fermentation nature, salts may or may not be added to the fermentation process. Salted vegetables that are fermented include sauerkraut, kimchi and gherkin (Breidt et al. 2013). There are several non-salted, and acidic fermented vegetables that include gundruk, khalpi, sinki, ingiangang, etc. (Tamang and Sarkar 1993, Ray and Panda 2007, Rhee et al. 2011, Montet et al. 2014).

3. Health and Nutritional Aspects of Lactic Acid Fermented Vegetables and Fruits

Lactic acid fermented vegetables have several health-related nutritional benefits besides possessing the ability to preserve. These are discussed below in brief:

3.1. Food Preservation

Nowadays, there are several methods of preserving fresh fruits and vegetables and these include drying, freezing, canning and pickling. But, many of these options are not suitable for small-scale or household use in under- developed and developing countries (Steinkraus 2002). For instance, small-scale canning of vegetables may cause food safety problems and contamination by food-borne pathogens, such as *Listeria monocytogenes* and *Escherichia coli* (Reina et al. 2005). The main aim of lactic fermentation is to preserve food, increase its shelf-life and improve food quality and palatability (Ray and Sivakumar 2009).

Strains of *Lactobacillus plantarum* and *Pediococcus pentosaceus*, which were isolated from various pickled vegetables, showed an antagonistic effect towards *Salmonella* invasion (Chiu et al. 2007). Similarly, strains of LAB isolated from gundruk, sinki, khalpi, inziangsang and fermented bamboo shoot products had the capacity to inhibit *Staphylococcus aureus* and to adhere to mucus secreting HT29 MTX cells (Tamang et al. 2009).

3.2. Removal of Anti-Nutritional Factors

Many vegetables and fruits contain naturally-occurring toxins and anti-nutritional compounds (Drewnowski and Gomez-Carneros 2000). For example, cassava roots contain two cyanogenic glucosides, linamarin and lotaustralin (Ray and Ward 2006). When the roots are naturally fermented by a mixed population of yeasts (*Saccharomyces cerevisiae* and *Candida* spp.) and LAB (*Lactobacillus*, *Leuconostoc* and *Pediococcus*), the cyanogen level is reduced drastically (Kostinek et al. 2005). Likewise, LAB reduce the toxic elements in African locust beans and leaves of *Cassava obtusifolia* during preparation of 'kawal', a Sudanese food (Dirar 1993). Strains of LAB isolated from ethnic fermented vegetables of the Himalayas (e.g., gundruk, sinki, khalpi, inziangsang and fermented bamboo shoot products) had the capacity to degrade anti-nutritive factors (Tamang et al. 2009).

3.3. Mineral and Vitamin Preservation

The micronutrient availability is enhanced in lactic fermented vegetables because of significant reduction in the phytase enzyme. Iron bioavailability was reported to be higher in lacto-pickled carrots (Rakin et al. 2004), beetroot (Yoon et al. 2004) and sweet potato (Panda et al. 2007).

Kimchi is the name given to various traditional fermented vegetables, which are emblematic of the Korean culture. Kimchi is mainly manufactured with Chinese cabbage (*Brassica pekinensis*) and radish, but other seasoning ingredients such as garlic, green onion, ginger, red pepper, mustard, parsley, fermented seafood (jeotgal), carrot and salt are also used (Jung et al. 2011). Due to its nutritional properties, kimchi was recently included in the list of the top five 'world's healthiest foods' (http://eating.health.com/2008/02/01/worlds-healthiest-foods-kimchi-korea/). These beneficial effects are attributed to the functional components (vitamins, minerals, fiber and phytochemicals) (Lee et al. 2011).

3.4. Improvement of Food Digestibility

LAB contain various intracellular and extracellular food digestive enzymes i.e. α-amylase (Panda and Ray 2016), pectinase (Kunji et al. 1996) and proteinase (Shurkhno et al. 2005). These enzymes aid in improving the digestibility of fermented fruits and vegetables. Proteases digest vegetable proteins during fermentation and some indigestible compounds, such as sulphur compounds (aliin, allicin, ajoene, allylpropl, trisulfide, sallylcysteine, vinyldithiine and S-allylmercaptocystein) in garlic or onion, improving their digestibility (Di Cagno et al. 2013). Single-culture lactic fermentation (*Lactobacillus casei*, *Lb. plantarum*) and sequential culture fermentation (*Saccharomyces boulardii* + *Lb. casei*; *S. boulardii* + *Lb. plantarum*) drastically reduced the content of phytic acid, polyphenols and trypsin inhibitors of a food mixture containing rice flour, green paste and tomato pulp, while significantly improving the *in vitro* digestibility of starch and protein (Sindhu and Khetarpal 2002).

3.5. Enhancement of Anti-oxidant Activity

Two strains (POM1 and C2) or LP09 of *Lb. plantarum*, which were previously isolated from tomatoes and carrots and another commercial strain of *Lb. plantarum* (LP09), were selected to singly ferment (30°C for 120 hours) pomegranate juice. Pomegranate juice was further stored at 4°C for 30 days. Filtered pomegranate juice, without added starters (un-started pomegranate juice) was used as the control. The concentration of total polyphenolic compounds and antioxidant activity was highest for started pomegranate juice, with some differences that depended on the starter used. Fermentation increased the concentration of ellagic acid and enhanced the anti-microbial activity. Fermented pomegranate juice scavenged the reactive oxygen species generated by H_2O_2 (hydrogen peroxide) and modulated the synthesis of immune-mediators from peripheral blood mononuclear cells. Un-started and fermented pomegranate juice inhibited the growth of K562 tumor cells (Filannino et al. 2013). Similar findings were reported by Di Cagno et al. (2011) with fermented red and green smoothies prepared from fruits like blackberries, prunes, kiwifruits, papaya and fennels with selective inoculation of LAB (*Weissella cibaria, Lb. plantarum, Lactobacillus* spp. and *Lb. pentosus*). Polyphenolic compounds especially ascorbic acid, were better preserved in started red and green smoothies compared to un-started samples. This reflected on the free radical scavenging activity.

3.6. Other Functional Properties

Other than the human gastrointestinal tract, raw fruits and some vegetables harbour functional probiotic LAB strains that possess resistance to gastric acidity and bile toxicity, and the capacity of adhesion to the gut epithelium, to hydrolyze nutritional constituents, which cannot be metabolized by the host (e.g. fructooligosaccharides, or FOS), and to synthesize anti-microbial substances (Di Cagno et al. 2013). In this context, these food matrices possess intrinsic chemical and physical parameters that, for some traits, mimic those of the human gastrointestinal tract, such as the extremely acidic environment, buffering capacity, high concentration of indigestible nutrients (fiber, inulin and FOS) and anti-nutritional factors (tannins and phenols) (Rossi et al. 2005, Di Cagno et al. 2013). In most cases, the autochthonous microbiota of fruits and vegetables colonizes and adheres to surfaces, inciting antagonistic activity towards spoilage and pathogenic microorganisms. A large number of autochthonous LAB, previously isolated from carrots, French beans, cauliflower, celery, tomatoes and pineapples, were assayed for functional properties (Vitali et al. 2012). Strains of *Lb. plantarum* particularly showed high survival under simulated gastric and intestinal conditions, stimulated a large number of immune-mediators by peripheral blood mononuclear cells, including cytokines with pro- and anti-inflammatory activities, strongly adhered to Caco-2 cell, used FOS as the only carbon source and inhibited pathogens from human sources. Similar results were found for autochthonous strains, which were isolated from kimchi and belonged to the species *Lb. plantarum* and *Lb. sakei* (Lee et al. 2011). For these strains, the capacity to low cholesterol was found. Functional properties were also attributed to LAB strains, which were isolated from Nozawana-zuke, a traditional Japanese pickle made with *Brassica campestris* L. var. *rapa* (Kawahara and Otani 2006).

3.7. Disadvantages of Lactic Acid Fermentation

Microbial decarboxylation of amino acids results in the formation of biogenic amines that can be found in fermented foods. These amino-compounds impart an unpleasant flavor to the product or can be toxic (Garcia-Ruiz et al. 2011). Ingestion of certain amines by humans can cause headache, fever and vomiting — symptoms similar to those due to microbial food poisoning. Histidine, ornithine, tyrosine and lysine are the main amino acids that can be decarboxylated into biogenic amines, such as histamine, putrescine, tyramine and cadaverine, respectively (Sahu et al. 2016). Some bacteria commonly involved in lactic acid fermentation of vegetables, such as *Leuconostoc mesenteroides*, have been found to contain the amino acid, decarboxylase, that induces biogenic amine formation (Arena et al. 1999). Conversely, other bacteria such as *Lb. plantarum* and *Pediococcus* can limit biogenic amines through the production of amine oxidase enzymes (Garcia-Ruiz et al. 2011).

The determination of biogenic amines in food products is critical to assessing the potential health risks before consumption. Biogenic amines are reported in various fermented vegetables, such as sauerkraut and kimchi (Montet et al. 2014). Fermentation of cabbage with certain lactic starters, such as *Lb. casei* subsp. *casei*, *Lb. plantarum*, and *Lb. curvatus* reduce the biogenic amine content of sauerkraut (Rabie et al. 2011).

4. Probiotic Role of LAB

Human gut microflora may contain 10^{13}–10^{14} microorganisms that play an important role on human health. In addition, some food products contain live bacteria that translate through the human gastrointestinal tract to impart beneficial effects on health (Ray et al. 2014). Species of *Lactobacillus* and *Bifidobacterium* are considered as probiotics and are claimed to provide benefits when consumed in plenty and regularly (Swain et al. 2014). Most acidic fermented vegetables are known to be reserves for obtaining probiotic LAB culture (Di Cagno et al. 2013, Swain et al. 2014). These probiotic strains play a crucial role as these bacteria can survive and colonize the gastrointestinal tract to confer functional properties and health benefits (Suskovic et al. 2010). Several strains of non-hemolytic LAB, such as *Lb. pentosus*, *Lb. plantarum*, and *Lb. paracasei* were isolated from fermented food products (Patel and Shah 2016). These LAB produce several bioactive compounds during the fermentation process, such as organic acids, bacteriocins, vitamins, exopolyssacharides, enzymes and so on that directly or indirectly improve the consumer's health in the long run. Some bioactive compounds instantly or directly impact the health status of the consumer (Patel and Shah 2016, Ramachandran 2016). As per the research findings, it not only improves the health status but also the immunity, curing several chronic diseases, such as cancer and hepatitis, etc. (Patel and

Shah 2014). Table 1 shows the functional activities of some autochthonous LAB (isolated *in situ* from vegetable matrices) (Di Cagno et al. 2013).

5. Bioactive Components in Fermented Vegetables

The following bioactive compounds are found in fermented vegetables:

5.1. *Organic Acids*

The LAB initiate rapid and adequate acidification of raw materials through the production of various organic acids in food during growth and cell multiplication from *in situ* carbohydrates. In fermented foods, the souring effect is primarily due to the fermentative conversion of carbohydrates to organic acids (lactic and acetic acids) with a concomitant lowering of the pH of food (Montet et al. 2014). This is an important characteristic that leads to an increased shelf-life and safety of the final product (Ray and Panda 2007, Swain et al. 2014). Apart from organic acids generated in food matrices, other metabolites that are produced are: ethanol, bacteriocins, aroma compounds, exopolysaccharides, enzymes such as amylases and pectinases, vitamins and minerals (Ray and Joshi 2014). Some of these metabolites lead to anti-microbial activity and inhibition of spoilage-causing and/or disease-causing bacteria, thus helping to maintain and preserve the nutritive qualities of foods for an extended shelf-life (O'Sullivan et al. 2002). Phenyl lactic acid is recognized as the major factor responsible for anti-fungal activity that prolongs the shelf-life of fermented foods. The inhibitory properties of phenyl lactic acid have been reported against moulds isolated from bakery products, flours and cereals, including some mycotoxigenic species such as *Aspergillus ochraceus*, *Penicillium verrucosum* and *P. citrinum*; and some bacterial contaminants, namely *Listeria* spp., *Staphylococcus aureus* and *Enterococcus faecalis* (Dieuleveux and Gueguen 1998, Lavermicocca et al. 2003).

5.2. *Bacteriocins*

Bacteriocins produced by LAB can be defined as biologically active proteins or protein complexes (protein aggregates, lipocarbohydrate proteins, glycoproteins, etc.) that display bactericidal action exclusively towards Gram-positive bacteria and particularly towards closely-related species. Bacteriocins are produced by strains of *Lactococcus*, *Lactobacillus*, *Pediococcus* (Klaenhammer, 1988, Piard and Desmazeaud 1992), *Leuconostoc*, *Carnobacterium*, *Streptococcus*, *Enterococcus* and *Bifidobacterium* (Meghrous et al. 1990). Their synthesis is ribosomally regulated in the cell; however, the precise mechanisms are still not clear. Most bacteriocins have a narrow-spectrum while some have wide-spectrum activities. Nisin is the most commercially used bacteriocin in the food industry. Other bacteriocins, including diplococcin, plantaricin, acidophilin and bulgaricin, have not

Table 1: Functional Activities and Bioactive Compounds of Autochthonous Lactic Acid Bacteria Isolated from Raw and Fermented Vegetables

Microorganisms	Food matrix	Bio-active compounds and Functionalactivities	References
Lb. fermentum	Nozawana-zuke; traditional Japanese pickle	Immuno-enhancing effect	Kawahara and Otani (2006)
Pd. pentosaceus MP12, *Lb. plantarum* LAP6	Pickled cabbage	Adhesion on mouse epithelial cells; inhibition on pathogenic bacteria (*Salmonella*)	Chiu et al. (2007)
Lb. plantarum IB2	Ethnic fermented vegetables from Asia (gundruk, sinki, khalpi, inziangsang, and fermented bamboo)	Anti-microbial activity towards *Staphylococcus aureus* and to adhere to mucus secreting HT29 MTX cells	Tamang et al. (2009)
Lb. plantarum	Various raw fruits and vegetables	Stimulation of immune-mediators, adhesion to human intestinal Caco-2 cells, inhibition of *Escherichia coli* and *Bacillus megaterium*	Vitali et al. (2012)
Lc. lactis subsp. *lactis*	Various raw and fermented vegetables	Growth in the presence of bile salts, removal of cholesterol from the growth media	Kimoto et al. (2004)
Lb. plantarum c16, c19	Olives	Growth in the presence of bile salts, adhesion to IPEC-J2 cells, inhibition of *E. coli*	Bevilacqua et al. (2010)
Lb. casei, *Lb. plantarum*, *S. boulardii* + *Lb. casei*, *S. boulardii* + *Lb. plantarum*	Tomato pulp + rice flour + green paste	Reduced content of phytic acid and trypsin inhibitors	Sindhu and Khetrapal, (2002)
Lb. plantarum	Pomegranate juice	Increase in polyphenols and anti-oxidant activity	Filannino et al. (2013)
W. cibaria, *Lb. plantarum*, *Lb. pentosus*	Smoothies prepared from black-berries, prunes, kiwifruits, papaya	Increase in anti-oxidant activity	Di Cagno et al. (2011)
Lb. plantarum	Inziangsang (fermented leafy vegetable)	Bacteriocin production	Tamang et al. (2009)
LAB	Swiss chard, spinach	Bacteriocin production	Ponce et al. (2008)
LAB	Green vegetables	Exo-polysaccharide production	Grosu- Tudor and Zamfir (2014)
Lb. plantarum MTCC 1407	Sweet potato	Enzyme (amylase) production	Panda et al. (2007, 2008, 2009b)

yet been utilized commercially (Patel and Shah 2016). It is notable that the majority of bacteriocins produced by the strains of *Lb. acidophilus* have a broad spectrum of activity against food-borne pathogens and spoilage bacteria. Reuterin, chemically identified as 3-hydroxy propionaldehyde, is an example of a non-protein bacteriocin produced by certain strains of *Lb. reuteri*. It is water-soluble, stable over a wide pH range and has a broad spectrum of activity against most bacteria, yeasts and moulds, thus making it a prospective food preservative (Rattanachaikunsopon and Phumkhachorn 2010).

Bacteriocins produced by LAB and isolated from dairy, meat and fish products are reported in literature (Campos et al. 2006), but studies on bacteriocin-producing LAB isolated from organic vegetables are very scarce. Ponce et al. (2008) report that bacteriocin-like compounds can be isolated from fermented leafy vegetables, such as swiss chard and spinach. In another study, Tamang et al. (2009) isolated 84 strains of LAB from ethnic fermented vegetables of the Himalayan region. Most LAB strains showed anti-microbial activities against the used indicator strains; however, only *Lb. plantarum* IB2 (BFE 948) isolated from inziangsang, a fermented leafy vegetable product, produced a bacteriocin against *Staphylococcus aureus* S1.

5.3. Vitamins

Vitamins are involved in many essential functions of the body, like cell metabolism, synthesis of nucleic acids and antioxidant activities. Inherently, vitamins cannot be synthesized by humans and animals through natural means. However, several species of bacteria, yeasts, fungi and algae may serve to produce folic acid, vitamin B_{12} or cobalamin, vitamin K_2 or menaquinone, riboflavin, thiamine and other essential vitamins (LeBlanc et al. 2011).

Folic acid is important for many metabolic reactions, such as the biosynthesis of DNA and RNA and the inter-conversions of amino acids. Also, folate possesses antioxidant competence that protects the genome by preventing free radical attack (LeBlanc et al. 2007). Dietary folate is essential for humans, since it cannot be synthesized by the mammalian cells. Folate may be found in legumes, leafy greens, some fruits and vegetables, in liver and fermented dairy products. Folate deficiencies are associated with a variety of disorders like osteoporosis, Alzheimer's disease, coronary heart disease and increased risk of breast and colorectal cancer as indicated from epidemiological studies (Rossi et al. 2011). Currently the food industry is focusing on the strategy to select and employ folate-producing probiotic strains for fermentation with an elevated amount of 'natural' folate for the inherent tendency to provide desire health benefits without increasing the production costs (Rossi et al. 2011). Among LAB, many *Lactobacillus* spp. and *Lactococcus* spp. including *Lb. plantarum*,

Lb. bulgaricus, Ln. lactis, Streptococcus thermophilus and *Enterococcus* spp. have the ability to produce folate. On the other hand, some lactobacilli (*Lb. gasseri, Lb. salivarius, Lb. acidophilus* and *Lb. johnsonii*) used as both starter cultures and probiotics, cannot synthesize folate because they lack few specific genes involved in the folate biosynthesis (LeBlanc et al. 2007). *Bifidobacteria* are dynamic for folate production though it is strain-specific and depends on the medium. Optimization and selection of suitable growth conditions can result in high levels of folate per unit of biomass. The extracellular production of folate, vitamin B_{12} and thiamine in cultures of LAB isolated from *nukazuke*, a traditional Japanese pickle, and the relationship between the vitamin production and such properties of LAB as tolerance to salts and ethanol, were studied (Masuda et al. 2012). Among the 180 isolates of LAB, two strains of *Lb. sakei* and a strain of *Lb. plantarum* extracellularly produced high levels of folate (about 100 µg/L). A strain of *Lb. coryniformis* and one of *Lb. plantarum* produced about 2 µg/L of vitamin B_{12}, although the level was not high.

Production of thiamin and nicotinic acid (vitamin B_3) was found to be strain-dependent as all tested strains of *Bifidobacterium bifidum, Bf. longum, Bf. infantis,* and some of *Bf. breve* were high producers, while *Bf. adolescentis, Bf. longum* and others lacked this ability (Deguchi et al. 1985). Vitamin B_{12}, popularly known as cobalamin, is required for the metabolism of fatty acids, amino acids, nucleic acids and carbohydrates (Quesada-Chanto et al. 1994, Taranto et al. 2003).

Normally, vitamin K is present as phylloquinone (Vitamin K_1) in green plants and as menaquinone (K_2) produced by some intestinal bacteria like LAB, especially by various strains of genera *Lactococcus, Lactobacillus, Enterococcus, Leuconostoc* and *Streptococcus* (O'Connor et al. 2005, Cooke et al. 2006). Vitamin K deficiency is found to be associated with some clinical disorders like intracranial hemorrhage in newborn infants and possible bone fracture resulting from osteoporosis (LeBlanc et al. 2011). In this regard, LAB-producing menoquinone could be useful in supplementing vitamin K in humans (Morishita et al. 1999).

5.4. Bioactive Peptides

Bioactive peptides are short-chain peptides generally consisting of three to 20 amino acids and these are formed by degradation of the original protein by the fermenting LAB (Szwajkowska et al. 2011). These peptides have several health benefits, such as anti-hypertensive, anti-lipemic, anti-oxidative, anti-microbial, immune-modulating, osteoprotective and opiate effective (Möller et al. 2008, Patel and Shah 2016).

Some vegetables harbour LAB as natural flora, and some of these can produce bioactive peptides (Hebert et al. 2008). It is implicit that with lactic acid fermentation, these peptides are also available in fermented vegetables, though no literature is available to support this hypothesis.

5.5. *Exo-polysaccharides* (*EPS*)

The EPS produced by LAB can be divided into two categories: homo-polysaccharides, containing only one type of monosaccharides and hetero-polysaccharides, composed of repeating units, varying in size from di- to hepta-saccharides that could also contain non-sugar molecules. The production of homo-polysaccharides requires the presence of a specific substrate, such as sucrose, and the assembly of monosaccharide units takes place outside the bacterial cell (Cerning 1990). Most EPS-producing LAB belong to the genera, *Streptococcus*, *Lactococcus*, *Leuconostoc*, *Pediococcus* and *Lactobacillus* (Patel and Prajapati 2013).

Grosu-Tudor and Zamfir (2014) provided the information that LAB strains isolated from fermented vegetables, were able to produce large amounts of EPS with the potential for application in food biotechnology (i.e. to improve the rheological properties of fermented products).

5.6. *Enzymes*

LAB synthesize diverse types of enzymes that impact the compositional, processing and organolyptic properties along with the overall quality of fermented products. The enzymatic activity of the LAB, such as amylase, phytase, phosphatase, etc, isolated from fermented vegetables, has been reported (Tamang et al. 2009, Panda and Ray 2016). Various species of *Lactobacillus*, *Lactococcus*, *Leuconostoc*, *Pediococcus*, *Weissella* and *Bifidobacterium* are found to produce carbohydrate-degrading enzymes, like glucosidases, amylases and xylanases (Novik et al. 2006). LAB were found to produce higher percentages of α- and/or β-galactosidase in comparison to bifidobacteria (Alazzeh et al. 2009). In a series of publications, our research group demonstrated the application of amylase-producing probiotic strain of *Lb. plantarum* MTCC 1407 in lactic acid fermentation of sweet potato (Mohapatra et al. 2007, Panda and Ray 2007, 2008, Panda et al. 2007, 2009a, b).

6. Future Perspective

Nowadays the consumer's demand for organic foods with the minimum use of chemical preservatives is increasing; hence, fermented vegetables and fruits are referred to as functional foods or health-enhancing foods due the presence of in several bioactive compounds. Fermentation of vegetables with LAB increases the safety aspects as they served as an inimitable source for developing novel products and applications, especially those that can satisfy the consumer's demands for natural products and health benefits. It is well established that certain LAB and bifidobacteria have the ability to synthesize vitamins and bioactive peptides in food matrices, but not much information is available regarding the amount of produced. Recent advances dealing with fermenting microflora, particularly LAB and their functional ingredients, propose that we have yet to realize their full potential. LAB fermentation also decreases the anti-nutritional factors and toxic metabolites present in some vegetables. In addition, increasing vegetarianism and consumer's demand for alternatives to dairy-based probiotic foods, due to their high fat and cholesterol content or lactose content, may lead to an increased demand for fermented vegetables, which are equally beneficial in terms of health.

Acknowledgement

The authors are grateful Dr Nadine Zakhia-Rozis, Food Safety Team, CIRAD-PERSYST, UMR Qualisud, Montpellier Cedex 5, France, for critically reading the manuscript and offering valuable suggestions.

References

Alazzeh, A.Y., S.A. Ibrahim, D. Song, A. Shahbazi and A.A. Abu Ghazaleh. 2009. Carbohydrate and protein sources influence the induction of α-and β-galactosidases in *Lactobacillus reuteri*. *Food Chemistry* 117: 654-659.

Arena, M.E., F.M. Saguir and N.C. Manca de Nadra. 1999. Arginine, citruline and ornithine metabolism by lactic acid bacteria from wine. *International Journal of Food Microbiology* 52: 155-161.

Bevilacqua, A., C. Altieri, M.R. Corbo, M. Sinigaglia and L.I. Ouoba. 2010. Characterization of lactic acid bacteria isolated from Italian Bella di Cerignola table olives: Selection of potential multifunctional starter cultures. *Journal of Food Science* 75: 536-544.

Breidt, F., R.F. McFeeters, I. Pérez-Díaz and C.H. Lee. 2013. Fermented vegetables. Pp. 841-855. *In*: M.P. Doyle and L.R. Beuchat (Eds.). *Food Microbiology: Fundamentals and Frontiers*, 4th edn. ASM Press, Washington, DC.

Campos, A., O. Rodriguez, P. Calo-Mata, M. Prado and J. Barros-Velazquez. 2006. Preliminary characterization of bacteriocins from *Lactococcus lactis*, *Enterococcus faecium* and *Enterococcus mundtii* strains isolated from turbot (*Psetta maxima*). *Food Research International* 39: 356-364.

Cerning, J. 1990. Exocellular polysaccharides produced by lactic acid bacteria. *FEMS Microbiology Reviews* 87: 113-130.

Chiu, H.H., C.C. Tsai, H.Y. Hsih and H.Y. Tsen. 2007. Screening from pickled vegetables the potential probiotic strains of lactic acid bacteria able to inhibit the Salmonella invasion in mice. *Journal of Applied Microbiology* 104: 605-612.

Cooke, G., J. Behan and M. Costello. 2006. Newly identified vitamin K-producing bacteria isolated from the neonatal faecal flora. *Microbial Ecology Health Disease* 18: 133-138.

Deguchi, Y., T. Morishita and M. Mutai. 1985. Comparative studies on synthesis of water-soluble vitamins among human species of bifidobacteria. *Agricultural and Biological Chemistry* 49: 13-19.

Di Cagno, R., R. Coda, M. De Angelis and M. Gobbetti. 2013. Exploitation of vegetables and fruits through lactic acid fermentation. *Food Microbiology* 33: 1-10.

Di Cagno, R., G. Minervini, C.G. Rizzello, M. De Angelis and M. Gobbetti. 2011. Effect of lactic acid fermentation on antioxidant, texture, color and sensory properties of red and green smoothies. *Food Microbiology* 28: 1062-1071.

Dieuleveux, V. and M. Guéguen. 1998. Anti-microbial effects of D-3-phenyllactic acid on *Listeria monocytogenes* in TSB-YE medium, milk and cheese. *Journal of Food Protection* 61: 1281-1285.

Dirar, M. 1993. The Indigenous Fermented Foods of the Sudan. *CAB International.* Wallingford, UK.

Drewnowski, A. and C. Gomez-Carneros. 2000. Bitter taste, phytonutrients and the consumers: A review. *The American Journal of Clinical Nutrition* 72(6): 1424-1435.

Filannino, P., L. Azzi, I. Cavoski., O. Vincentini, G. Carlo, S. Rizzello, M. Gobbetti and R. Di Cagno. 2013. Exploitation of the health-promoting and sensory properties of organic pomegranate (*Punica granatum* L.) juice through lactic acid fermentation. *International Journal of Food Microbiology* 163: 184-192.

García-Ruiz, A., E.M. González-Rompinelli, B. Bartolomé and M.V. Moreno-Arribas. 2011. Potential of wine-associated lactic acid bacteria to degrade biogenic amines. *International Journal of Food Microbiology* 148: 115-120.

Grosu-Tudor, S.-S. and M. Zamfir. 2014. Exopolysaccharide production by selected lactic acid bacteria isolated from fermented vegetables. *Scientific Bulletin.* Series F. *Biotechnologies,* Vol. XVIII, 2014 ISSN 2285-1364, CD-ROM ISSN 2285-5521, ISSN Online 2285-1372, ISSN-L

Hebert, E.M., G. Mamone, G. Picariello, R.R. Raya, G. Savoy, P. Ferranti and F. Addeo. 2008. Characterization of the pattern of αs1- and β-casein breakdown and release of a bioactive peptide by a cell envelope proteinase from *Lactobacillus delbrueckii* subsp. lactis CRL 581. *Applied and Environmental Microbiology* 74: 3682-3689.

Holzapfel, W.H. 2002. Appropriate starter culture technologies for small-scale fermentation in developing countries. *International Journal of Food Microbiology* 75:197-212.

Jung, J.Y., S.H. Lee, J.M. Kim, M.S. Park, J.W. Bae, Y. Hahn, E.L. Madsen and C.O. Jeon. 2011. Metagenomic analysis of kimchi, a traditional Korean fermented food. *Applied and Environmental Microbiology* 77: 2264-2274.

Kawahara, T. and H. Otani. 2006. Stimulatory effect of lactic acid bacteria from commercially available Nozawana-zuke pickle on cytokine expression by mouse spleen cells. *Bioscience Biotechnology and Biochemistry* 70: 411-417.

Kimoto, H., M. Nomura, M. Kobayashi, T. Okamoto and S. Ohmomo. 2004. Identification and probiotic characteristic of *Lactococcus* strains from plant materials. *Japanese Agricultural Research Quarterly* 38 : 111-117.

Klaenhammer, T.R. 1988. Bacteriocins of lactic acid bacteria. *Biochemistry* 70: 337-379.

Kostinek, M., I. Specht, V.A. Edward, U. Schillinger, C. Hertel, W.H. Holzapfel and C.M.A.P. Franz. 2005. Diversity and technological properties of predominant lactic acid bacteria from fermented cassava used for the preparation of gari, a traditional African food. *Systematic and Applied Microbiology* 28(6): 527-540.

Kunji, E.R.S., I. Mierau, A. Hagting, B. Poolman and W.N. Kinings. 1996. The proteolytic systems of lactic acid bacteria. *Antonie Van Leeuwenhoek* 70: 181-221.

Lavermicocca, P., F. Valerio and A. Visconti. 2003. Anti-fungal activity of phenyllactic acid against moulds isolated from bakery products. *Applied and Environment Microbiology* 69: 634-640.

LeBlanc, J.G., G.S. de Giori, E.J. Smid, J. Hugenholtz and F. Sesma. 2007. Folate production by lactic acid bacteria and other food-grade microorganisms. *Communicating Current Research Education Topics and Trends in Applied Microbiology* 1: 329-339.

LeBlanc, J.G., J.E. Laino, M. Juarez del Valle, V. Vannini, D. van Sinderen, M.P. Taranto, G. Font de Valdez, G. Savoy de Giori and F. Sesma. 2011. B-group vitamin production by lactic acid bacteria – Current knowledge and potential applications. *Journal of Applied Microbiology* 111: 1297-1309.

Lee, H., H. Yoon, Y. Ji, H. Kim, H. Park and J. Lee. 2011. Functional properties of *Lactobacillus* strains from kimchi. *International Journal of Food Microbiology* 145: 155-161.

Mas, A., J.M. Guillamon, M.J. Torija, G. Beltran, A.B. Cerezo, A.M. Troncoso and M.C. Garcia-Parrilla. 2014. Bioactive compounds derived from the yeast metabolism of aromatic amino acids during alcoholic fermentation. *Biomedical Research International* 2014: 898045.

Masuda, M., M. Ide, H. Utsumi, T. Niiro, Y. Shimamura and M. Murata. 2012. Production potency of folate, vitamin B_{12} and thiamine by lactic acid bacteria isolated from Japanese pickles. *Bioscience, Biotechnology, and Biochemistry* 76: 2061-2067.

Meghrous, J., P. Eulogy, A. Junelles, J. Ballongue and H. Petitdemange. 1990. Screening of bifidobacteria strains for bacteriocin production. *Biotechnology Letters* 12: 575-580.

Mohapatra, S., S.H. Panda, S.K. Sahoo, P.S. Shiv Kumar and R.C. Ray. 2007. β- Carotene-rich sweet potato curd: Production, nutritional and proximate composition. *International Journal of Food Science and Technology* 42: 1305-1314.

Möller, N.P., K.E. Scholz-Ahrens, N. Roos and J. Schrezenmeir. 2008. Bioactive peptides and proteins from foods: Indication for health effects. *European Journal of Nutrition* 47: 171-182.

Montet, D., R.C. Ray and N. Zakhia-Rozis. 2014. Lactic acid fermentation of vegetables and fruits. Pp. 108-140. *In*: R.C. Ray and D. Montet (Eds.). *Microorganisms and Fermentation of Traditional Foods*. CRC Press, Boca Raton, Florida, USA.

Morishita, T., N. Tamura, T. Makino and S. Kudo. 1999. Production of menaquinones by lactic acid bacteria. *Journal of Dairy Sciences* 82: 1897-1903.

Novik G., A. Gamian J. Francisco and E.S. Dey. 2006. A novel procedure of the isolation of glycolipids from bifidobacterium adolescentis 94 BIM using super-critical carbon dioxide. *Journal of Biotechnology* 121: 555-562.

O' Sullivan, L., R.P. Ross and C. Hill. 2002. Potential of bacteriocin-producing lactic acid bacteria for improvements in food safety and quality. *Biochemistry* 84: 593-604.

O'Connor E.B., E. Barrett, G. Fitzgerald, C. Hill, C. Stanton and R.P. Ross. 2005. Production of vitamins, exopolysaccharides and bacteriocins by probiotic bacteria. *In*: A.Y. Tamime (Ed.). *Probiotic Dairy Products*. Blackwell Publishing, Oxford, UK.

Panda, S.H., M. Paramanick and R.C. Ray. 2007. Lactic acid fermentation of sweet potato (*Ipomoea batatas* L.) into pickles. *Journal of Food Processing and Preservation* 37(1): 83-101.

Panda, S.H. and R.C. Ray. 2007. Lactic acid fermentation of β-carotene rich sweet potato (*Ipomoea batatus* L.) into lacto-juice. *Plant Foods for Human Nutrition* 62: 65-70.

Panda, S.H. and R.C. Ray. 2008. Direct conversion of raw starch to lactic acid by *Lactobacillus plantarum* MTCC 1407 in semi-solid fermentation using sweet potato (*Ipomoea batatas* L.) flour. *Journal of Scientific and Industrial Research*, India 67: 531-537.

Panda, S.H. and R.C. Ray. 2016. Amylolytic lactic acid bacteria: Microbiology and technological interventions in food fermentation. Pp. 148-165. *In*: D. Montet and R.C. Ray (Eds.). *Fermented Foods: Biochemistry and Biotechnology*. CRC Press, Boca Raton, Florida, USA.

Panda, S.H., S.K. Naskar, P.S. Shivakumar and R.C. Ray. 2009b. Lactic acid fermentation of anthocyanin- rich sweet potato (*Ipomoea batatus* L.) into lacto-juice. *International Journal of Food Science and Technology* 44: 288-296.

Panda, S.H., S. Panda, P.S. Shiva Kumar and R.C. Ray. 2009a. Anthocyanin-rich sweet potato lacto-pickle: Production, nutritional and proximate composition. *International Journal of Food Science and Technology* 44: 445-455.

Patel, A. and J.B. Prajapati. 2013. Food and health applications of exopolysaccharides produced by lactic acid bacteria. *Advances in Dairy Research* 1: 107.

Patel, A. and N. Shah. 2014. Recent advances in anti-microbial compounds produced by food grade bacteria in relation to enhance food safety and quality. *Journal of Innovative Biology* 1(4): 189-194.

Patel, A. and N. Shah. 2016. Bio-active compounds of fermented foods. Pp. 283-303. *In*: D. Montet and R.C. Ray (Eds.). *Fermented Foods: Biochemistry and Biotechnology*. CRC Press, Boca Raton, Florida, USA.

Piard, J.C. and M. Desmazeaud. 1992. Inhibiting factors produced by lactic acid bacteria. 2. Bacteriocins and other anti-microbial substances. *Lait* 72: 113-142.

Ponce, A.G., M.R. Moreira, C.E. del Valle and S.I. Roura. 2008. Preliminary characterization of bacteriocin-like substances from lactic acid bacteria isolated from organic leafy vegetables. *LWT-Food Science and Technology* 41: 432-441.

Quesada-Chanto, A., A.S. Afschar and F. Wagner. 1994. Microbial production of propionic acid and vitamin B12 using molasses or sugar. *Applied Microbiology Biotechnology* 41: 378-383.

Rabie, M.A, H. Siliha, S. El-Saidy, A.A. El-Badawy and F.X. Malcata. 2011. Reduced biogenic amine contents in sauerkraut via addition of selected lactic acid bacteria. *Food Chemistry* 129: 1778-1782.

Rakin, M., J. Baras, M. Vukasinovic and M. Milan. 2004. The examination of parameters for lactic acid fermentation and nutritive value of fermented juice of beetroot, carrot and brewer's yeast autolysate. *Journal of the Serbian Chemical Society* 69(8-9): 625-634.

Ramachandran, L. 2016. Nutritional and therapeutic significance of protein-based bioactive compounds liberated by fermentation. Pp. 304-317. *In*: D. Montet and R.C. Ray (Eds.). *Fermented Foods: Biochemistry and Biotechnology*. CRC Press, Boca Raton, Florida, USA.

Rattanachaikunsopon, P. and P. Phumkhachorn. 2010. Potential of cinnamon (*Cinnamomum verum*) oil to control *Streptococcus iniae* infection in tilapia (*Oreochromis niloticus*). *Fish Sciences* 76: 287-293.

Ray, R.C. and V.K. Joshi. 2014. Fermented Foods: Past, present and future scenario. Pp. 1-36. *In*: R.C. Ray and D. Montet (Eds.). *Microorganisms and Fermentation of Traditional Foods*. CRC Press, Boca Raton, Florida, USA.

Ray, R.C. and S.H. Panda. 2007. Lactic acid fermented fruits and vegetables: An overview. Pp. 155-188. *In*: M.V. Palino (Ed.). *Food Microbiology Research Trends*. Nova Science Publishers Inc., Hauppauge, New York, USA.

Ray, R.C. and P.S. Sivakumar. 2009. Traditional and novel fermented foods and beverages from tropical root and tuber crops: Review. *International Journal of Food Science and Technology* 44: 1073-1087.

Ray, R.C. and O.P. Ward. 2006. Post-harvest microbial biotechnology of topical root and tuber crops. Pp. 345-396. *In*: R.C. Ray and O.P. Ward (Eds.). *Microbial Biotechnology in Horticulture*, vol. 1. Science Publishers, New Hampshire, USA.

Ray, R.C., A.F. El Sheikha and R. Sashikumar. 2014. Oriental fermented functional (probiotics) foods. Pp. 283-311. *In*: R.C. Ray and D. Montet (Eds.). *Microorganisms and Fermentation of Traditional Foods*. CRC Press, Boca Raton, Florida, USA.

Reina, L.D., F. Breidt, H.P. Fleming and S. Kathariou. 2005. Isolation and selection of lactic acid bacteria as biocontrol agents for non-acidified refrigerated pickles. *Journal of Food Science* 70(1): 7-11.

Rhee, S.J., J.E. Lee and C.H. Lee. 2011. Importance of lactic acid bacteria in Asian fermented foods. *Microbial Cell Factory* 10 (Suppl. 1): 1-13.

Rossi, M., A. Amaretti and S. Raimondi. 2011. Folate production by probiotic bacteria. *Nutrition* 3: 118-134.

Rossi, M., C. Corradini, A. Amaretti, M. Nicolini, A. Pompei, S. Zanoni and D. Matteuzzi. 2005. Fermentation of fructooligosaccharides and inulin by Bifidobacteria: A comparative study of pure and fecal cultures. *Applied and Environmental Microbiology* 71: 6150-6158.

Sahu, L., S.K. Panda, S. Paramithiotis, N. Zdolec and R.C. Ray. 2016. Biogenic amines in fermented foods. Pp. 318-332. *In*: D. Montet and R.C. Ray (Eds.). *Fermented Foods: Biochemistry and Biotechnology*. CRC Press, Boca Raton, Florida, USA.

Shurkhno, R.A., R.G. Gareev, A.G. Abul'khanov, S.Z. Validov, A.M. Boronin and R.P. Naumova. 2005. Fermentation of a high-protein plant biomass by introduction of lactic acid bacteria. *Prikladnaia Biohimia I Mikrobiologiia* 41(1): 79-89.

Sindhu, S.C. and N. Khetarpal. 2002. Effect of probiotic fermentation on anti-nutrients and *in vitro* protein and starch digestibilities of indigenously developed RWGT food mixture. *Nutrition and Health* 16: 173-181.

Steinkraus, K.H. 2002. Fermentations in world food processing. *Comprehensive Reviews in Food Science and Food Safety* 1: 23-32.

Suskovic, J., B. Kos, J. Beganovic, A.L. Pavunc, K. Habjanic and S. Matosic. 2010. Anti-microbial activity — The most important property of probiotic and starter lactic acid bacteria. *Food Technology and Biotechnology* 48: 296-307.

Swain M. R., M. Anandharaj, R.C. Ray and R.P. Rani. 2014. Fermented fruits and vegetables of Asia: A potential source of probiotics. *Biotechnology Research International* 2014: 250424.

Szwajkowska, M., A. Wolanciuk, J. Barłowska, J. Król and Z. Litwińczuk. 2011. Bovine milk proteins as the source of bioactive peptides influencing the consumers' immune system—A review. *Animal Science Papers and Reports* 29: 269-280.

Tamang, J.P. and P.K. Sarkar. 1993. Sinki: A traditional lactic acid fermented radish taproot product. *Journal of General and Applied Microbiology* 39: 395-408.

Tamang, J.P., B. Tamang, U. Schillinger, C. Guigas and W.H. Holzapfel. 2009. Functional properties of lactic acid bacteria isolated from ethnic fermented vegetables of the Himalayas. *International Journal of Food Microbiology* 135: 28-33.

Taranto, M.P., J.L. Vera, J. Hugenholtz, G.F. De Valdez and F. Sesma. 2003. *Lactobacillus reuteri* CRL1098 produces cobalamin. *Journal of Bacteriology* 185: 5653-5647.

Vermeirssen, V., J. Van Camp and W. Verstraete. 2014. Bioavailability of angiotensin converting-enzyme-inhibitory peptides. *British Journal of Nutrition* 92: 357-366.

Vitali, B., G. Minervini, C.G. Rizzello, E. Spisni, S. Maccaferri, P. Brigidi, M. Gobbetti and R. Di Cagno. 2012. Novel probiotic candidates for humans isolated from raw fruits and vegetables. *Food Microbiology* 31: 116-125.

Yoon, K.Y., E.E. Woodams and Hang Y.D. 2004. Probiotication of tomato juice by lactic acid bacteria. *Journal of Microbiology* 42(4): 315-318.

4

Safety of Lactic Acid Fermented Vegetables

A. Jagannath[1]*

1. Introduction

Fermentation has been used since time immemorial primarily to improve the storage quality and safety of foods, enabling generations to survive through periods of food shortage and climatic excesses. Apart from this primary aim, depending on specific products, fermentation improves palatability, sensory properties, functionality of foods and nutritional value while reducing toxicity. In fact while all other forms of food preservation involve a certain loss of nutrients, fermentation is a process which ensures an increase in the nutrient content (Jagannath et al. 2014, Jagannath et al. 2012a) and improvement in chemical and microbiological qualities (Jagannath et al. 2012b).

What started as an art to conserve food for the lean season has developed into a modern science today so much so that carefully selected, desirable microorganisms containing high levels of viable cells are intentionally added to various foods to initiate, accelerate and accomplish the desired fermentation and end products. These microorganisms, called starter cultures, can be either moulds, yeasts or bacteria and their substrates milk, meat, fruits, vegetables, cereals, legumes, root, tubers, etc. Among bacteria, lactic acid bacteria are the most predominantly used organisms.

2. Spontaneous and Controlled Fermentation

The safety of lactic acid fermentation of vegetables is largely influenced by whether the fermentation is spontaneous or controlled. In spontaneous fermentation, a particular species or class of bacteria which has the

[1] Department of Fruit and Vegetable Technology, Defence Food Research Laboratory, Siddartha Nagar, Mysore: 570 011. Karnataka, India.

* Correspondent author: E-mail: jagann10@yahoo.com

highest growth rate and is best adapted to the prevailing conditions grow, dominate and alter the nature of food favorably, ultimately producing certain metabolites which create conditions inimical to its own growth, but benefit another class or species of organisms which in turn grow and repeat the process. In short, spontaneous fermentation is characterized by a definite microbial succession which ultimately stabilizes the product for long-term storage. In majority of cases spontaneous fermentation is employed for stabilizing the product.

Controlled fermentation, on the other hand, involves using starter cultures. Steinkraus (1983) lists three low-cost starter technologies — natural inoculation, transfer of an old batch of fermented product to a new batch (back slopping) and indigenously derived starter cultures as the major ways to initiate fermentation. Use of starter cultures can shorten the time required for fermentation while ensuring a better control of the process.

3.   Safety of Lactic Acid Fermented Vegetables

Fermentation, whether spontaneous or controlled, ensures rapid colonization of desirable microorganisms, thus preventing the invasion, survival and contamination of vegetables by either pathogenic or spoilage organisms. This process, called competitive exclusion, is the essence of most vegetable fermentations.

When starters are used, to ensure safety of lactic fermentation of vegetables, certain key factors, like the ability of the starter culture to inhibit or eliminate food-borne pathogens and degradation of anti-nutrition factors or microbial metabolites, like mycotoxin present in the fermenting substrate, need to be taken into consideration.

Lactic acid fermented vegetables are characterized by a rapid drop in pH to the level of 4.0 or less, which prevents the growth of pathogens. Apart from this, metabolites like acetic acid, hydrogen peroxide, bacteriocins also contribute towards safety of lactic acid fermented vegetables (Holzapfel et al. 1995). This mode of pathogen control depends on factors like initial level of contamination and levels of hygiene and sanitation which in turn, depend on local conditions and on the degree of acidification (Nout and Motarjemi 1997). Hence lactic acid fermentation of vegetables can under no circumstances be used as a substitute for the clean and hygienic methods of production, handling and processing because pathogens, that are capable of surviving and growing in these adverse conditions, do exist. The importance of good manufacturing practices and the principles of Hazard Analysis Critical Control Points are also equally important in ensuring a safe product.

Proper fermentation temperature is important in ensuring the nature of the microflora which predominates. Lower temperatures in the range of 68°–75°F ensure a good fermentation in three to four weeks' time while temperatures in the range of 60°–75°F result in a longer fermentation time of five to six weeks. Fermentation at more than 75°F generally spoils the product.

The safety of lactic acid fermented vegetables can primarily be addressed under two heads, viz. biological and chemical safety (Nout and Motarjemi 1997). However, such a clear demarcation seldom exits in reality as some of the biological entities like microorganisms do produce chemical metabolites, which in principle are largely responsible for exerting anti-bacterial effect that contributes to the safety of fermented vegetables. Further, plant derived biomolecules have antibacterial activity (Kyung and Fleming 1994, Tolonen et al. 2002), which can also influence the type of microflora that can predominate and be of food safety concern.

3.1. Biological Safety

Lactic acid fermentation is an established form of biopreservation wherein natural occurring or carefully selected microorganisms or their anti-microbials are used for extending the shelf-life of food. The antagonistic activity of these bacteria against spoilage and pathogenic organisms makes these fermented vegetables relatively safe. Consequently very few instances are reported where fermented vegetables are implicated as vehicles of food poisoning. Nevertheless, the fermentation process, if improperly done, can lead to food poisoning outbreaks or spoilage due to secondary fermentation by yeast. Improper fermentation can also result from using one type of culture suitable for one vegetable for other vegetables. For example, sauerkraut culture may not be suitable for fermentation of carrots and vice versa. Inoculation pack studies, wherein target organisms inoculated into specific fermented foods studied for their ability to survive in fermented vegetables, are widely reported in scientific literature.

Listeria monocytogenes, a food-borne pathogen has been isolated from improperly fermented vegetables as this organism can tolerate low pH and high salt concentrations. Insufficient acid production and anaerobic conditions can favour the growth and proliferation of *Clostridium botulinum* which can produce a harmful neurotoxin. *E. coli* O157:H7 is another emerging pathogen which is of concern in fermented vegetables as it can survive hostile acid environments.

Bacteriophages inhibit starter cultures and permit growth of other lactic acid bacteria likely to be present, thereby influening microbial succession.

3.2. Chemical Safety

The chemical safety of lactic acid fermented vegetables is studied with reference to preformed toxins, mycotoxins, anti-nutritional factors, salt and biogenic amines. The presence of lactic acid bacteria can prevent the production of mycotoxins or toxic secondary fungal metabolites, formed during the growth and proliferation of certain molds. However, lactic acid fermentation of fruits and vegetables offers no protection against preformed toxins, be it of fungal origin, like mycotoxins or bacterial toxins. Hence a final cooking of the fermented vegetable may overcome this problem but at the cost of losses in vitamins and minerals. Lactic acid fermentation also has a role in ensuring food safety due to its effect on naturally occurring toxins in vegetables which can either be inherent in the raw material or microbially derived (Westby et al. 1997). A classical example is the fermentation of soaked roots of cassava wherein microbial growth is essential for a reduction in cyanogenic glucosides, like linamarin and lotaustralin (Westby and Choo 1994). The effect of fermentation on the naturally occurring toxins inherently present in cassava has been reviewed by Westby et al. 1997. Members of *Leuconostoc, Lactobacillus, Alcaligenes* and possibly *Corynebacterium spp.* and *Candida* spp. are reported to be associated with this mixed-strain spontaneous fermentation (Onyekwere et al. 1989).

The other main anti-nutritional compounds in vegetables are phytate, tannins, saponins, oxalates, lectins and enzyme inhibitors. However, studies on the effect of lactic acid fermentation on the levels of these anti-nutritional compounds are limited. Metabolites of lactic acid bacteria, like organic acids, lactic acid, acetic acid, low molecular weight metabolites like diacetyl, acetaldehyde, fatty acids, hydrogen peroxide, bacteriocins etc., can inhibit spoilage or pathogenic organisms and ensure safety of lactic acid fermented vegetables. The protonated forms of organic acids are uncharged, hence, can cross biological membranes and acidify the cytoplasm of target organisms, exerting their antagonistic effects and reducing the risks of food-borne diseases.

Bacteriocins produced by lactic acid bacteria can either have a narrow spectrum of activity (e.g. lactacin B, F, and lactocin 27) or a broad spectrum of activity (Gram-positive organisms, like nisin, pediocin A and pediocin PA-1).

The isomers of lactic acid bacteria also have a bearing on the safety of lactic acid fermented vegetables. Lactic acid bacteria produce either D (–) lactic acid or L (+) lactic acid. The D (–) isomer is not hydrolyzed by the lactate dehydrogenase isoenzymes of the body and is known to cause acidosis, disturbing the acid-alkali balance in the blood. The L (+) isomer is specific to some genera, like *Streptococci, Lactococci, Enterococci* and *Carnobacteria* while for *Lactobacilli* and *Pediococci* the isomers are specific to a particular species (Holzapfel 1997).

3.2.1. Nitrites and nitrates: All vegetables are classified into different categories based on their nitrate (Bryan and Hord 2010) and nitrite contents. Beetroot, celery, lettuce, rocket, spinach have above 2500 mg/kg nitrate or 40 mmol and are classified as vegetables with very high nitrate content. Chinese cabbage, celeriac, endive, leek, parsley, kohlrabi have a nitrate content of about 1000–2500 mg/kg or 18–40 mmol and classified as vegetables with high nitrate content, while cabbage, dill, turnips, carrot juice have moderate nitrate content (500–1000 mg/kg or 9–18 mmol). Broccoli, carrot, cauliflower, cucumber, pumpkin come under low-nitrate containing vegetables (200–500 mg/kg or 3–9 mmol) while asparagus, artichoke, broad beans, green beans, peas, capsicum, tomato, watermelon, sweet potato, potato, garlic, onion, eggplants and mushroom are classified under very low category i.e. <200 mg/kg or <3 mmol.

Inorganic nitrates have undergone a radical change in their perception *vis a vis* their presence in food. Since the 8th century, Chinese medicine which used potassium nitrite and nitrates to mediate hypotensive actions to the implication during the mid 20th century, of excessive nitrate consumption from drinking water in infants, to methemoglobinemia (Majumdar 2003) and to human carcinogenesis from processed meat, the role of nitrate and nitrites seems to be a perpetual topic of active research (Jagannath et al. 2014). Very recently nitrate in beetroot juice is reported to decrease the systolic blood pressure (Hobbs et al. 2012) and oxygen consumption during walking and running (Lansley et al. 2011), thus bringing the research a full circle, on the physiological role of nitrates and nitrites.

Consequent to these changes, methods and means to reduce or remove nitrates from vegetables during cultivation, processing and post-processing treatment have been tried and some of these are microbiological methods employing lactic acid bacteria (Hybenova et al. 1995, Oh et al. 2004, Yan et al. 2008). Infact utilization and reduction of nitrates has long been used as a criterion for selection of lactic acid bacteria in fermentation of vegetables. (Hybenova et al. 1995) and many reports exist of vegetables being subjected to lactic acid fermentation with the intention of reducing their nitrate and nitrite contents (Oh et al. 2004, Yan et al. 2008). Certain lactic acid bacteria isolated from the Korean fermented kimchi show reduction in the sodium nitrite concentration (Oh et al. 2004), thereby rendering the product free from nitrites. Such a selection of starters could help in reducing the anti-nutrients associated with fermented vegetables. Fermentation employing starter cultures is more effective in lowering the nitrite concentrations and biogenic amines as compared to spontaneous fermentation (Yang et al. 2003). Research shows that although spontaneous lactic acid fermentation can lower the nitrates and nitrites because of diverse microflora, controlled fermentation does not result in any appreciable reduction in these anti-nutritional factors (Jagannath et al. 2015). Nevertheless lactic acid fermentation of vegetables reduces the nitrate and nitrite contents (Preiss et al. 2002, Feng-Di et al. 2009, Yan et al. 2008).

3.2.2. Oxalates: Oxalates are common in plants belonging to amaranth, aroid/arum, goosefoot, Beta, Spinacia, wood sorrel, buckwheat and purlane families where they occur in the range of 400-900 mg/100g (Noonan and Savage 1999). High oxalate diets lead to problems ranging from corrosion in the mouth to gastric haemorrhage and renal failure (Noonan and Savage 1999). High oxalate content in foods also exert a negative effect on calcium and iron absorption. Hence foods are classified with respect to oxalate: calcium ratios into those greater than 2, approximately 1 or with a ratio lesser than 1. Free oxalate and calcium can precipitate in the urine and form kidney stones. Processing of raw produce will reduce the oxalate content to varying degrees (Savage et al. 2000, Judprasong et al. 2006, Lisiewska et al. 2011, Akhtar et al. 2011). Fermentation in certain studies are reported to reduce the oxalate content (Antai and Obong 1992) with spontaneous fermentation showing much greater loss compared to controlled fermentation (Jagannath et al. 2015).

3.2.3. Safety of Salt: Salt plays an important role in the stabilization of a variety of fermented vegetables. Salt governs the type and extent of microbial activity that occurs and helps prevent softening of tissue. The concentration of salt is usually in the range of 2–3 per cent in Sauerkraut but can be as high as 5–6 per cent in kimchi. Salt extracts water from high-moisture vegetables, suppresses the growth of undesirable bacteria and creates conditions required for the growth of beneficial organisms. Salt also contributes to the flavour of the finished product by maintaining a balance between the acid and salt. Dry salting involves sprinkling salt over vegetables like cauliflower, red bell, pimiento peppers, salt-cured ripe olives, cabbage, etc. (Vaughn 1982), while brine salting involves preparation of a salt solution ranging from 5–10 per cent and is applied to cucumbers, olives, etc. Irrespective of the nature of salting, the final fermented product has relatively high salt concentrations which has a direct implication on the health of the consumer. Attempts to replace sodium chloride with potassium chloride (Choi et al. 1994) and calcium chloride (McFeeters and Perez-Diaz 2010) in high-salt-containing lactic acid fermented vegetables have been made. Use of select organisms in order to reduce the salt concentration in fermented vegetables is also being pursued (Johanningsmeier et al. 2007).

Apart from health safety of high-salt-containing vegetables, waste disposal of brine solutions left after fermentation also poses environmental challenges and concerns.

3.2.4. Biogenic amines: The safety of lactic acid fermented foods in general and vegetables in particular is influenced by the formation of certain organic compounds, called biogenic amines. Decarboxylation of amino acids, amination and transamination of aldehydes and ketones present in vegetables can result in formation of nitrogenous compounds, collectively called biogenic amines which can be aliphatic, aromatic or heterocyclic. Putrescine, cadaverine, spermine, spermidine are aliphatic, tyramine and phenylethylamine are aromatic while histamine and tryptamine are classified as heterocyclic biogenic amines. Karovicova and Kohajdova (2005) reviewed various aspects of biogenic amines in foods. The enzymes required for transformation of amino acids to biogenic amines are derived from the microflora present in a spontaneous fermentation or intentionally added as starter organisms in a controlled fermentation. These amines in vegetables are formed largely by lactic acid bacteria belonging to the genera *Lactobacillus, Enterococcus, Carnobacterium, Pediococcus, Lactococcus* and *Leuconostoc* as a part of the osmotic stress response to the hostile environment of sodium chloride concentrations, lack of oxygen, etc. likely to be present in fermented vegetables. For the production of amines, these microbes adopt the strategy to survive in an acidic environment or to open up alternate metabolic energy-yielding pathways to offset suboptimal substrate concentrations (Cotter and Hill 2003).

3.2.4.1. Conditions for the formation of biogenic amines: Many factors influence the formation of these biogenic amines. Availability of free amino acids which act as precursors for the synthesis of biogenic amines, presence of decarboxylase-positive micro organisms, conditions that favor the growth of these organisms or the synthesis and activity of these enzymes are some of them. Temperatures in the range of 20–37°C are a prerequisite for the growth of most bacteria-containing decarboxylases (Maijala 1993).

The amino acid decarboxylase activity has an optimum pH in the range of 4.0–5.5 (Halaszet al. 1994) and is influenced by the presence of sugars with a reported optimum concentration of 0.5–2.0 per cent in the case of glucose (Bardócz 1995). Sodium chloride (Santos 1996), oxygen supply and redox potential also affect the production of biogenic amine and the activity of decarboxylase enzyme.

3.2.4.2. Toxicology: Although biogenic amines are involved in natural biological processes such as synaptic transmission, blood pressure control, allergic response and cellular growth control (Zeisberger 1998, Singewald and Philippu 1996, Toninello et al. 2004), there exists a threshold level beyond which these very compounds affect the health and well-being of humans. Biogenic amines can react with nitrites to form nitrosamines which are reported to be carcinogenic.

The most frequently implicated amine in food-borne intoxication is histamine (Shalaby 1996, Lehane and Olley 2000) where 1000 mg is reported to be highly toxic, while as low as 100 mg has medium toxicity and 10 mg is the tolerable limit. Histamine acts as a neurotransmitter and vasodilator on the central nervous and cardiovascular systems respectively. High levels of histamine can induce migraine, headaches (Akerman et al. 2002), vertigo, vomiting, hypotension, arrhythmia and anaphylaxia (Maintz and Novak 2007). Tyramine, tryptamine and B-phenylethylamine are vasoactive amines that cause hypertensive effects and increase in blood sugar content (Shalaby 1996). Based on scientific research European Union-related surveys, reports and consumption data, a qualitative risk assessment of biogenic amines (BA) in fermented foods revealed histamine and tyramine as the most toxic and food-safety-relevant biogenic amines (EFSA, 2011).

3.2.4.3. Methods to prevent and reduce biogenic amines in fermented vegetables: Conditions which favor the growth of amine-producing organisms and activity of their enzymes should be avoided. Low temperature is a must for extended periods of storage so as to inhibit the proteolytic and decarboxylase activities of bacteria. Most of the risk management options highlighted for fish and fishery products in a joint FAO/WHO expert meeting on the public health risks of histamine and other biogenic amines products can also be applied for the safety of lactic acid-fermented vegetables (FAO/WHO 2012). High temperatures can effectively reduce the causative organism for biogenic amine production but are less effective in detoxifying preformed biogenic amines in fermented foods. Naila et al. (2010) reviewed the existing and emerging approaches for the control of biogenic amines in food and can serve as an excellent starting point for future research in ensuring the safety of fermented products. Some of the emerging methods are starter selection, application of high hydrostatic pressures, irradiation, packaging, using food additives and preservatives.

A careful selection of starters which do not produce biogenic amines during the fermentation process is an important preventive measure. The use of specific starters with either negative decarboxylase activity or those possesing amino oxidase activity to carry out fermentation is an effective method for suppressing the growth of biogenic amine-producing bacteria (Rabie et al. 2011, Leuschner and Hammes 1998). The use of amino-oxidase producing organisms has the potential of preventing or reducing the accumulation of biogenic amines in fruits and vegetables (Leuschner et al. 1998). Apart from these, the enzyme diamine oxidase (DAO) can also be used to decontaminate foods containing biogenic amines.

Preservatives like sodium and potassium sorbates, sodium hexametaphosphate, citric acid, succinic acid, D-sorbitol, malic acid, ascorbic acid, glycine and sugar limit the formation of biogenic amines. The use of spice-active ingredients like curcumin, capsaicin, piperine, gingerol, allicin, eugenol and cinnamic acid derived from turmeric, chillies, pepper, ginger, garlic, clove, cinnamon respectively can also inhibit formation of biogenic amines.

High hydrostatic pressures in the range of >300 MPa and irradiation at >10 kGy can affect both the organism and enzymes produced and is an area calling for research.

Active packaging techniques, like removal of gases (vacuum packaging), reducing (controlled atmosphere packaging) or replacing oxygen by other gases (modified atmosphere packaging, MAP) too restrict biogenic amine-forming bacteria or enzyme activity. In other words, control of biogenic amines may require a combination of methods to effectively address the public health concerns but often their very presence is an indicator of severe food-safety concerns as a result of improper hygiene.

4. Conclusion

Lactic acid fermentation is often projected as an alternate technology to safe guard food (Nout and Motarjemi 1997) where cold storage and processing is not possible. In fruits and vegetables it ensures food security through improved food preservation, increasing the range of raw materials that can be used to produce edible food products and removing anti-nutritional factors to make food safe. Further, the process requires no sophisticated equipment, is a cheap and energy-efficient means of preserving perishable raw materials and has practical utility both in cold temperate and humid tropics. Many byproducts of lactic fermentation trigger anti-microbial activities which keep the product free from spoilage or pathogenic organisms. In spite of this, there are instances where fermented foods have been implicated in food-borne illnesses (Nout 1994). Ensuring the safety of lactic acid fermented foods is further complicated by the nature of the industry. The majority of fermented foods are prepared at household level or in small industries where observation of food hygiene and sanitation is not ensured. Documenting traditional knowledge, improving the understanding of fermented products, developing a scientific understanding of the microbial processes, refining the fermentation technique, disseminating the improved methods and creating a supportive policy are some of the measures which can give fermentation its due place in ensuring food security for the millions (Battcock and Azam Ali 1998).

References

Akerman, S., D.J. Wlliamson, H. Kaube and P.J. Goadsby. 2002. The role of histamine in dural vessel dilation. *Brain Research* 956: 96-102.

Akhtar, M.S., B. Israr, N. Bhatty and A. Ali. 2011. Effect of cooking on soluble and insoluble oxalate contents in selected Pakistani vegetables and beans. *International Journal of Food Properties* 14: 241-249.

Antai, S.P. and U.S. Obong. 1992. The effect of fermentation on the nutrient status and some toxic components of *Icaciniamanni*. *Plant Foods for Human Nutrition* 42: 219-224.

Bardócz, S. 1995. Polyamines in food and their consequences for food quality and human health. *Trends in Food Science and Technology* 6 (10): 341-346.

Battcock, M. and S. Azam-Ali. 1998. Fermented fruits and vegetables: A global perspective. *FAO Agricultural Services Bulletin* No. 134. FAO Rome.

Bryan, N.S. and N.G. Hord. 2010. Dietary Nitrates and nitrites. *In:* Bryan N. (Ed.), *Food Nutrition and the Nitric Oxide Pathway*. Destech Pub Inc: Lancaster, PA. Pp. 59-77.

Choi, S.Y., L.R. Beuchat., L.M. Perkins and T. Nakayama. 1994. Fermentation and sensory characteristics of kimchi containing potassium chloride as a partial replacement for sodium chloride. *International Journal of Food Microbiology* 21(4): 335-340.

Cotter, P.D. and C. Hill. 2003. Surviving the acid test: responses of gram-positive bacteria to low pH. *Microbiology and Molecular Biology Reviews* 67: 429-445.

EFSA. 2011. Scientific opinion on risk based control of biogenic amine formation in fermented foods. EFSA Panel on Biological Hazards (BIOHAZ) *EFSA Journal* 2011; 9(10): 2393. European Food Safety Authority (EFSA), Parma, Italy.

FAO/WHO. 2012. Joint FAO/WHO Expert Meeting on the Public Health Risks of Histamine and Other Biogenic Amines from Fish and Fishery Products, 23-27 July 2012 FAO Headquarters, Rome Italy.

Feng-Di J., J. Bao-Ping, L. Bo and L. Fei. 2009. Effect of fermentation on nitrate, nitrite and organic acids in traditional pickled Chinese cabbage. *Journal of Food Processing and Preservation* 33(1): 175-186.

Halasz, A., A. Barath, L. Simon-Sarkadi and W. Holzapfel. 1994. Biogenic amines and their production by microorganisms in food. *Trends in Food Science and Technology* 5(2): 42-49.

Hobbs, D., A.N. Kaffa, T.W. George, L. Methven and J.A. Lovegrove. 2012. Blood pressure-lowering effects of beetroot juice and novel beetroot-enriched bread products in normotensive male subjects. *British Journal of Nutrition* 108(11): 2066-2074.

Holzapfel, W.H. 1997. Use of starter cultures in fermentation on a household scale. *Food Control* 8(5/6): 241-258.

Holzapfel, W.H., R. Geisen and U. Schillinger. 1995. Biological preservation of foods with reference to protective cultures, bacteriocins and food-grade enzymes. *International Journal of Food Microbiology* 24: 343-362.

Hybenova, E., M. Drdak., R. Guoth and J. Gracak. 1995. Utilization of nitrates — A decisive criterion in the selection of *Lactobacilli* for bioconversion of vegetables. *Zeitschrift-fuer-Lebensmittel-Untersuchung-und-Forschung*. 200: 213-216.

Jagannath, A., Manoranjan Kumar and P.S. Raju. 2014. Fermentative stabilization of betanin content in beetroot and its loss during processing and refrigerated storage. *Journal of Food Processing and Preservation* 39: 606-613.

Jagannath, A., Manoranjan Kumar and P.S. Raju. 2015. The recalcitrance of oxalate, nitrate and nitrites during the controlled lactic fermentation of commonly consumed green leafy vegetables. *Nutrition and Food Science* 45(2): 336-346.

Jagannath, A., P.S. Raju and A.S. Bawa. 2012a. Controlled lactic fermentative stabilization of ascorbic acid in amaranthus paste. *LWT-Food Science and Technology* 48: 297-301.

Jagannath, A., P.S. Raju and A.S. Bawa. 2012b. A two-step controlled lactic fermentation of cabbage for improved chemical and microbiological qualities. *Journal of Food Quality* 35: 13-20.

Johanningsmeier, S., R.F. McFeeters., H.P. Fleming and R.L. Thompson. 2007. Effects of *Leuconostoc mesenteroides* starter culture on fermentation of cabbage with reduced salt concentrations. *Journal of Food Science* 72(5): M166-M172.

Judprasong, K., S. Charoenkiatkul, P. Sungpuag, K. Vasanachitt and Y. Nakjamanong. 2006. Total and soluble oxalate contents in Thai vegetables, cereal grains and legume seeds and their changes after cooking. *Journal of Food Composition and Analysis* 19(4): 340-347.

Karovicova, J. and Z. Kohajdova. 2005. Biogenic amines in food. *Chemical Papers* 59(1): 70-79.

Kyung, K.H and H.P. Fleming. 1994. Anti-bacterial activity of cabbage juice against lactic acid bacteria. *Journal of Food Science* 59: 125-129.

Lansley, K.E., P.G. Winyard, J. Fulford, A. Vanhatalo, S.J. Bailey, J.R. Blackwell, F.J. DiMenna, M.Gilchrist, N. Benjamin and A.M. Jones. 2011. Dietary nitrate supplementation reduces the O_2 cost of walking and running: A placebo-controlled study. *Journal of Applied Physiology* 110(3): 591-600.

Lehane, L. and Olley, J. 2000. Histamine fish poisoning revisited. *International Journal of Food Microbiology* 58: 1-37.

Leuschner, R.G. and W.P. Hammes. 1998. Degradation of histamine and tyramine by *Brevibacterium* linens during surface ripening of Munster cheese. *Journal of Food Protection* 61(7): 874-878.

Leuschner, R.G., M. Heidel and W.P. Hammes. 1998. Histamine and tyramine degradation by food-fermenting microorganisms. *International Journal of Food Microbiology* 39: 1-10.

Lisiewska, Z., P. Gebczynski, J. Shupski and K. Kur. 2011. Effect of processing and cooking on total and soluble oxalate content in frozen root vegetables prepared for consumption. *Agricultural and Food Science* 20: 305-314.

Maijala, R.L. 1993. Formation of histamine and tyramine by some lactic acid bacteria in MRS-broth and modified decarboxylation agar. *Letters in Applied Microbiology* 17: 40-43.

Maintz, L. and N. Novak. 2007. Histamine and histamine intolerance. *American Journal of Clinical Nutrition* 85: 1185-1196.

Majumdar, D. 2003. The blue baby syndrome: nitrite poisoning humans. *Resonance* 8(10): 20-30.

McFeeters, R.F. and I. Perez-Diaz. 2010. Fermentation of cucumbers brined with calcium chloride instead of sodium chloride. *Journal of Food Science* 75(3): C291-296.

Naila, A., S. Flint, G. Fletcher, P. Bremer and G. Meerdink. 2010. Control of biogenic amines in food—existing and emerging approaches. *Journal of Food Science* 75(7): R139-R150.

Noonan, S.C. and G.P. Savage. 1999. Oxalate content of foods and its effect on humans. *Asia-Pacific Journal of Clinical Nutrition* 8(1): 64-74.

Nout, M.J.R. 1994. Fermented foods and food safety. *Food Research International* 27: 291-298.

Nout, M.J.R. and Y. Motarjemi. 1997. Assessment of fermentation as a household technology for improving food safety: A joint FAO/WHO workshop. *Food Control* 8(5/6): 221-226.

Oh, C.K., M.C. Oh and S.H. Kim. 2004. The depletion of sodium nitrite by lactic acid bacteria isolated from kimchi. *Journal of Medicinal Food* 7(1): 38-44.

Onyekwere, O.O., I.A. Akinrele, O.A. Koleoso and G. Heys. 1989. Industrialization of Gari fermentation. Pp. 363-410. *In*: K.H. Steinkraus (Ed.). *Industrialization of Indigenous Fermented Foods*. Marcel Dekker, New York.

Preiss, U., M. Koeniger and R. Arnold. 2002. Effect of fermentation on components of vegetables. *Deutsche-Lebensmittel-Rundschau* 98 (11): 400-405.

Rabie, M.A., S. Hassan., S. El-Saidy, A.A. El-Badawy and F.X. Malcata. 2011. Reduced biogenic amine contents in sauerkraut via addition of selected lactic acid bacteria. *Food Chemistry* 129: 1778-1782.

Santos, M.H.S. 1996. Biogenic amines: Their importance in foods. *International Journal of Food Microbiology* 29(2-3): 213-231.

Savage, G.P., L. Vanhanen., S.M. Mason and A.B. Ross. 2000. Effect of cooking on the soluble and insoluble oxalate content of some New Zealand foods. *Journal of Food Composition and Analysis* 13: 201-206.

Shalaby, A.R. 1996. Significance of biogenic amines in food safety and human health. *Food Research International* 29: 675-690.

Singewald, N. and A. Philippu. 1996. Involvement of biogenic amines and amino acids in the central regulation of cardiovascular homeostasis. *Trends in Pharmacological Sciences* 17: 356-363.

Steinkraus, K.H. 1983. *Handbook of Indigenous Fermented Foods*. Marcel Dekker. New York.

Tolonen, M., M. Taipale, B. Viander, J.H. Pihlava, H. Korhonen and E.L. Ryhanen. 2002. Plant derived biomolecules in fermented cabbage. *Journal of Agricultural and Food Chemistry* 50: 6798-6803.

Toninello, A., M. Salvi., P. Pietrangeli and B. Mondovì. 2004. Biogenic amines and apoptosis: mini review article. *Amino Acids* 26: 339-343.

Vaughn, R.H. 1982. Lactic acid fermentation of cabbage, cucumbers, olives and other produce. *In*: Prescott and Dunns *Industrial Microbiology*, fourth edition, Gerald Reed (Ed.). CBS Publishers and Distributors, Delhi, India.

Westby, A. and B.K. Choo. 1994. Cyanogen reduction during the lactic fermentation of cassava. *Acta Horticulturae* 375: 209-215.

Westby, A., A. Reilly and Z. Bainbridge. 1997. Review of the effect of fermentation on naturally occurring toxins. *Food Control* 8 (5/6): 329-339.

Yan, P.M., W.T. Xue, S.S. Tan, H. Zhang and X.H. Chang. 2008. Effect of inoculating lactic acid bacteria starter cultures on the nitrite concentration of fermenting Chinese paocai. *Food Control* 19: 50-55.

Yang, X.M., Q.M. Liu, L.F. Xi and X.Y. Xu. 2003. Effect of fermentation inculcated *Lactobacillus* on quality and nitrite content of Chinese sauerkraut. *Journal of Zhejiang Agriculture University* 29(3): 291-294.

Zeisberger, E. 1998. Biogenic amines and thermoregulatory changes. *Progress in Brain Research* 115: 159-176.

5

Sauerkraut Fermentation

Britta Wiander[1]*

1. Background

Fermented foods produced from plant materials are an accepted and essential part of the diet in most parts of the world. They include a wide diversity of raw materials, using technology from very simple to the most advanced and achieving a huge range of sensory and textural qualities in the final products. Nearly all vegetables can be fermented by lactic acid bacteria (LAB). While most fermentations are at village or household level, others have achieved commercial application and play an important role in national economies. Almost all food fermentations have been or remain native to a country or culture and most were developed before recorded history. How these fermentations first developed is a question that can never be completely answered, but was undoubtedly closely related to a need for preservation and for improved organoleptic and textural qualities. Preservation of plant material by lactic acid fermentation has been documented as far back as 1000–1500 B.C. (Woodford 1985). However, it is believed that the origin of fermentation of plant material is in China and was discovered by the Chinese during the 3rd century B.C. (Pederson 1979, Buckenhüskes1993). Historically, preservation of perishable animal and plant products has been of great importance, for example, cabbage can be stored only for a rather short time, but fermented cabbage can, however, be stored for months.

Fermented foods are delicious products that are prepared from raw or heated raw materials resulting in characteristic properties (e.g. taste, aroma, visual appearance, texture, consistency, shelf-life, hygienic safety) by a process in which microorganisms are involved. In certain products, the endogenous enzymes of the raw material play an important role (Hammes 1990).

[1] Putkitehtaantie 191, 32270 Metsämaa, Finland.
* Correspondent author: E-mail: britta.wiander@evira.fi

Lactic acid bacteria have an essential role in the preservation and production of healthy foods. Lactic acid bacteria are generally recognized as safe (GRAS) microorganisms (Jay 1996). Generally the lactic acid fermentations are low-cost and often require little or no heat in their preparation. Thus, they are fuel-efficient. Lactic acid fermented foods play an important role in feeding the world's population on every continent today. As the world population rises, lactic acid fermentation will acquire more importance in preserving fresh vegetables, fruits, cereals and legumes for feeding the vast humanity. In underdeveloped countries lactic acid fermentation is an important way to preserve food because of the low energy requirements (Daeschel et al. 1987).

Fermentation of cabbage is an ancient preservation method used in different parts of the world. Even today fermented cabbage, also called sauerkraut, is a popular food and consumed in many countries. Sauerkraut is produced in both households in small scale and in industrial scale. Producing sauerkraut is a simple process which can be performed in households with simple equipment, but on the other hand, sauerkraut fermentation is a very complex process involving many phases.

Salt (NaCl) is traditionally used in the fermentation of sauerkraut. The highest reported amount of salt used has been approximately 3% (w/w). However, today consumers prefer a lower salt content in their food, mainly because of health reasons.

2. Lactic Acid Bacteria

Lactic acid bacteria describe a group of bacteria drawn from several genera to produce lactic acid. Although lactic acid bacteria co-exist with plants, their role on the plant surface is still unknown. It is believed that they protect plants from pathogenic microorganisms by producing antagonistic compounds, such as acids, bacteriocins and anti-fungal agents (Visser et al. 1986).

Lactic acid bacteria form a phylogenetically ancient group that differentiated from a common clostridial type of ancestor when the oxygen atmosphere was forming about two billion years ago. They are usually non-motile and catalase negative organisms. Except for some streptococci, lactic acid bacteria are rarely pathogenic, but some reports have associated some strains with human diseases. Furthermore, lactic acid bacteria are gram-positive, aerotolerant and generally non-sporulating, non-respiring rods or cocci. The major fermentation product is lactic acid (von Wright 2011).

According to the taxonomic classification of today, lactic acid bacteria belong to Phylum Firmicutes, Class Bacilli and Order Lactobacillales. Lactic acid bacteria are divided into families: *Aerococcaceae*, *Carnobacteriaceae*, *Enterococcaceae*, *Lactobacillaceae*, *Leuconostocaceae* and *Streptococcaceae* (http://www.uniprot.org/taxonomy/186826).

Lactic acid bacteria are virtually non-proteolytic and their ability to attack amino acids is also limited. Regardless of the presence or absence of oxygen, they generate ATP by a fundamentally anaerobic process. Those species that ferment glucose primarily to lactic acid are termed homofermentative and those that produce a mixture of products are termed heterofermentative. The term homolactic could be used to refer to those lactic acid bacteria that contain aldolase, but not transketolase.

3. Natural Fermentation

3.1. Fermentation Process

Natural (spontaneous) lactic acid fermentation of cabbage into sauerkraut is an ancient method to preserve cabbage. Sauerkraut can be stored for long periods, whereas fresh cabbage has a short shelf-life. Fermentation is an important method to preserve highly perishable foods. Lactic acid fermentation of cabbage and other vegetables is a common way of preserving fresh vegetables in the Western world, in China and in Korea. The traditional way to produce sauerkraut is by natural fermentation. This method is still used today. The natural fermentation process itself does not necessarily demand high investments.

In the sauerkraut fermentation process, salt is mixed with sliced cabbage, from which the core has been removed, and the cabbage mix is tightly pressed together in an appropriate vat. The fermentation vat can in household use be for example a few litres but at the industrial scale, the amounts of cabbage to be fermented are on a much higher level. The fermentation of cabbage proceeds under anaerobic conditions. The cabbage mix can be covered by food-grade plastic upon which water is poured to inhibit air from entering the fermentation vat. Different fermentation temperatures can be used.

Fermentation is a simple way of preserving food as the raw vegetables only need to be sliced and salt added. The salt extracts liquid from the sliced vegetables which serve as a substrate for growth of the lactic acid bacteria. Conditions should be maintained as anaerobic as possible to prevent growth of microorganisms that might spoil the sauerkraut.

Sauerkraut is in one way a simple product to manufacture, but on the other hand is the sauerkraut fermentation a complex process and the manufacture of sauerkraut of uniform high quality is not simple (Fleming 1987). The spontaneous lactic acid fermentation is a complex microbial process in which the lactic acid bacteria finally dominate the microbial flora. The original number of lactic acid bacteria in the raw material is normally very low (Daeschel et al. 1987). Therefore, there is always a risk that the fermentation will not proceed in the desired way and that the spoilage microbes naturally occurring in the raw material will take over.

Sauerkraut is traditionally produced by spontaneous fermentation, which depends on the naturally-occurring lactic acid bacteria present in the raw material. Normally the number of lactic acid bacteria in the cabbage is very low. The most critical point of fermentation process is in the beginning of fermentation as the lactic acid bacteria present in the raw material to be fermented must be able to increase radically in number to rapidly drop the pH of the sliced cabbage. When pH level fail to drop rapidly enough, then the spoilage bacteria, normally present in the cabbage, take over and fermentation will not proceed in the desired way, resulting in an end product which has neither good microbiological nor sensory quality.

3.2. The Role of Lactic Acid Bacteria in the Fermentation Process

Raw cabbage contains a very low number of lactic acid bacteria which initiate the fermentation process in sliced cabbage. The beginning of fermentation is a critical point in the fermentation process because the success of the fermentation depends greatly upon how rapidly the pH of the sliced cabbage begins to decrease.

Pederson determined in 1930 the sequence of microorganisms developing in typical sauerkraut fermentation. In the proceeding of the spontaneous sauerkraut fermentation, one lactic acid bacterium gives way to a more acid-tolerant lactic acid bacterium. *Ln. mesenteroides* initiates growth in shredded cabbage over a wide range of temperatures and salt concentrations, producing carbon dioxide and lactic acid as well as acetic acid which quickly lowers the pH. The rapid drop of pH inhibits the growth of undesirable microorganisms that could have a negative impact on the fermentation process. The carbon dioxide produced replaces the air and facilitates anaerobiosis required for fermentation. The fermentation is completed in sequence by *Lb. brevis* and *Lb. plantarum*. *Lb. plantarum* is responsible for the high acidity. *P. cerevisiae* can also contribute to acid production (Pederson and Albury 1969).

The heterofermentative lactic acid bacteria dominate the microflora in the beginning of the sauerkraut fermentation. At a later stage of fermentation, the heterofermentative lactic acid bacteria are replaced by the more acid-tolerant homofermentative species which produce only lactic acid (Pederson and Albury 1969). Typically, *Ln. mesenteroides* grows early during fermentation and *Lb. plantarum* terminates fermentation (Neish 1952, Pederson and Albury 1961).

Lactic acid bacteria are inhibitory to many microorganisms when they are cultured together and this is the basis of extended shelf-life and improved microbiological safety of lactic acid fermented products. This effect consists of many factors. The most important are the organic acids, mainly lactic and acetic acid that are produced. It is seen that acetic acid has a greater anti-microbial effect than lactic acid (Eklund 1989), as the acetic acid has a greater pK_a value (4.75) than lactic acid (3.86). It is therefore greatly important in natural fermentations that acetic acid is produced in the early stage of fermentation (Adams and Hall 1988). Heterofermentative lactic acid bacteria produce significant quantities of acetic as well as lactic acid.

Vegetable fermentation, such as sauerkraut production, shows remarkable stability, even though starter cultures are rarely used and the raw material is not pasteurized. This stability may result from the growth of heterofermentative lactic acid bacteria in the initial phase of the fermentation. Production of a mixture of lactic and acetic acids by the lactic acid bacteria seems to restrict the growth of competing microorganisms at the critical stage of the process, besides making an important contribution to the organoleptic properties of the product (Adams and Hall 1988).

Rapid pH decrease in the beginning of fermentation signifies success of the entire fermentation process, as rapid increase in acidity minimises the influence of spoilage bacteria. If the pH decrease in the beginning of fermentation is delayed, the spontaneous fermentation process most probably will not proceed in the desired way.

Most cabbage intended for sauerkraut fermentation is shredded or chopped before tanking, although a small quantity is brined whole or cut into chunks. Thus, nutrients from cabbage are immediately available for the lactic acid bacteria (Fleming et al. 1985). Fermentable carbohydrates in cabbage include sucrose, fructose and glucose. All the other essential nutrients for growth of lactic acid bacteria are present in the raw material. The availability of these nutrients for the growth of lactic acid bacteria depends upon treatment of the raw material before salting or brining. Nutrients are immediately available in cabbage which has been shredded (Fleming et al. 1985).

Acidity is an important component flavor in sauerkraut. The level of acidity is dependent upon the concentration of fermentable sugars in fresh cabbage and the extent to which these sugars are converted into acids. Fermentation continues until all fermentable sugars are used or until the product becomes so acidic that the growth of lactic acid bacteria is inhibited (Fleming 1987).

The sauerkraut fermentation is highly dynamic with numerous physical, chemical and microbiological changes influencing the quality of the final product. Fermentation may be divided into two phases — the first is gaseous and the second is non-gaseous. During the gaseous phase, CO_2 (carbon dioxide) is produced (Fleming 1987). The gas-forming lactic acid bacteria are of the heterofermentative type and the non-gas-forming are the homofermentative (Fleming 1987). Lactic acid, acetic acid, mannitol and ethanol are typical products formed during sauerkraut fermentation.

After fermentation is completed, no major chemical change occurs if the fermentation has resulted in the preservative levels of acid and the air is eliminated from sauerkraut. Entry of air can cause oxidative deterioration of the product quality. The problem increases if the brine level is below the level of sauerkraut, as oxygen diffusion is much more rapid in gas than in liquid. Oxidative reactions can result in darkening and disflavoring of the sauerkraut. Ascorbic acid present in the product can act as an antioxidant to delay these changes, but decreases rapidly in the presence of oxygen (Fleming 1987).

3.3. Problems Related to the Spontaneous Sauerkraut Fermentation

The production of sauerkraut of uniformly high quality is not simple. In spontaneous fermentation, the lactic acid bacteria naturally occurring in cabbage are responsible for the sauerkraut fermentation. The number of lactic acid bacteria in cabbage is usually very low, but it is under good fermentation conditions, usually enough to start the fermentation process. In the spontaneous lactic acid fermentation process, the lactic acid bacteria dominate the microbial flora. There are, however, some risks related to spontaneous sauerkraut fermentation.

As soon as the cabbage has been sliced, the number of microorganisms increases in the cabbage mix. The nutrients from the cabbage will, because of cutting, be easily available for the microorganisms in the cabbage mix and this leads to growth of the microflora. The fermentation of the cabbage mix should therefore start and the pH decrease as soon as possible to inhibit the growth of pathogens and other undesirable microbes.

Fermentation conditions should be maintained as anaerobic as possible to prevent growth of microorganisms that might spoil the sauerkraut. If air enters the fermentation vat, it will result in lowered sensory and microbiological quality. Many defects in sauerkraut are caused by oxygen. Discoloration on the sauerkraut surface can occur by an autochemical reaction involving oxygen. Flavours and odours result from metabolic compounds, especially organic acids like propionic and butyric acids which are produced by aerobic yeasts and moulds (Oberg and Brown 1993).

Slimy or ropy sauerkraut is generally the result of dextran formation caused by *Ln. mesenteroides* and is a transitory problem since the dextrans are utilized by the other lactic acid bacteria. Slimy sauerkraut caused by pectinolytic activity is permanent in effect (Vaughn 1985). Pectinases, produced by yeast, cause slimy sauerkraut and growth of the red-pigmented yeast *Rhodotorula*, resulting in pink-coloured sauerkraut. Also *Lb. brevis* is able to produce a red pigment if the pH does not decrease properly (Oberg and Brown 1993).

Spontaneous fermentation depends on the naturally-occurring lactic acid bacteria present in the raw material. The most critical point in the fermentation process is in the beginning of fermentation because the lactic acid bacteria originally present in the cabbage must be able to increase radically in number to rapidly drop the pH of the cabbage mix. When pH fails to drop rapidly enough, spoilage bacteria, also normally present in the cabbage, will take over and the fermentation will not proceed as it should, resulting in an end product that has neither good microbiological nor sensory quality.

It is of great importance that the nutrients in cabbage are available to the naturally-occurring lactic acid bacteria in cabbage. The nutrients from sliced cabbage are released during the cutting and further by pressing the sliced cabbage tightly to produce brine. The release of the nutrients from cabbage enables the growth of lactic acid bacteria. As a result, the pH decreases rapidly caused by lactic and acetic acids produced by lactic acid bacteria. If pathogens and other undesired microorganisms in the sliced cabbage grow more rapidly than the lactic acid bacteria, it leads to unsuccessful sauerkraut fermentation and the fermentation product will have neither good microbiological nor sensory quality. Thus care should be taken in proper handling of the raw cabbage material and all equipment coming in contact with raw material should be properly cleaned.

3.4. Salt Concentration in Sauerkraut Fermentation

Salt concentration in sauerkraut fermentation has traditionally been quite high. Concentrations of 2–3% (w/w) have usually been used. The nutrients released by salt from sliced cabbage are used by the lactic acid bacteria, which in turn produce lactic and acetic acids. The production of these acids results in a decrease of the pH in the cabbage mix. The rapid growth of lactic acid bacteria and production of lactic and acetic acids lower the pH and ensure a desired proceeding of the sauerkraut fermentation, resulting in sauerkraut with good microbiological and sensory qualities.

Today, however, consumers are more aware of the importance of eating healthy food. The general trend in industrialized countries is to reduce the salt level of foods to prevent cardiovascular diseases. Therefore attention is paid to lower the salt concentration in sauerkraut fermentation. The use of salt also creates a waste problem, encouraging the reduced use of salt in sauerkraut fermentation. Research has been made in evaluating whether the salt concentration could successfully be lowered without loss in microbiological safety and sensory quality.

Fermentation of cabbage into sauerkraut traditionally proceeds in the presence of NaCl (sodium chloride) and the highest reported amount of salt used is around 3% (w/w). Attention is paid to the use of sodium in production of sauerkraut and different salt concentrations were tried in sauerkraut fermentation (Delanoe and Emard 1971, Gangopadhyay and Mukherjee 1971, Mayer et al. 1973, Niven 1980, Peñas et al. 2010). Consumers nowadays wish to lower their sodium intake, which has led to research in trying to partially replace NaCl with KCl (potassium chloride) in foods. In sauerkraut production, the NaCl content can be over 2% (w/w). Sauerkraut containing various amounts of NaCl were investigated and the lowest percentage added was 0.5% (Pederson 1940, Fleming and McFeeters 1985, Trail et al. 1996, Viander et al. 2003). Sauerkraut was prepared utilizing hydrolyzed protein in combination with the salt present at 1.0–4.5% (Hsu et al. 1984, Wedral et al. 1985). A patent was worked out for making sauerkraut where part of the salt normally added is replaced by an alcohol/acid mixture (Owades 1991). Reduced NaCl concentrations in combination with lactic acid bacteria starter cultures were investigated where the used NaCl concentrations were 1% (Delclos 1992) and 0.5%, 1.0% and 2.0% (Johanningsmeier et al. 2007). Various salt concentrations of 0.5–2.0% were used in several studies (Petäjä et al. 2000, Viander et al. 2003, Johanningsmeier et al. 2007, Peñas et al. 2010), showing that it is possible to lower the salt concentration without loss in either microbiological safety or sensory quality. Mineral salt was also used in sauerkraut fermentation and studies showed that sauerkraut and sauerkraut juice can be successfully produced by using mineral salt in low concentrations (Wiander and Palva 2011).

The fermented product, kimchi, is also traditionally produced with NaCl and efforts were made to replace part of the NaCl with KCl. It was found that kimchi with acceptable sensory qualities can be produced using brines containing up to 50% KCl instead of NaCl. This provides a mechanism for substantially reducing the sodium intake in individuals who frequently consume kimchi (Choi et al. 1994).

3.5. Use of Starters in Sauerkraut Fermentation

Traditionally sauerkraut is produced by spontaneous fermentation in which the lactic acid bacteria originally existing in the cabbage raw material are responsible for the fermentation process. The number of lactic acid bacteria in the raw material is very low, but it is feasible enough to induce the fermentation process. However, there are risks connected with the spontaneous sauerkraut fermentation as the fermentation process does not always result in a product with good microbiological and sensory quality.

A conscious dealing with bacteria in starter cultures was not possible until microorganisms had been discovered and generally accepted as agents of fermentation and the role of bacteria in the various fermentation processes had been recognized (Hammes 1990).

Although sauerkraut is widely produced by spontaneous fermentation, starter cultures are also used to ensure high and uniform quality of the fermentation product. Starter cultures are defined as preparations which contain living microorganisms, which are applied with the intention of making use of their microbial metabolism (Hammes 1990). Pure starter cultures can be used as single-strain or multiple-strain cultures. Furthermore, undefined cultures consisting of fermenting substrate taken from a selected process which has resulted in high quality end product are also used. In addition, 'back slopping' is used in fermentation processes, where continuous inoculation from previous batches has taken place. Selection criteria for lactic acid bacteria to be used in vegetable and vegetable juice fermentations were summarised in 1993 by Buckenhüskes

(Daeschel et al. 1987, Lücke et al. 1990, Buckenhüskes 1992) and the selection of lactic acid bacteria for use in vegetable fermentations was discussed by Daeschel and Fleming (1984). The use of starter cultures was discussed and studied with increasing interest (Frank 1973, Fleming et al. 1985, Buckenhüskes et al. 1986, Delclos 1992, Harris et al. 1992, Breidt et al. 1993, Breidt et al. 1995, Halasz et al. 1999, Wiander and Ryhänen 2005).

Although commercial starter cultures are available on the market, they are not being largely used in sauerkraut production. The most common reasons why producers do not want to use commercial starter cultures have been discussed by authors Lücke et al. (1990) and Hammes (1991). The benefits of using commercial starter cultures were listed by Lücke et al. (1990) and one benefit is that the use of defined starter cultures in the fermentation process ensures accurate proceeding of the process. Losses in seasonal vegetables can also be prevented by using controlled fermentation as enough cold storage facilities are not available for fresh vegetables (Desai and Sheth 1997). If the starter culture functions properly, the growth of both spoilage microorganisms and pathogens gets retarded during manufacture and storage of the fermented food product.

Lactic acid bacteria strains were isolated from spontaneous sauerkraut fermentations (Stetter and Stetter 1980, Valdez et al. 1990, Harris et al. 1992) and isolated strains have further been used as starters in fermentation of sauerkraut (Breidt et al. 1993, Desai and Sheth 1997, Petäjä et al. 2000, Wiander and Ryhänen 2008). Lactic acid bacteria strains isolated from sauerkraut were used in fermentation of sausage (Vogel et al. 1993). Furthermore, lactic acid bacteria strains were isolated from fermented Chinese cabbage and further used as starters in producing sour Chinese cabbage (Han et al. 2014, Xiong et al. 2014).

Improvement in overall quality is a continuing task for manufacturers of all food products, including sauerkraut. Perhaps of equal or greater importance for the sauerkraut industry is the need for improvement in uniformity of product quality, as it can influence retail sales of sauerkraut, and may be asked for by more demanding buyers (Fleming 1987).

In the starter-induced fermentation starter culture is added and mixed with sliced cabbage before the cabbage mix is pressed tightly. The cabbage mix should be properly covered to enable anaerobic fermentation conditions.

Various starter cultures for use in sauerkraut fermentation are available in the market. The starter cultures can be bought as 'ready to use' and added to the cabbage mix without further enrichment of the starter culture before use. Starter cultures can also be produced and enriched by the producer, but right equipment and premises are needed to ensure proper handling of starter cultures. This option requires investments in equipment and premises and also special training of the staff.

3.6. Summary of Benefits Related to Use of Starter Cultures

The use of starter cultures is a good alternative to spontaneous fermentation of sauerkraut. By using starter cultures, the fermentation process can be more controlled in comparison to spontaneous fermentation. Starter cultures can be added to the sliced cabbage mix as single or multiple-starter cultures. Single-starter cultures contain only one lactic acid bacteria strain, whereas multiple-starter cultures can contain two to several lactic acid bacteria strains. By using specific starters, the same desired taste of sauerkraut can be achieved repeatedly from one time to another. Variation in the used starter cultures results in differently tasting products.

The added starter culture enables rapid production of lactic and acetic acids which results in a quick decrease of pH. The rapid drop of the pH decreases significantly the negative effects caused by possible pathogens and other undesirable microorganisms present in the raw material. As the impact of undesirable microorganisms is reduced, the accurate proceeding of the fermentation process is guaranteed and the sauerkraut will have good microbiological and sensory quality.

Starter cultures can be used to optimize the sauerkraut fermentation and to lower the risks which can lead to failure in the fermentation process. The use of starter cultures increases the cost of sauerkraut production, but the accurate proceeding of the fermentation process is better guaranteed due to use of starters. The percentage of waste produced decreases by using starter cultures, which results, on the other hand, in better profit for the producer. In addition, the use of starter cultures speeds up the fermentation process and sauerkraut can be produced in a much shorter period as compared to spontaneous fermentation.

4. Sauerkraut Juice

Sauerkraut juice can be produced from fermented sauerkraut. The fermented sauerkraut is pressed by using convenient pressing equipment to get the juice extracted from the sauerkraut. This sauerkraut juice can be used with ordinary meals or used in small portions daily as wished. The sauerkraut juice can be stored at the same temperatures as sauerkraut for long periods.

4.1. Sauerkraut Juice Produced from Spontaneously Fermented Sauerkraut

Sauerkraut juice can be produced from spontaneously fermented sauerkraut. The juice is extracted from sauerkraut after the fermentation process has reached a pH below 4.2. The microbiological and sensory quality of the pressed sauerkraut juice depends on the quality of fermented sauerkraut itself. The microbiological flora of the cabbage and fermentation conditions in general have a great impact on how the fermentation process proceeds and whether the fermented sauerkraut has the desired taste and microbiological quality.

Pressed sauerkraut juice can be consumed directly after being produced or it can be stored at the same temperatures as fermented sauerkraut. The sauerkraut juice can also be pasteurized. If the heat treatment is done at higher temperatures (120–140°C) the juice is sterilized and can be stored for very long periods, even years. If the sauerkraut juice is heat treated, it will not contain live lactic acid bacteria.

Sauerkraut juice produced from spontaneously fermented sauerkraut does not always have the same taste, even though by following the same procedure. The difference can be caused by variation in the chemical composition and microbiological flora of the cabbage species. Differences can exist, for example, in the sugar content, number of lactic acid bacteria, pathogens and the overall microbiological flora.

4.2. Sauerkraut Juice Produced from Induced Sauerkraut Fermentation

To ensure a more uniform sensory and microbiological quality of the pressed sauerkraut juice, starter cultures can be used in sauerkraut fermentation. By using starter-induced sauerkraut fermentations for producing juice, it is possible to produce juices with the same taste repeatedly. This will also make it easier for the producer to market the product.

The added starter culture(s) produces more effective lactic and acetic acids which rapidly decrease the pH in the sliced cabbage mix, thereby lowering the influence of other microbes naturally existing on the leaves of the cabbage. This guarantees a better microbiological quality of the produced sauerkraut juice.

The taste of sauerkraut juice can be modified by using specific starter culture combinations in inducing sauerkraut fermentation. The starters can be single or multiple-starter cultures, containing two or more lactic acid bacteria strains. By using different starter cultures or changing the amount or proportion between the starter cultures, the taste of sauerkraut juice can be modified in the desired manner. The taste of sauerkraut juice can also be modified by using different salt types and salt concentrations.

Studies were made to evaluate the impact of different starter cultures on the sensory quality of the sauerkraut juice. The use of *Ln. mesenteroides* isolated from spontaneous sauerkraut fermentation showed that sauerkraut juice with very smooth taste can be produced. All consumers do not like the harsh taste of sauerkraut or sauerkraut juice, so the alternative smoother taste of sauerkraut products is a good option for the market. Also the combination of isolated *Ln. mesenteroides* and *P. dextrinicus*, both isolated from spontaneous sauerkraut fermentation, has resulted in a smoother taste compared to some other starter cultures (Wiander and Ryhänen 2008).

By using different supplements in sauerkraut fermentation, the taste of sauerkraut juice can be modified. The supplements can, for example, be different herbs and spices which are added to the sliced cabbage mix before the fermentation starts. In fact there are an endless number of possibilities to produce sauerkraut juice with different tastes.

5. Hygienic Conditions

Great focus should be put on the production conditions. Even though spontaneous sauerkraut fermentation seems a simple process, there are many critical points in the proceeding of the fermentation.

The cabbage raw material should be properly harvested and if not instantly used, it should be stored in a clean storage facility. The raw cabbage should be properly handled at every stage. The cabbage and other raw materials and ingredients used in fermentation should be of good microbiological quality. Cabbage heads can be washed with water to prevent soil and dirt from entering the fermentation process. The outer leaves of the cabbage are often removed as well as the core. The chopping of cabbage should be done in a clean environment. Furthermore, besides clean working clothes, the staff should know how to handle the raw material used in the fermentation process. With a trained and motivated staff, the production risks are lowered and the waste percentage minimized.

All equipment coming in touch with the raw material used in fermentation should be cleaned to eliminate all possible contamination risks. Soil and dirt on the raw material and improperly cleaned equipment increase the contamination risk, lowering the microbiological quality of the fermented sauerkraut. Furthermore, the sensory quality of the fermented product probably suffers or may even turn out to be unacceptable.

When the sauerkraut fermentation is complete, attention should be paid to handling of the product to avoid any post contamination. The products should be stored at the right temperatures and further suitably stored in a clean storage to maintain their good microbiological and sensory quality.

It is of great importance to pay attention to the working conditions. All equipment should be regularly and thoroughly cleaned and the raw material properly harvested, handled and stored. Critical points where contamination of the raw material could occur should be recognized and eliminated. Without clean equipment and high quality raw material, sauerkraut of good microbiological and sensory quality cannot be successfully produced.

5.1. Heat Treatment of Sauerkraut

Sauerkraut can, after fermentation, be refrigerated and sold as fresh sauerkraut, without heat treatment. It can be stored for months before being placed on the market. Storage temperatures for fresh sauerkraut range between 2–4°C. At these temperatures, the metabolic activities in the sauerkraut are low and good microbiological and sensory quality maintained. Sauerkraut should be stored in an anaerobic environment without the risk of post-contaminating the product. The pH of sauerkraut should be below 4.2 if stored without heat treatment.

However, sauerkraut can be heat treated to stop the function of enzymes and eliminate possible pathogens by heating to 70–80°C. The heat treatment stabilizes the product and results in a stable pH, which enables a longer shelf-life. The heat treatment also destroys gas-producing microorganisms, thus eliminating the risk for packages to burst. As a result of pasteurization, the product can be stored at room temperature. The heat-treated sauerkraut does not, on the other hand, contain live lactic acid bacteria as they are destroyed during the process.

Sauerkraut is also heat treated by using high temperatures up to 120–140°C. The heat treated sauerkraut is stored at room temperature for a long time, upto several years. The sauerkraut does not contain any live microorganisms after the heat treatment. The heat treatment may reduce the sensory quality, making the sauerkraut taste as if boiled. The heat treatment extends its shelf-life, but the heat treatment results in loss of live lactic acid bacteria and the consequent loss in some health-beneficial characteristics.

5.2. Packaging

Sauerkraut can be packed in plastic bags, in glass jars and in metal cans. The choice of packaging material depends largely on whether it would be heat treated or stored as refrigerated. If the sauerkraut is heat treated, the packaging material is either glass or metal; if refrigerated, then plastic packaging material can be chosen. Glass and metal cans are totally airtight and are therefore very good packaging materials. At high temperatures (120–140°C) heat-treated sauerkraut packed in glass jars or metal cans is stored for years. Plastic materials can also be used as such or in combination with other materials which improve the characteristics of the plastic material.

6. Lactic Acid Bacteria and Health Beneficial Aspects

We know that sauerkraut fermentation has been used as a preservation method for centuries. In addition we know that sauerkraut is a healthy food for several reasons. Not until rather recently we discovered the health-beneficial properties of sauerkraut. Nowadays much focus is paid on the functions of the gastrointestinal tract and the role of lactic acid bacteria in this environment. Lactic acid bacteria enable the fermentation of cabbage to sauerkraut and sauerkraut has often, on the basis of consumers' experiences, considered to improve the function of the stomach. Sauerkraut which has not been heat treated contains usually a high number of live lactic acid bacteria.

Many lactic acid bacteria strains are known to have a positive impact on human and animal health (Gilliland 1989, Hose and Sozzi 1990, Kalantzopoulos 1997). Lactic acid bacteria were used as probiotics in cases of intestinal disorders, for example lactose intolerance, acute gastro-enteritis caused by rotavirus and other enteric pathogens, adverse effects of pelvic radiotherapy, constipation, inflammatory bowel disease and food allergy (Salminen et al. 1988, Gilliland 1990, Isolauri et al. 1991, Salminen et al. 1996, Masood et al. 2011, van Baarlen et al. 2013, Park et al. 2014). Lactic acid bacteria can inhibit growth of pathogens due to their ability to produce anti-bacterial agents, for example acidic compounds and bacteriocins (Dalie et al. 2010, Mills et al. 2011). Low-salted sauerkraut produced with *Ln. mesenteroides* provides high beneficial antioxidant and anti-carcinogenic compounds in addition to possessing low sodium content (Peñas et al. 2010).

References

Adams, M.R. and C.J. Hall. 1988. Growth inhibition of food-borne pathogens by lactic and acetic acids and their mixtures. *International Journal of Food Science and Technology* 23: 287-292.

Breidt, F., K.A. Crowley and H.P. Fleming. 1993. Isolation and characterization of nisin-resistant *Leuconostoc mesenteroides* for use in cabbage fermentations. *Applied and Environmental Microbiology* 59, 11: 3778-3783.

Breidt, F., K.A. Crowley and H.P. Fleming. 1995. Controlling cabbage fermentations with nisin and nisin-resistant *Leuconostoc mesenteroides*. *Food Microbiology* 12, 2: 109-116.

Buckenhüskes, H., M. Schneider and W.P. Hammes. 1986. *Die milchsaure Vergaerung pflanzlicher Rohware unter besonderer Beruecksichtigung der Herstellung von Sauerkraut. Chemie, Mikrobiologie und Technologie der Lebensmittel* 10: 42-53.

Buckenhüskes, H.J. 1992. Advances in vegetable fermentation. *In: Les BactériesLactiques. Actes du Colloque LACTIC 91. Adria Normandie. Centre de Publication de I'Université de Caen,* France.

Buckenhüskes, H.J. 1993. Selection criteria for lactic acid bacteria to be used as starter cultures for various food commodities. *FEMS Microbiology Reviews* 12: 253-272.

Choi, S.Y., L.R. Beuchat, L.M. Perkins and T. Nakayama. 1994. Fermentation and sensory characteristics of kimchi containing potassium chloride as a partial replacement for sodium chloride. *International Journal of Food Microbiology* 21: 335-340.

Daeschel, M.A. and H.P. Fleming. 1984. Selection of lactic acid bacteria for use in vegetable fermentations. *Food Microbiology* 1, 4: 303-313.

Daeschel, M.A., R.E. Andersson and H.P. Fleming. 1987. Microbial ecology of fermenting plant materials. *FEMS Microbiology Reviews* 46: 357-367.

Dalie, D.K.D., A.M. Deschamps and F. Richard-Forget. 2010. Lactic acid bacteria — Potential for control of mould growth and mycotoxins: A review. *Food Control* 21: 370-380.

Delanoe, R. and L.O. Emard. 1971. Experimental manufacture of sauerkraut in Quebec. *Quebec Laitieret Alimentaire* 30, 7: 11-14.

Delclos, M. 1992. Vegetable preservation by a mixed organic acid fermentation. *Dissertation Abstracts International-B*, 52, 9: 4537.

Desai, P. and T. Sheth. 1997. Controlled fermentation of vegetables using mixed inoculum of lactic cultures. *Journal of Food Science and Technology* 34, 2: 155-158.

Eklund, T., 1989. *In*: Gould, G.W. (Ed.). *Mechanisms of Action of Food Preservation Procedures* Pp. 161-199.

Fleming, H.P. 1987. *Considerations for the Controlled Fermentation and Storage of Sauerkraut*. Pp. 26-32.

Fleming, H.P. and R.F. McFeeters. 1985. Residual sugars and fermentation products in raw and finished commercial sauerkraut. *New York State Agricultural Experiment Station Special Report* 56: 25-29.

Fleming, H.P., R.F. McFeeters and M.A. Daeschel. 1985. *In*: Gilliland, S.E. (Ed.). *Bacterial Starter Cultures for Foods*, Pp. 97-124.

Frank, H.K. 1973. *Starterkulturen in der Lebensmitteltechnik. Chemie, Mikrobiologie und Technologie der Lebensmittel* 2, 2: 52-56.

Gangopadhyay, H. and S. Mukherjee. 1971. Effect of different salt concentrations on the microflora and physico-chemical changes in sauerkraut fermentation. *Journal of Food Science and Technology* 8, 3: 127-131.

Gilliland, S.E. 1989. Acidophilus milk products: a review of potential benefits to consumers. *Journal of Dairy Science* 72: 2483.

Gilliland, S.E. 1990. Health and nutritional benefits from lactic acid bacteria. *FEMS Microbiology Reviews* 87: 175-188.

Halasz, A., A. Barath and W.H. Holzapfel. 1999. The influence of starter culture selection on sauerkraut fermentation. *European Food Research and Technology* 208: 434-438.

Hammes, W.P. 1990. Bacterial starter cultures in food production. *Food Biotechnology* 4, 1: 383-397.

Hammes, W.P. 1991. Fermentation of non-dairy foods. *Food Biotechnology* 5, 3: 293-303.

Han, X., H. Yi, L. Zhang, W. Huang, Y. Zhang, L. Zhang and M. Du. 2014. Improvement of fermented Chinese cabbage characteristics by selected starter cultures. *Journal of Food Science* 79: M1387-M1392.

Harris, L.J., H.P. Fleming and T.R. Klaenhammer. 1992. Characterization of two nisin-producing *Lactococcuslactis* subsp. lactis strains isolated from a commercial sauerkraut fermentation. *Applied and Environmental Microbiology* 58, 5: 1477-1483.

Harris, L.J., H.P. Fleming and T.R. Klaenhammer. 1992. Novel paired starter culture system for sauerkraut, consisting of a nisin-resistant *Leuconostoc mesenteroides* strain and a nisin-producing *Lactococcuslactis* strain. *Applied and Environmental Microbiology* 58, 5: 1484-1489.

Hose, H. and T. Sozzi. 1990. *In*: *Processing and Quality of Foods*. vol. 2. *Food Biotechnology*: *Avenues to Healthy and Nutritious Products*.

Hsu, J.Y., E.R. Wedral and W.J. Klinker. 1984. Preparation of sauerkraut utilizing hydrolyzed protein. *United States Patent US4428968*.

Isolauri, E., M. Juntunen, T. Rautanen, P. Sillanaukee and T. Koivula. 1991. A human *Lactobacillus* strain (*Lactobacillus* GG) promotes recovery from acute diarrhoea in children. *Pediatrics* 88: 90-97.

Jay, J. 1996. *Modern Food Microbiology*, 5th ed. Chapman and Hall, New York.

Johanningsmeier, S., R.F. McFeeters, H.P. Fleming and R.L. Thompson. 2007. Effects of *Leuconostoc mesenteroides* starter culture on fermentation of cabbage with reduced salt concentrations. *Journal of Food Science* 72, 5: M166-172.

Kalantzopoulos, G. 1997. *Fermented Products with Probiotic Qualities*. Anaerobe 3: 185-190.

Lücke, F.K., J.M. Brümmer, H.J. Buckenhüskes, A. Garrido Fernandez, M. Rodrigo and J.E. Smith. 1990. Starter culture development. *In*: Zeuthen, P., Cheftel, J.C., Eriksson, C., Gormley, T.R., Linko, P. and Paulus, K. (Eds.). *Processing and Quality of Foods*. vol. 2. *Food Biotechnology: Avenues to Healthy and Nutritious Products*, Elsevier Applied Science, London and New York.

Masood, M.I., M.I. Qadir, J.H. Shirazi and I.U. Khan. 2011. Beneficial effects of lactic acid bacteria on human beings. *Critical Reviews in Microbiology* 37: 91-98.

Mayer, K., G. Pause and U. Vetsch.1973. *Bildung biogener Amine während der Sauerkrautgärung. IndustrielleObst und Gemüseverwertung* 58, 11: 307-309.

Mills, S., C. Stanton, C. Hill and R.P. Ross. 2011. New developments and applications of bacteriocins and peptides in foods. *Annual Review Food Science and Technology* 2: 299-329.

Neish A.C. 1952. Analytical methods for bacterial fermentations. *Nat. Res. Counc. Can. Bull.* 46-8-3.

Niven, C.F. 1980. Technology of sodium in processed foods: General bacteriological principles, with emphasis on canned fruits and vegetables, and dairy foods. *In:* American Medical Association. Sodium and potassium in foods and drugs. *Na & K Symposium.* USA. Pp. 45-48.

Oberg, C.J. and R.J. Brown. 1993. Focusing on the chemistry and microbiology of vegetables in preservation by fermentation. *Products of Chemistry.* George B. Kauffman (Ed.). California State University, Fresno.

Owades, J.L. 1991. Method of making salt-free sauerkraut. *United States Patent US5064662.*

Park, KY., J.K. Jeong, Y.E. Lee and J.W. 3rd Daily. 2014. Health benefits of kimchi (Korean fermented vegetables) as a probiotic food. *Journal of Medicinal Food* 17, 1: 6-20.

Pederson C.S. and M.N. Albury. 1961. The effect of pure culture inoculation on fermentation of cucumbers. *Food Technology* 15: 351-354.

Pederson C.S. and M.N. Albury. 1969. The sauerkraut fermentation. *New York State Agricultural Experiment Station Technical Bulletin No. 824.* Cornell University, Geneva, N.Y.

Pederson, C.S. 1940. The relation between quality and chemical composition of canned sauerkraut. *New York State Agricultural Experiment Station Bulletin No. 693:* 1-15.

Pederson, C.S. 1979. *Microbiology of Food Fermentations,* second edition. Pp. 153-209, Avi, Westport, CI.

Peñas, E., J. Frias, B. Sidro and C. Vidal-Valverde. 2010. Chemical evaluation and sensory quality of sauerkrauts obtained by natural and induced fermentations at different NaCl levels from brassica oleracea Var. capitata Cv. Bronco Grown in eastern Spain: Effect of storage. *Journal of Agricultural and Food Chemistry,* 58, 6: 3549-3557.

Petäjä, E., P. Myllyniemi, P. Petäjä, V. Ollilainen and V. Piironen. 2000. Use of inoculated lactic acid bacteria in fermenting sour cabbage. *Agricultural and Food Science in Finland* 9: 37-48.

Salminen, E., I. Elomaa, J. Minkkinen, H. Vapaatalo and S. Salminen. 1988. Preservation of intestinal integrity during radiotherapy using live *Lactobacillus acidophilus* cultures. *Clinical Radiology* 39: 435-437.

Salminen, S., E. Isolauri and E. Salminen. 1996. Clinical uses of probiotics for stabilizing the gut mucosal barrier: Successful strains and future challenges. *Anthonie van Leeuwenhoek* 70: 347-358.

Stetter, H. and K.O. Stetter.1980. *Lactobacillus bavaricus* sp. nov., a new species of the subgenus *Streptobacterium. Zentralblatt für Bakteriologie* - 1C, 1, 1: 70-74.

Trail, A.C., H.P. Fleming, C.T. Young and R.F. McFeeters. 1996. Chemical and sensory characterization of commercial sauerkraut. *Journal of Food Quality* 19: 15-30.

Valdez de, G. F., G.S. de Giori, M. Garro, F. Mozzi and G. Oliver. 1990. Lactic acid bacteria from naturally fermented vegetables. *Microbiologie - Aliments - Nutrition* 8: 175-179.

van Baarlen, P., J.M. Wells and M. Kleerebezem. 2013. Regulation of intestinal homeostasis and immunity with probiotic lactobacilli. *Trends in Immunology* 34: 208-215.

Vaughn, R.H. 1985. The microbiology of vegetable fermentations. *In: Microbiology of Fermented Foods,* Wood, B.J.B., (Ed.). Elsevier Applied Science, New York. Pp. 49-109.

Viander, B., M. Mäki and A. Palva. 2003. Impact of low salt concentration, salt quality on natural large-scale sauerkraut fermentation. *Food Microbiology* 20: 391-395.

Visser, R., W.H. Holzapfel, J.J. Bezuidenhout and J.M. Kotze. 1986. Antagonism of lactic acid bacteria against phytopathogenic bacteria. *Applied and Environonmental Microbiology* 52: 552-555.

Vogel, R. F., M. Lohmann, M. Nguyen, A.N. Weller and W.P. Hammes. 1993. Molecular characterization of *Lactobacillus curvatus* and *Lact. sake* isolated from sauerkraut and their application in sausage fermentations. *Journal of Applied Bacteriology* 74: 295-300.

von Wright, A. 2011. *Lactic Acid Bacteria*. Boca Raton, London, New York: CRC Press, Taylor & Francis Group.

Wedral, E.R., W.J. Klinker and J.Y. Hsu. 1985. Flavouring process. *European Patent EP0106236B1*.

Wiander, B. and A. Palva. 2011. Sauerkraut and sauerkraut juice fermented spontaneously using mineral salt, garlic and algae. *Agricultural and Food Science* 20: 169-175.

Wiander, B. and E.L. Ryhänen. 2005. Laboratory and large-scale fermentation of white cabbage into sauerkraut and sauerkraut juice by using starters in combination with mineral salt with a low NaCl content. *European Food Research & Technology* 220: 191-195.

Wiander, B. and E.L. Ryhänen. 2008. Identification of lactic acid bacteria strains isolated from spontaneously fermented sauerkraut and their use in fermentation of sauerkraut and sauerkraut juice in combination with a low NaCl content. *Milchwissenschaft* 63: 386-389.

Woodford, M.M. 1985. The silage fermentation. *Microbiology of Fermented Foods*, vol. 2. Wood, B.J.B., (Ed.). 85-112, Elsevier, New York.

Xiong, T., X. Li, Q.Q. Guan, F. Peng and M.Y. Xie. 2014. Starter culture fermentation of Chinese sauerkraut: Growth, acidification and metabolic analyses. *Food Control* 41: 122-127.

6

Kimchi: A Well-known Korean Traditional Fermented Food

Jayanta Kumar Patra[1,2]*, Gitishree Das[1] and Spiros Paramithiotis[3]

1. Introduction

The Korean peninsula is surrounded by ocean on three sides — east, west, and south — and by rugged mountains in the north. This isolation of Korea from its neighboring countries allowed the ancient Korean people to develop a unique culture and distinct ethnicity. Probably the most long lasting cultural property of the early Korean people is their language and their food habit (Fig. 1). Koreans are renowned in the world for their skills in preparation of fermented foods for more than 1500 years (Han et al. 1998, Surh et al. 2008). Thus Korean food culture was developed due to the necessity of preserving food during the long, harsh winters when the temperature dipped to –20°C. In such adverse climatic condition, foods like fish, salted beans and vegetables were preserved by the fermentation process in large clay pots for use throughout the winters (Oh et al. 2014). These fermented foods and beverages are based on the cultural preferences of different geographical areas and the heterogeneity of tradition where they are produced.

In general, the fermentation is a slow decomposition process of organic substances that is induced by a group of microorganisms or enzymes that help in the conversion of carbohydrates into organic acids or alcohols (FAO 1998). The fermentation products vary according to the raw materials and preparation techniques (Surh et al. 2008) used. In many cases, the method of production of different traditional fermented foods was not known but was passed down to subsequent generations as a family tradition and technique (Swain et al. 2014). Normally, drying and salting were the most

[1] School of Biotechnology, Yeungnam University, Gyeongsan, Republic of Korea.
[2] Research Institute of Biotechnology & Medical Converged Science, Dongguk University, Gyeonggi-do 10326, Republic of Korea
[3] Department of Food Science and Human Nutrition, Agricultural University of Athens, Athens, Greece.
* Corresponding author: jkpatra@ynu.ac.kr

Fig. 1: 'Bibimbap', a Popular Korean Food Served with Different Types of Kimchi

common practices employed in the old food preservation technique. The fermentation process was evolved in order to preserve different types of fruits and vegetables by organic acid and alcohols during their harvesting season for use at the time of scarcity. This process of food preservation also adds desirable flavor, texture, reduces toxicity and decreases cooking time (Rolle and Satin 2002).

World Health Organization (WHO) and the Food and Agriculture Organization (FAO) have recommended the intake of a specific dose of fruits and vegetables in the daily food to prevent chronic pathologies, such as coronary heart problems, hypertension, and the risk of strokes (Swain et al. 2014). Today's consumer prefers fresh foods and beverages that are highly nutritional, beneficial to health and ready to consume (Endrizzi et al. 2009) due to the awareness created on maintenance of health. Lactic acid (LA) fermentation of vegetables and fruits is a common practice for maintenance and improvement of the nutritional and sensory features of food commodities (Demir et al 2006, di Cagno et al. 2013). A wide variety of lactic acid bacteria (LAB), such as *Lactobacillus brevis, Lb. casei, Lb. fermentum, Lb. kimchii, Lb. plantarum, Lb. pentosus, Leuconostoc mesenteroides, Ln. fallax, Weissella confusa, W. cibaria, W. koreensis,* and *Pediococcus pentosaceus* used in the fermentation process are isolated from various traditional naturally-fermented foods (Jung et al. 2014). Availability of specific nutrients, such as vitamins and minerals, as well as the acidic nature of fruits and vegetables, provide a favorable medium for fermentation by LAB.

Numerous fermented foods and beverages are consumed by the Koreans as well as many people of the world (Fig. 2). The major fermented food items, except for alcoholic beverages that are consumed nowadays in Korea, are basically divided into three categories (Surh et al. 2008). The first category is kimchi, which is most widely and popularly consumed. It is prepared from napa cabbage or the Chinese cabbage and/or has radish as its main ingredient, along with different kinds of vegetables (Surh et al. 2008). The fermentation process is completed within a short period of time. The second type of fermented food that is popularly consumed are the soy-based products, that include *chongkukjang* (quick-fermented soybean paste), *doenjang* (soybean paste), *ganjang* (soy sauce) and, *gochujang* (hot pepper-soybean paste) (Surh et al. 2008). Traditionally, such fermented products are prepared once in a year and consumed the year round. The fermented products prepared from fish and shellfish constitute the third category. These products are used in the preparation of kimchi or consumed as they are (Surh et al. 2008). Although different types of fermented foods are consumed in Korea and other East Asian countries, like Japan and China, scientific research regarding preparation, processing and health benefits are mainly concentrated on kimchi. In the present chapter, we focus on kimchi, its mode of preparation, processing and its medicinal potential.

Fig. 2: Different Types of Fermented Foods Available in the Republic of Korea

2. About Kimchi

Kimchi is one of the important traditional fermented foods consumed in Korea and other East Asian countries, like Japan and China. There is information on kimchi in the ancient Korean book *Samkuksaki*, which was published in A.D. 1145 and according to it, it is thought to be a simple fermentation product made by preserving vegetable in brine stored in a stone jar (Cheigh and Park 1994). Since then, kimchi has been developed into different types of products, using different vegetables, such as *baechu* (napa cabbage; *Brassica rapa* L. ssp. *Pekinensis* [Lour.] Han), cucumber, radish, or scallion as the major ingredient by adopting different types of traditional methods of preparation (Surh et al. 2008). The most important step in kimchi is the salting of raw vegetables; draining of excess of salt; then mixing with other minor ingredients, such as garlic, ginger, red pepper, fermented fish, salt, sugar and scallion, before storing the mixture for the fermentation to take place (CAC 2001, Surh et al. 2008, Kim et al. 2012). The different stages of the fermentation process are described in Fig. 3.

The average composition of different raw materials for the preparation of kimchi commercially (baechu-kimchi) includes baechu, 85.9 per cent; radish, 2.8 per cent; red pepper powder, 2.9 per cent; salt, 2.5 per cent; garlic, 1.4 per cent; ginger, 0.7 per cent scallion, 1.5 per cent; fermented fish, 1.9 per cent; sugar, 0.8 per cent; and flavor enhancers, 0.3 per cent (Park et al. 1994). The addition of various components to the kimchi preparation results in increase of the nutritive value with different functions. The different compounds, such as ascorbic acid, carotenoids, chlorophylls, capsaicin, sulfur-containing compounds, polyphenols, fibers and compounds generated in the fermentation process, such as lactic acid, glycoproteins, bacteriocin as well as lactic acid bacteria obtained from the raw ingredients result in higher laxation with increased antibacterial, antioxidant, anti-aging, antiobesity and anti-tumor properties (Song 2004).

Traditionally, kimchi was prepared and stored at home for personal consumption. However, the commercialization of kimchi has increased sharply in countries like Korea, China, Japan and the United States due to their popularity among the people (Kim et al. 2012). Kimchi is usually stored in two ways, either for three weeks at 4°C, for ripening or for only three to four days at room temperature. The raw kimchi can be eaten either as salad mixed with sesame seeds, sesame seed oil and sugar, while the over-ripened kimchi is usually boiled with meat (*jigae*). It is reported that the average consumption of kimchi by the Korean population on a daily basis is 124.3 g, with a maximum of 154.5 g consumed by the 30–49 year olds (Surh et al. 2008). The most important problem in the commercialization of packaged kimchi is the continuous fermentation process by lactic acid bacteria during distribution and storage that eventually decreases significantly the quality of kimchi. Over-fermentation of kimchi results in excessive acidification (sour taste) due to the production of acid by LAB with softening of its texture and diffusion of color (Cheigh and Park 1994, Lim et al. 2001). Thus, the extension of kimchi shelf-life and maintaining of its quality by minimizing LAB growth, is a major concern for the kimchi industry (Swain et al. 2014). Different types of kimchi are available in the local market of the Korean peninsula (Table 1, Fig. 4).

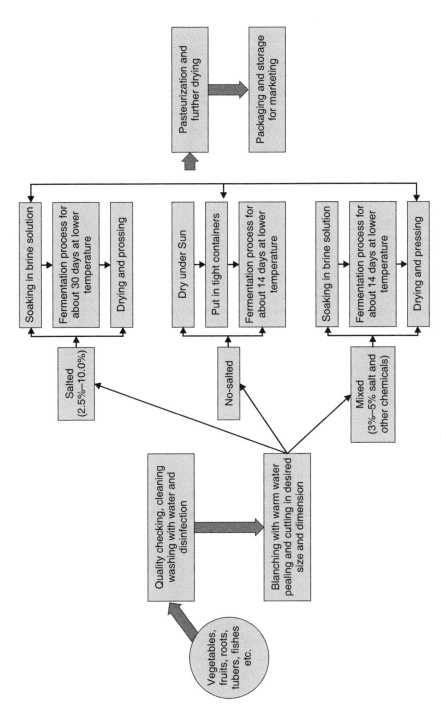

Fig. 3: Different Stages of the Fermentation Process

Table 1: Different Types of Kimchi Found in Republic of Korea

Varieties	Type
Chinese cabbage	Chinese cabbage kimchi (most popular), white kimchi (with no chili powder), *bossam* kimchi (whole heads of cabbages are wrapped with large cabbage leaves), *gotchorri* (kimchi without fermentation)
Ggagduki	Radish orcucumberis cut into small cubes and mixed with other ingredients. *Ggagduki*, oyster *ggagduki*, and shrimp *ggagduki* fall into this category
Kimchi with meat, shell fish, or lavers	Chicken kimchi, pheasant kimchi, ear shell kimchi, green laver kimchi, seafood kimchi
Kimchi with vegetables, as their base which are usually used as added materials for other kimchi	Leek kimchi, scallion kimchi, garlic kimchi
Nabak kimchi	Sliced vegetables are sliced and used for kimchi preparation. *Nabak* kimchi (cabbage and radish), *sokbakji* (with seafood), and soy sauce kimchi (salty soy sauce is used instead of salt)
Radish	*Dongchimi* (with plenty of liquid in it), *chonggak* kimchi (young radishes with leaves are used), *beeneul* kimchi (with big chunks of radish)
Vegetable Fruits	Cucumber *sobagi* (stuffed), cucumber pickle, gourd kimchi, eggplant kimchi, pumpkin kimchi, burdock kimchi
Vegetables with leaves or long leafstalks	Leaf mustard kimchi, sweet potato leafstalk kimchi, young radish kimchi, dropwort kimchi, kimchi with various wild grasses, lettuce *gotchorri*

3. Ingredients for Kimchi Preparation

The basic composition of kimchi includes vegetables like napa cabbage, cucumber, radish and other ingredients such as garlic, ginger, red pepper, fermented fish, salt, sugar and scallion. Then the whole mixture is stored for the fermentation to occur (CAC 2001, Surh et al. 2008, Kim et al. 2012). The ingredients in kimchi are well described by Cheigh and Park (1994). The raw materials used in the preparation of kimchi are categorized into four different groups: (a) major raw materials, (b) different types of spices, (c) seasonings and (d) other essential additional materials (Kim et al. 2012). The major raw materials group includes about 30 different types of vegetables such as, Chinese cabbage or napa cabbage, radish, young Oriental radish, ponytail radish and cucumber. The different types of spices used for preparation of kimchi include red pepper, green onion, ginger, garlic, onion, mustard, black pepper, and cinnamon. For seasoning of the kimchi, salt, salt-pickled seafood, sesame seed, monosodium glutamate, soybean sauce and corn syrup are the materials used. Apart

Fig. 4: Different Types of Kimchi Available in Local Markets of Republic of Korea

from these, the additional ingredients included in kimchi preparation include various types of vegetables (carrot, water cress, mushrooms), cereals (barley, rice), fruits (apple, pear), seafoods (oyster, shrimp), meats (pork, beef), etc. (Cheigh and Park 1994).

Certain types of kimchi are prepared without red chili peppers and are soaked in a tasty liquid that is less spicy and suitable for children, whereas others are made spicier by adding of red chili pepper powder and are consumed by the aged. However, garlic is always used in kimchi to add to its flavor (Oh et al. 2014). To make kimchi, the napa cabbage is mixed well with different kinds of vegetables, seasoning fish and salt, before being packed into jars. The ambient bacteria in the environment are dominated by *Lactobacillus, Leuconostoc,* and *Weissella* genera, resulting in the fermentation procedure and release of by-products that immensely contribute to the pungent flavor of kimchi which is a rich source of beneficial bioactive compounds. Only after the completion of the fermentation process can the product be called kimchi (Kim et al. 2012, Oh et al. 2014). Thus, it is different from the Chinese or Japanese salted vegetables that are only salted without fermentation. This is the reason why kimchi is most widely eaten throughout the world and is considered one of the best inventions so far in the food sector (Oh et al. 2014).

The addition of red chili pepper to kimchi increases the antioxidant capacity due to the high levels of vitamin C and rich capsaicin contents. Apart from these, the addition of red pepper to kimchi inhibits the purtrefactive microorganisms and prevents the food from spoiling. The Korean-made kimchi requires less quantity of salt due to the use of red pepper powder (Oh et al. 2014). As a result, kimchi can be stored for long periods of time and consumed even if not very salty. The use of red pepper powder to kimchi dates back to the history when the Korean people suffered greatly due to the lack of food until mid-1960s. Since then, the amazing growth of the South Korean economy has raised the annual income per capita to $20,000 and the people have become richer and gained access to an abundant variety of food for their livelihood without becoming overweight. Many people believe that since the Koreans eat a lot of kimchi that contains a high proportion of fiber (from the vegetables), capsaicin (from the chili peppers), appropriate quantities of high-grade protein (from fish and seafood), generous amounts of calcium (from the salted seafood) and an abundance of vitamins and minerals (from vegetables), a proper body weight and growth is maintained (Oh et al. 2014). However, the Korean nutritionists are worried about the increasing incidence of obesity among the young generation, who tend to skip kimchi consumption but are attracted to the imported instant foods (Oh et al. 2014).

4. Production Procedure

The Chinese cabbage kimchi, which is popularly known as 'Baechu', is one of the most important types of kimchi that are available in Korea. The basic method of its preparation is described in Fig. 5.

First, all the raw materials are collected, then the baechu cabbage is trimmed to small pieces and washed thoroughly with tap water. The excess water is drained out and a small amount of common salt is added and kept for two to three hours; — the process is known as brining. Then the excess amount of water is drained from the baechu cabbage and kept aside for preparation of kimchi. All other raw materials, such as, radish, spring onion and carrot are washed thoroughly with tap water followed by grading and cutting (Cheigh and Park 1994, Park et al. 2014). The premixture of chopped and sliced sub-ingredients including garlic, red pepper powder, salt-fermented fish (anchovy, etc.), and other

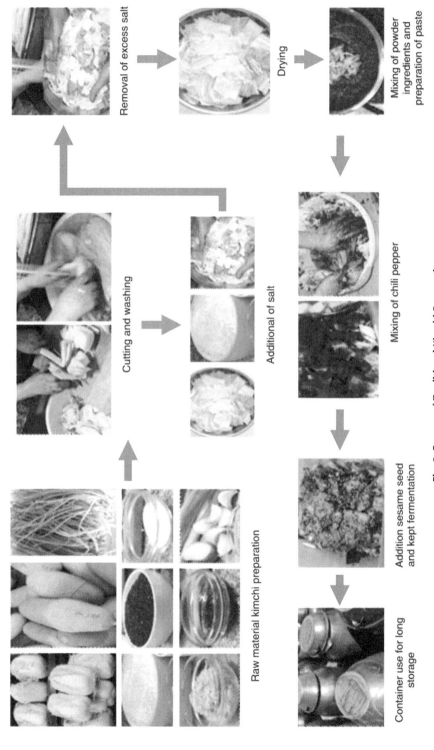

Fig. 5: Process of Traditional Kimchi Preparation

vegetables are mixed and stuffed between the leaves of the Chinese cabbage. The normal standardized proportion of the baechu kimchi composition includes ingredients such as brined baechu cabbage (100 per cent); 13 per cent of sliced radish; 3.5 per cent of red pepper powder; 1.4 per cent of chopped garlic; 0.6 per cent of chopped ginger; 1.0 per cent sugar; 2 per cent green onion and; a final salt level of 2.5 per cent; 2.2 per cent of fermented anchovy juice (optional) (Cho 1999). Apart from these essential elements, various other materials like watercress, mustard, pear, apple, pine nut, chestnut, gingko nut, cereals, fishes, crabs, etc. are also added to the premixture of kimchi, depending on the family tradition or economic status or availability of seasonal materials. After addition all the ingredients, the baechu kimchi is packaged properly and then fermented, depending upon its use. If it needs to be stored for a long time, then it is fermented at low temperature (5°C); if eaten within a short period, then the fermentation of kimchi takes place at room temperature. The recipe and other fermentation conditions are adjusted depending upon the requirement (Kil 2004).

5. Physicochemical and Microbiological Changes During Kimchi Fermentation

Based on the acidity, the fermentation of kimchi is generally divided into four stages: initial stage with acidity < 0.2 per cent; the immature stage with acidity between 0.2–0.4 per cent; the optimum-ripening stage with acidity between 0.4–0.9 per cent and the over-ripening or rancid stage with acidity > 0.9 per cent (Codex 2001). The type and quantity of lactic acid bacteria involved in the kimchi fermentation vary with the pH and acidity of the medium. Presence of *Ln. mesenteroides* is observed initially during early fermentation (pH 5.64–4.27 and acidity of 0.48–0.89 per cent). However, towards the latter part of fermentation, *Lb. sakei* dominates the fermentation process (pH 4.15 and acidity 0.98 per cent). Furthermore, the kimchi fermentation process is also influenced by the environment as well as the surrounding temperature (Lee et al. 2008a). The kimchi lactic acid bacteria reduce the pH value by producing organic acids from carbohydrates during fermentation, which causes striking changes in the bacterial community, depending on the fermentation stage (Jung et al. 2014).

The involvement of different types of microorganisms in the kimchi fermentation process was studied for the first time in Korea in 1939 and since then, many lactic acid bacteria have been isolated and thoroughly characterized (Lee et al. 1997). Due to the limitation of different culturing methods, an increasing number of studies depend on culture-independent methods, like 16S rDNA-based PCR analysis to sequence and identify the bacterial flora present in kimchi and other fermented foods (Lee et al. 2005a). Denaturing gradient gel electrophoresis (DGGE) and PCR pyro-sequencing approaches are popularly applied for the sequencing and identification of various microorganisms involved in the kimchi fermentation (Jung et al. 2013, Jeong et al. 2013a, Park et al. 2014).

6. Microbial Community Structure in Kimchi

Lactic acid bacteria are present in very low quantity and number in the different types of raw materials (vegetables and seasonings) used in kimchi preparation. However, their number gradually increases and they rapidly dominate the fermentation process owing to the favorable conditions that facilitate their growth, such as low temperature, anaerobic conditions and the presence of salt (1.5–4.0 per cent). In the early studies on kimchi microbial community, traditional methods were used that were based on the morphological and phenotypic identification of bacterial species grown on agar media, although they were often unsuccessful (Shin et al. 1996). The conventional procedure of identification of bacterial isolates from kimchi revealed the presence of two dominant species viz. *Ln. mesenteroides* and *Lb. plantarum as* that are abundantly present in kimchi products (Lim et al. 1989). However, by the molecular diagnostic technology, different species belonging to *Leuconostoc* and *Lactobacillus* species, including *Ln. gasicomitatum*, *Ln. citreum*, *Ln. gelidum* and *Lb. brevis* were identified in kimchi samples (Kim et al. 2000, Cho et al 2006). Various culture-independent methods have confirmed the presence of these microflora in kimchi products and also confirmed that majority of bacteria in kimchi are culturable (Kim and Chun, 2005, Lee et al. 2005a). Kim and Chun (2005) reported the presence of *W. koreensis* as the predominant species in commercially-available kimchi as identified by culture independent methods. Likewise, Lee et al. (2005a), using culture-independent molecular techniques, found that *W. confusa, Ln. citreum, Lb. sakei* and *Lb. curvatus* predominate during kimchi fermentation. From these results, it was concluded that lactic acid bacteria that contribute immensely in kimchi fermentation include *Lb. brevis, Lb. curvatus, Lb. plantarum, Lb. sakei, Lactococcus lactis, Ln. citreum, Ln. gasicomitatum, Ln. mesenteroides, Pd. pentosaceus, W. koreensis* and *W. confusa* (Table 2) (Choi et al. 2002, Cho et al. 2006, Kim et al. 2012, Lee et al. 2012a).

Nowadays, a wide variety of biochemical, phenotypic and immunological methods are available for detection, identification and differentiation of bacteria (Settanni and Corsetti 2007). Recently, Park et al. (2014) studied the microbial communities of 13 types of Korean commercial kimchi at pH 4.2–4.4 (optimally ripened kimchi) using the pyro-sequencing technique and reported that this microecosystem consisted of *W. koreensis* (27 per cent) that dominated and followed by *Lb. sakei* and *Lb. graminis* (14 per cent), *W. cibaria* (9 per cent), *Ln. mesenteroides* (8 per cent), *Lb. gelidum* (6 per cent), *Ln. inhae* and *Ln. gasicomitatum* (1 per cent), *Ln. kimchii* and *W. confusa* (0.3 per cent), and other bacteria (16 per cent). At the genus level, *Weissella* sp. prevailed the microcommunity with 44 per cent, followed by

Lactobacillus sp. with 38 per cent, and *Leuconostoc* sp. with 17 per cent (Park et al. 2014). Of late, different types of probiotic lactic acid bacteria are used as starter culture in kimchi fermentation.

Table 2: List of Predominant Kimchi Lactic Acid Bacteria

Lactobacillus **spp.**	*Leuconostoc* **spp.**	*Weissella* **spp.**
Lb. alimentarius	Ln. carnosum	W. cibaria
Lb. brevis	Ln. citreum	W. confusa
Lb. curvatus	Ln. dextranicum	W. koreensis
Lb. fermentum	Ln. gasicomitatum	W. paramesenteroides
Lb. fructosus	Ln. gelidum	W. soli
Lb. homohiochii	Ln. hozapfelii	
Lb. kimchii	Ln. inhae	
Lb. leichimannii	Ln. kimchii	
Lb. mali	Ln. lactis	
Lb. minor	Ln. mesenteroides	
Lb. paraplantarum	Ln. paramesenteroides	
Lb. pentosus	Ln. pseudomesenteroides	
Lb. plantarum		
Lb. sakei		

It was also reported that apart from the lactic acid bacteria, some yeasts also play a significant role in the fermentation of kimchi. Possibly some yeasts cause the off-flavor or food spoilage during kimchi fermentation (Chang et al. 2008). Diverse yeasts, including *Candida, Pichia, Saccharomyces,* and *Kluyveromyces* were isolated from kimchi during their late fermentation stages since they grow well at low pH (Chang et al. 2008, Jeong et al. 2013a). The *Saccharomyces* spp. consume free sugars and produce ethanol as a major product and glycerol as a by-product after approximately 30 days of kimchi fermentation (Jeong et al.2013a). Apart from these, with the decrease of kimchi lactic acid bacteria and *Saccharomyces* spp. population, growth of *Candida* spp., that are suspected to be responsible for spoilage during kimchi fermentation, is usually observed at the later stage of fermentation (Chang et al. 2008, Jeong et al. 2013a).

Furthermore, the infections caused by bacteriophages constitute a major problem in industrial food fermentation and a number of protective measures were employed for their control (Moineau and Levesque 2005, Kleppen et al. 2012). It was reported that bacteriophages affect the bacterial biota in different fermented foods and are responsible for the variability that is frequently observed in vegetable fermentation (Kleppen et al. 2011, Lu et al. 2012). In a recent study, a high amount of bacteriophage DNA was found in kimchi, which indicates the effect of bacteriophages in kimchi fermentation (Jung et al. 2011, Park et al. 2011).

7. Factors Affecting the Microbiota of Kimchi

There are many factors that are responsible for the dynamics of the microbial load during kimchi fermentation. More specifically, kimchi varieties, raw materials (vegetable type, harvesting area and season and cultivar), ingredients (seasonings), salt concentration, fermentation temperature and starter cultures used are the most important factors that affect the kimchi microbiota. The fermentation temperature influences the kimchi fermentation (Cho et al. 2006). The ingredients (seasonings) that are used in kimchi preparation greatly affect its type and quantity of microbiota (Lee et al. 2008b). When kimchi contained more garlic, the population of lactic acid bacteria significantly increased, while that of other aerobic bacteria decreased (Lee et al. 2008b). The latter finding indicated that the source of lactic acid bacteria might be garlic or the nutrient content of garlic that promoted the growth of lactic acid bacteria. The use of red pepper powder delayed the kimchi fermentation progress, including the growth of lactic acid bacteria and the production of metabolites during the initial stage of fermentation (Jeong et al. 2013b); the proportion of *Weissella* sp. was higher in kimchi with red pepper powder than in kimchi without red pepper powder, while the proportion of *Leuconostoc* spp. and *Lactobacillus* spp. was lower in kimchi with red pepper powder (Jung et al. 2014).

The concentration of salt is another important factor affecting the microbiota of kimchi. Kimchi is generally fermented at lower salt concentrations (2–3 per cent), which is favorable to the growth of lactic acid bacteria. It was reported that at low salt concentration (<2 per cent), acidification of kimchi occurred rapidly with quick growth of lactic acid bacteria, whereas at high salt concentration (>3 per cent), fermentation and ripening of kimchi were delayed (Mheen and Kwon 1984). Kimchi is generally fermented at low temperatures (2–6°C) for proper ripening and preservation, but sometimes it is fermented at higher temperatures (15–25°C) to expedite the fermentation process. The growth of lactic acid bacteria in kimchi occurs more rapidly at elevated temperatures, which results in a more efficient decrease in the pH value during fermentation (Mheen and Kwon 1984).

8. Metabolism of Kimchi Lactic Acid Bacteria

The fermentation of kimchi is generally processed by obligate aerobic microorganisms belonging to *Leuconostoc* spp. and *Weissella* spp.; facultative anaerobic microorganisms such as *Lb. curvatus*, *Lb. plantarum* and *Lb. sakei* as well as different types of hetero fermentative lactic acid bacteria. These hetero fermentative group of lactic acid bacteria produce carbon dioxide, lactic acid and ethanol from glucose through the 6-phosphogluconate/phosphoketolase pathway under anaerobic conditions. Glucose and fructose were found in kimchi as major free sugars (Jeong et al. 2013a, b,

Jung et al. 2011). Similarly, mannitol, that imparts a refreshing taste to kimchi, is also observed during kimchi fermentation (Jeong et al. 2013a, b, Jung et al. 2011), resulting from the metabolic activities of *Leuconostoc* spp. that is known as the major mannitol-producing lactic acid bacteria and are dominant in the microbiota of kimchi.

Besides lactate and acetate as the major organic acids, other diverse group of organic acids, such as butanoic acid, 2-methylpropionic acid, propionic acid, fumaric acid, tartaric acid and succinic acid are produced as minor components during fermentation (Shim et al. 2012). The amino acids, which play an important role in imparting taste to kimchi, are accumulated during fermentation (Jeong et al. 2013a). Although kimchi has high vitamin content, there are no reports on the production of vitamins by any of the lactic acid bacteria involved in kimchi fermentation. However, genes that are involved in the synthesis of vitamin B group (folate and riboflavin) were present in the genome of *Ln. mesenteroides* and *Lb. sakei* (Jung et al. 2012), indicating that lactic acid bacteria might be involved in the production of vitamins in kimchi during the fermentation process (Jung et al. 2014).

9. Health Benefits of Kimchi

The traditional food culture of Korea is very close to Nature and thus they have very simple procedures of food preparation besides having simple food habits. Most Korean do not require any special health supplements as food, since the natural food they consume is full of all types of health supplements. Since time immemorial, the Korean philosophy emphasizes the importance of food for life and a simple food habit; food and medicine are believed to originate from the same source, so maintenance of a simple and normal food habit is important for preserving of health and curing diseases. During the spring season, availability of newly sprouted mugwort and fresh vegetables provided vitamins to the Korean, which were not available during the long winter season and the fresh vegetables transferred their nutrients to the people consuming them. Likewise, the fresh fragrances and aromatic flavors of spring vegetables helped in restoring the lost appetites and energy of the people (Oh et al. 2014).

Kimchi (baechu kimchi) is a low-calorie food containing high amounts of vitamins such as vitamin C, B-carotene, vitamin B complex, etc.; apart from minerals such as sodium, calcium, potassium, iron and phosphorus; dietary fiber and other functional components such as allyl compounds, gingerol, capsaicin, isothiocyanate and chlorophyll. Phytochemicals, such as indole compounds, b-sitosterol, benzyl isothiocyanate and thiocyanate are the important active compounds in kimchi reported to possess medicinal potential against obesity and cancer besides having antioxidant and anti-atherosclerotic properties (Park and Rhee 2005, Ahn 2007) (Table 3).

Table 3: Medicinal Potential of Kimchi

Medicinal potential	References
Anti-obesity	Yoon et al. (2004), Kim et al. (2011a)
Anti-cancer	Kil (2004), Kim et al. (2007), Han et al. (2011)
Anti-xidative	Lee et al. (2005b), Park et al. (2014)
Anti-bacterial	Kil et al. (2004)
Cholesterol-lowering	Kim and Lee (1997), Kil et al. (2004)
Immuno potential	Kim et al. (1997), Wu et al. (2000)
Functional foods	Park and Rhee (2005)

9.1. Antiobesity Potential

Red pepper powder is an important kimchi ingredient that is rich in capsaicin which causes loss of body fat by stimulating the spinal nerves and activating the release of catecholamines in the adrenal glands of the body (Park et al. 2014). This compound increases the body metabolism, thus decreasing the fat content. Various researchers have demonstrated that consumption of kimchi in rats led to lower body fat (Yoon et al. 2004). Moreover, Kim et al. (2011a) suggested that consumption of fermented kimchi may affect obesity, lipid metabolism and inflammatory processes in human beings as well; more accurately, if using fermented kimchi in a four-week diet period. Significant differences in the blood pressure, both systolic and diastolic, body fat, fasting glucose as well as total cholesterol were observed when compared to the group using fresh kimchi.

9.2. Anti-cancer Potential

The anti-cancer potential of kimchi is mainly due to the addition of different ingredients, such as Chinese pepper, mustard leaf and Korean mistletoe extracts to the kimchi during preparation (Kil 2004, Kim et al. 2007, Han et al. 2011). It was reported that kimchi supplemented with Korean mistletoe extract had greater anti-cancer potential (80 per cent inhibition ratio) than kimchi without Korean mistletoe extract (62 per cent) against the HT-29 human colon carcinoma cells (Kil 2004). Different kinds of salt, natural sea-salt (NS) without bittern or baked salt are also used to increase the taste and are reported to increase the anti-cancer potential of kimchi (Han et al. 2009).

9.3. Anti-oxidative Potential

The addition of different ingredients such as Chinese cabbage, spring onion, garlic, ginger, red pepper etc. to kimchi, and which are rich in chlorophyll, phenolic compounds, carotenoids, vitamins, minerals and other secondary metabolites and phytochemicals play a vital role in the increased antioxidant potential of kimchi (Park et al. 2014).

9.4. Anti-aging Potential

The anti-aging potential of kimchi in the brain of senescence-accelerated mouse was reported by Kim et al. (2002). It was concluded that kimchi has an important role in retarding aging through the reduction of free radical production and increase in anti-oxidative enzyme activities. The anti-aging activity of kimchi was also evaluated using stress-induced premature senescence of WI-38 human fibroblasts challenged with hydrogen peroxide (Kim et al. 2011b). It was reported that optimally ripened kimchi indeed delayed the aging process by regulating the inflammation mechanisms.

9.5. Anti-bacterial Potential

The antibacterial potential of kimchi has been attributed to the presence of sulfur containing compounds and various lactic acid bacteria. It has been reported that uptake of kimchi helps to reduce the effect of *Helicobacter pylori* infection in the digestive tract (Kil et al. 2004).

9.6. Cholesterol-Lowering Activity

The cholesterol-lowering effect of kimchi was studied by various researchers who indicated that consumption of kimchi on a regular basis helps to decrease the level of cholesterol in the body (Kim and Lee 1997, Kil et al. 2004).

9.7. Immuno Potential

The immune function of kimchi was also reported. Various researchers have proved that consumption of kimchi on a regular basis helps to increase the immune cell development and growth (Kim et al. 1997, Wu et al. 2000).

9.8. Functional Properties

Kimchi and its various products that are rich sources of hetero fermentative lactic acid bacteria obtained from a mixture of ingredients, such as vegetables and seasonings, are regarded as functional foods with higher concentrations of probiotic lactic acid bacteria, antioxidants, vitamins, minerals and dietary fibers (Park and Rhee 2005). The functional properties of several *Lb. sakei* and *Lb. plantarum* strains was studied by Lee et al. (2011). Their ability to survive conditions simulating the gastrointestinal tract, to inhibit growth of several food-borne pathogens, to lower cholesterol levels as well as adhere *in vitro* to human cell lines makes them potential candidates for application as multifunctional strains.

10. Industrial Production of Kimchi

Kimchi is an important globally popular food because of its taste and health benefits, increasing its global market worldwide (Jung et al. 2014). The kimchi companies in different parts of the Republic of Korea are now concentrating in production of uniform quality of kimchi. However, since kimchi is generally fermented by different naturally-occurring lactic acid bacteria derived from various types of raw materials, its quality also varies on the basis of the type of raw materials used, even when processed under controlled environmental conditions (Jung et al. 2014). In order to overcome this problem, the use of starter cultures was suggested as an alternative to maintain the quality during kimchi production (Jung et al. 2014).

Presently, different types of kimchi lactic acid bacteria (*Ln. mesenteroides*, *Ln. citreum*, and *Lb. plantarum*) with medicinal potential such as mannitol production, anti-microbial activity, acid and bile tolerances are being used as the starter culture in the preparation of kimchi (Kim et al. 2012, Ryu et al. 2012), though it is not always successful as it very often fails to result in products of high quality. Apart from the use of starter culture and their adaptability to kimchi fermentation conditions, the production of major organic acids (acetate and lactate) and mannitol as well as other components, such as minor amino acids, organic acids, flavoring compounds, biogenic amines, vitamins and bacteriocins are taken into consideration for the development of starter cultures for kimchi fermentation, since these substances also affect the taste and health-promoting and nutritional properties of kimchi products (Jung et al. 2014).

11. Safety Concerns

The increase in concern and awareness among the consumers about the type, quality and the raw materials used and their safety is increasing day by day. This change among the consumers is accelerated viathe media, which makes it possible for people to obtain sufficient information on processed foods and their safety. Recently, kimchi was presented in the *Health Magazine* of 2006 as one of the world's five healthiest foods (Lee et al. 2012b). The superior quality of the traditional Korean foods is primarily represented by fermented foods, such as kimchi and fermented soybean products (Doenjang, Gochujang, Kanjang), which are good sources of natural nutrients, such as essential amino acids, vitamins, and minerals (Lee et al. 2012b). However, there are many factors that need to be addressed before globalizing traditional foods, like kimchi. One of them is the ability of food-borne pathogens to persist in the final product originating either from the raw materials or surviving the fermentation process or through cross-contamination. Inatsu et al. (2004) studied the survival of *Escherichia coli* O157:H7, *Salmonella* Enteritidis, *Staphylococcus aureus* and *Listeria monocytogenes* in both commercial and laboratory-prepared kimchi. It was reported that all pathogens could survive at 10°C for seven days. Upon prolongation of the incubation, *S. aureus* reached enumeration limit after 12 days, whereas *S.* Enteritidis and *L. monocytogenes* reached that level after 16 days, while the *E. coli* O157:H7 population increased and remained at high levels throughout the 24 days of the experiment. Regarding the product prepared in the laboratory, *E. coli* O157:H7 and *S.* Enteritidis populations remained at high level while *S. aureus* and *L. monocytogenes* reached enumeration limit after 12 and 20 days, respectively. This issue was also assessed by Cho et al. (2011). The initial mixture of *soongchimchae*, i.e. a type of kimchi that combines fermented vegetables and meat, was spiked with *E. coli* O157:H7 and *L. monocytogenes* and allowed to ferment at 4°C for 15 days. It was reported that the population of *E. coli* O157:H7 and *L. monocytogenes* gradually decreased during fermentation and was below detection limit after 14 and 15, post-fermentation days, respectively.

11.1. Aflatoxin

Some researchers during the end of 1960's suggested that the high incidence of stomach cancer among Koreans and other Far Eastern people might be related to aflatoxins produced in the preparation of soybean paste (Crane et al. 1970). Since aflatoxin is one of the most powerful carcinogenic substance, it raised concern worldwide and researchers in Korea and Japan are investigating into the depth of such concept. Some researchers have shown that patients with gastric cancer showed significantly higher intake of kimchi and soybean paste than those who consumed less fermented products (Nan et al. 2005). This result confirmed that kimchi and soybean pastes are risk factors in gastric cancer since salts, together with other chemicals from the food, might be playing an important role in gastric cancer. However, some researchers who studied meju (a fermented soybean loaf) concluded that the fermented food consuming gastric cancer induced-rats had a suppressive effect on cancer development (Kim et al. 1985). Another group of researchers extensively studied kimchi and other fermented soybean products in Korea and demonstrated that aflatoxin B_1 and G_1 gradually degraded by 80–90 per cent after two months of fermentation and 100 per cent after three months of the fermentation process (Park et al. 2003). These results confirmed that the aflatoxin in soybean products and kimchi were not responsible for causing gastric cancer.

11.2. Nitrate, Nitrite, Amines

There are several controversial debates on the safety of kimchi related to nitrates, nitrites, secondary and biogenic amines. The formation of nitrite and secondary amines during kimchi fermentation was of great concern to researchers on the safety of kimchi consumption. However, they found that the amount of nitrite and secondary amines production is much below the permissible limits in food and thus it does not affect the consumers (Mah et al. 2004).

11.3. Reduction Policy of Hazardous Material in Kimchi

In order to minimize the hazardous effects of kimchi consumption, three safety measures can be followed. Firstly, use of good starter culture can be applied with more safe microorganisms. Secondly, the fermentation conditions may be changed as per suitability so that no toxic byproducts develop and lastly, various ingredients can be used in kimchi preparation to increase its nutritional and neutraceutical potentials (Mah et al. 2003).

12. Food Safety Management System in Korea

Two government agencies, Korea Food and Drug Administration (KFDA) and Ministry of Food, Agriculture and Forestry (MIFAFF) are engaged in the assessment of food safety in the Korean market. The MIFAFF is responsible for the safety of raw materials used in processed food and the KFDA checks imported/processed foods from different countries (Lee et al. 2012b).

13. Conclusion

Kimchi is one of the most popular fermented products prepared from Chinese cabbage along with other ingredients and potentially probiotic lactic acid bacteria. It has various nutritional and nutraceutical potential. The quality of kimchi is enhanced by manipulating the different kinds and amounts of ingredients and fermentation conditions. Although there are a number of challenges in kimchi processing and production, however this fermented food, that is a part of the normal diet in countries like Korea and Japan since many generations, could play a vital role in the global food industry. The health-promoting, probiotic and nutraceutical potential of kimchi can be optimized by manipulating its ingredients, and using appropriate probiotic starters and preparation methods, including fermentation. This kimchi can serve as one of the best healthy foods available in the world.

Acknowledgements

We are grateful to the local people of Gyeongsan area, Gyeongsan, Republic of Korea for allowing photographs to be taken of different types of kimchi sold in the local markets.

References

Ahn, S.J. 2007. The effect of kimchi powder supplement on the body-weight reduction of obese adult women [MS thesis]. Pusan National University, Busan, Korea.

CAC, 2001. *Codex Standard for Kimchi (CODEX STAN 223-2001)*. Codex Alimentarius Commission, Rome, Italy.

Chang, H.W., K.H. Kim, Y.D. Nam, S.W. Roh, M.S. Kim, C.O. Jeon, H.M. Oh and J.W. Bae. 2008. Analysis of yeast and archaeal population dynamics in kimchi, using denaturing gradient gel electrophoresis. *International Journal of Food Microbiology* 126: 159-166.

Cheigh, H.S. and K.Y. Park. 1994. Biochemical, microbiological and nutritional aspects of kimchi (Korean fermented vegetable products). *Critical Reviews in Food Science and Nutrition* 34: 175-203.

Cho, E.J. 1999. Standardization and cancer chemopreventive activities of Chinese cabbage kimchi [PhD thesis]. Pusan National University, Busan, Korea.

Cho, G.Y., M.H. Lee and C. Choi 2011. Survival of *Escherichia coli* O157:H7 and *Listeria monocytogenes* during kimchi fermentation supplemented with raw pork meat. *Food Control* 22: 1253-1260.

Cho, J.H., D.Y. Lee, C.N. Yang, J.I. Jeon, J.H. Kim and H.U. Han. 2006. Microbial population dynamics of kimchi, a fermented cabbage product. *FEMS Microbiology Letters* 257: 262-267.

Choi, H.J., C.I. Cheigh, S.B. Kim, J.C. Lee, D.W. Lee, S.W. Choi, J.M. Park and Y.R. Pyun. 2002. *Weissella kimchii* sp. nov., a novel lactic acid bacterium from kimchi. *International Journal of Systematic and Evolutionary Microbiology* 52: 507-511.

Codex Alimentarius Commission (Codex). 2001. Codex standard for kimchi. *Codex Standard* 223. Food and Agriculture Organization of the United Nations, Rome, Italy.

Crane, P.S., S.U. Rhee, and D.J. Seel. 1970. Experience with 1079 cases of cancer of the stomach seen in Korea from 1962 to 1968. *The American Journal of Surgery* 120: 747-751.

Demir, N., K.S. Bahceci and J. Acar. 2006. The effects of different initial *Lactobacillus plantarum* concentrations on some properties of fermented carrot juice. *Journal of Food Processing and Preservation* 30: 352-363.

di Cagno, R., R. Coda, M.D. Angelis, M. Gobbetti. 2013. Exploitation of vegetables and fruits through lactic acid fermentation. *Food Microbiology* 33: 1-10.

Endrizzi, I., G. Pirretti, D.G. Calo and F. Gasperi 2009. A consumer study of fresh juices containing berry fruits. *Journal of the Science of Food and Agriculture* 89: 1227-1235.

FAO. 1998. Fermented Fruits and Vegetables — A Global Perspective. vol. 134, *FAO Agricultural Services Bulletin*, Rome, Italy.

Han, G.J., A.R. Son, S.M. Lee, J.K. Jung, S.H. Kim and K.Y. Park. 2009. Improved quality and increased in vitro anti-cancer effect of kimchi by using natural sea salt without bittern and baked (Guwun) salt. *Journal of the Korean Society of Food Science and Nutrition* 38: 996-1002.

Han, W., W. Hu and Y.M. Lee. 2011. Anti-cancer activity of human colon cancer (HT-29) cell line from different fraction of *Zanthoxylum schnifolium* fruits. *Korean Journal of Pharmacognosy* 42: 282-287.

Han, B.J., B.R. Han and H.S. Whang. 1998. Kanjang and Doenjang, in one hundred Korean foods that Koreans must know. *Hyun Am Sa*, Seoul, 500-507.

Inatsu Y., M.L. Bari, S. Kawasaki and K. Isshiki. 2004. Survival of *Escherichia coli* O157:H7, *Salmonella* Enteritidis, *Staphylococcus aureus* and *Listeria monocytogenes* in Kimchi. *Journal of Food Protection* 67: 1497-1500.

Jeong, S.H., J.Y. Jung, S.H. Lee, H.M. Jin and C.O. Jeon. 2013a. Microbial succession and metabolite changes during fermentation of dongchimi, traditional Korean watery kimchi. *International Journal of Food Microbiology* 164: 46-53.

Jeong, S.H., H.J. Lee, J.Y. Jung, S.H. Lee, H.Y. Seo, W.S. Park and C.O. Jeon. 2013b. Effects of red pepper powder on microbial communities and metabolites during kimchi fermentation. *International Journal of Food Microbiology* 160: 252-259.

Jung J.Y., S.H. Lee and C.O. Jeon. 2014. Kimchi microflora: history, current status, and perspectives for industrial kimchi production. *Applied Microbiology and Biotechnology* 98: 2385-2393.

Jung, J.Y., S.H. Lee, H.M. Jin, Y. Hahn, E.L. Madsenand C.O. Jeon.2013. Metatranscriptomic analysis of lactic acid bacterial gene expression during kimchi fermentation. *International Journal of Food Microbiology* 163: 171-179.

Jung, J.Y., S.H. Lee, J.M. Kim, M.S. Park, J. Bae, Y. Hahn, E.L. Madsenand C.O. Jeon. 2011. Metagenomic analysis of kimchi, a traditional Korean fermented food. *Applied and Environmental Microbiology* 77: 2264-2274.

Jung, J.Y., S.H. Lee, S.H. Leeand C.O. Jeon. 2012. Complete genome sequence of *Leuconostoc mesenteroides* subsp. *mesenteroides* strain J18, isolated from kimchi. *Journal of Bacteriology* 194: 730.

Kil, J.H. 2004. Studies on development of cancer-preventive and anti-cancer kimchi and its anti-cancer mechanism [PhD thesis]. Pusan National University, Busan, Korea.

Kil, J.H., K.O. Jung, H.S. Lee, I.K. Hwang, Y.J. Kim and K.Y. Park. 2004. Effects of kimchi on stomach and colon health of *Helicobacter pylori*-infected volunteers. *Journal of Food Science and Nutrition* 9: 161-166.

Kim, B.J., J. Chun and H.U. Han. 2000. *Leuconostoc kimchii* sp. nov., a new species from kimchi. *International Journal of Systematic and Evolutionary Microbiology* 50: 1915-1919.

Kim, B.K., K.Y. Park, H.Y. Kim, S.C. Ahn and E.J. Cho. 2011b. Anti-aging effects and mechanisms of kimchi during fermentation under stress-induced premature senescence cellular system. *Food Science and Biotechnology* 20: 643-649.

Kim, E.K., S.Y. An, M.S. Lee, T.H. Kim, H.K. Lee, W.S. Hwang, S.J. Choe, T.Y. Kim, S.J. Han, H.J. Kim, D.J. Kim and K.W. Lee. 2011a. Fermented kimchi reduced body weight and improves metabolic parameters in overweight and obese patients. *Nutrition Research* 31: 436-443.

Kim, J.H., J.D. Ryu, H.G. Lee and J.H. Park. 2002. The effect of kimchi on production of free radicals and anti-oxidative enzyme activities in the brain of SAM. *Journal of the Korean Society of Food Science and Nutrition* 31: 117-123.

Kim J., J. Bang, L.R. Beuchat, H. Kim and J.H. Ryu. 2012. Controlled fermentation of kimchi using naturally occurring anti-microbial agents. *Food Microbiology* 32: 20-31.

Kim, M.J. and J.S. Chun. 2005. Bacterial community structure in kimchi, a Korean fermented vegetable food, as revealed by 16S rRNA gene analysis. *International Journal of Food Microbiology* 103: 91-96.

Kim, M.J., M.J. Kwon, Y.O. Song, E.K. Lee, H.J. Yoon and Y.S. Song. 1997. The effects of kimchi on hematological and immunological parameters *in vivo* and *in vitro*. *Journal of the Korean Society of Food Science and Nutrition* 26: 1208-1214.

Kim, J.Y. and Y.S. Lee. 1997. The effects of kimchi intake on lipid contents of body and mitogen response of spleen lymphocytes in rats. *Journal of the Korean Society of Food Science and Nutrition* 26: 1200-1207.

Kim, Y.T., B.K. Kim and K.Y. Park. 2007. Anti-mutagenic and anti-cancer effects of leaf mustard and leaf mustard kimchi. *Journal of Food Science and Nutrition* 12: 84-88.

Kim, J.P., J.G. Park, M.D. Lee, M.D. Han, S.T. Park, B.H. Lee and S.E. Jung. 1985. Co-carcinogenic effects of several Korean foods ongastric cancer induced by N-methyl-N-nitro-N-nitrosoguanidine in rats. *The Japanese Journal of Surgery* 15: 427-437.

Kim, Y.K.L., E. Koh, H.J. Chung and H. Kwon. 2000. Determination of ethyl carbamate in some fermented Korean foods and beverage. *Food Additives and Contaminants* 17: 469-475.

Kleppen, H.P., T. Bang, I.F. Nes and H. Holo. 2011. Bacteriophages in milk fermentations: Diversity fluctuations of normal and failed fermentations. *International Dairy Journal* 21: 592-600.

Kleppen, H.P., H. Holo, S.R. Jeon, I.F. Nes and S.S. Yoon. 2012. Novel Podoviridae family bacteriophage infecting *Weissella cibaria* isolated from kimchi. *Applied and Environmental Microbiology* 78: 7299-7308.

Lee, D.Y., S.J. Kim, J.H. Cho and J.H. Kim. 2008a. Microbial population dynamics and temperature changes during fermentation of kimjang kimchi. *Journal of Microbiology* 46: 590-593.

Lee, G.I., H.M. Lee and C.H. Lee. 2012b. Food safety issues in industrialization of traditional Korean foods. *Food Control* 24: 1-5

Lee H., H. Yoon, Y. Ji, H. Kim, H. Park, J. Lee, H. Shin and W. Holzapfel. 2011. Functional properties of *Lactobacillus* strains isolated from kimchi. *International Journal of Food Microbiology*, 145: 155-161.

Lee, J., K.T. Hwang, M.S. Heo, J.H. Lee and K.Y. Park. 2005b. Resistance of *Lactobacillus plantarum* KCTC 3099 from kimchi to oxidative stress. *Journal of Medicinal Food* 8: 299-304.

Lee, J.S., C.O. Chun, M.C. Jung, W.S. Kim, H.J. Kim, M. Hector, S.B. Kim, C.S. Park, J.S. Ahn, Y.H. Park and T.I. Mheen. 1997. Classification of isolates originating from kimchi using carbon-source utilization patterns. *Journal of Microbiology and Biotechnology* 7: 68-74.

Lee, J.S., G.Y. Heo, J.W. Lee, Y.J. Oh, J.A. Park, Y.H. Park, Y.R. Pyun and J.S. Ahn. 2005a. Analysis of kimchi microflora using denaturing gradient gel electrophoresis. *International Journal of Food Microbiology* 102: 143-150.

Lee, J.Y., M.K. Choi and K.H. Kyung. 2008b. Reappraisal of stimulatory effect of garlic on kimchi fermentation. *Korean Journal of Food Science and Technology* 40: 479-484.

Lee, S.H., M.S. Park, J.Y. Jung and C.O. Jeon. 2012a. *Leuconostoc miyukkimchii* sp. nov., isolated from brown algae (*Undaria pinnatifida*) kimchi. *International Journal of Systematic and Evolutionary Microbiology* 62: 1098-1103.

Lim, C.R., H.K. Park and H.U. Han. 1989. Reevaluation of isolation and identification of gram positive bacteria in kimchi. *The Korean Journal of Microbiology* 27: 404-414.

Lim, Y.S., K.N. Park, M.J. Bae and S.H. Lee. 2001. Anti-microbial effects of ethanol extracts of *Pinusdensiflora* Sieb. et Zucc on lactic acid bacteria. *Journal of the Korean Society of Food Science and Nutrition* 30: 1158-1163.

Lu, Z., I.M. Perez-Diaz, J.S. Hayes and F. Breidt. 2012. Bacteriophage ecology in a commercial cucumber fermentation. *Applied and Environonmental Microbiology* 78: 8571-8578.

Mah, J.H., J.B. Ahn, J.H. Park, H.C. Sung and H.J. Hwang. 2003. Characterization of biogenic amine-producing microorganisms isolated from Myeolchi-Jeot, Korean salted and fermented anchovy. *Journal of Microbiology and Biotechnology* 13: 692-699.

Mah, J.H., Y.J. Kim, H.K. No and H.J. Hwang. 2004. Determination of biogenic amines in kimchi, Korean traditional fermented vegetable products. *Food Science and Biotechnology* 13: 826-829.

Mheen, T.I. and T.W. Kwon. 1984. Effect of temperature and salt concentration on kimchi fermentation. *Korean Journal of Food Science and Technology* 16: 443-450.

Moineau, S. and C. Levesque. 2005. Control of bacteriophages in industrial fermentations. Pp. 286-296. *In*: E. Kutter and A. Sulakvelidze (Eds.). *Bacteriophages: Biology and Applications*. CRC Press, Boca Raton.

Nan, H.M., J.W. Park, Y.J. Song, H.Y. Yun, J.S. Park, T. Hyun, S.J. Youn, Y.D. Kim, J.W. Kang and H. Kim. 2005. Kimchi and soybean pastes are risk factors of gastric cancer. *World Journal of Gastroenterology* 11: 3175-3181.

Oh, S.H., K.W. Park, J.W. Daily and Y.E. Lee. 2014. Preserving the legacy of healthy Korean food. *Journal of Medicinal Food* 17: 1-5.

Park, E.J., K.H. Kim, G.C.J. Abell, M.S. Kim, S.W. Roh and J.W. Bae. 2011. Metagenomic analysis of the viral communities in fermented foods. *Applied and Environmental Microbiology* 77: 1284-1291.

Park, K.Y., J.K. Jeong, Y.E. Lee and J.W. Daily. 2014. Health benefits of kimchi (Korean fermented vegetables) as a probiotic food. *Journal of Medicinal Food* 17: 6-20.

Park, K.Y. and S.H. Rhee. 2005. Functional foods from fermented vegetable products: Kimchi (Korean fermented vegetables) and functionality. Pp. 341-380. *In*: J. Shi, C.T. Ho and F. Shahidi (Eds.). *Asian Functional Foods*. CRC Press, Inc., Boca Raton, FL, USA.

Park, K.Y., K.O. Jung, S.H. Rhee and Y.H. Choi. 2003. Anti-mutagenic effects of doenjang (Korean fermented soypaste) and its active compounds. *Mutation Research – Fundamental and Molecular Mechanisms of Mutagenesis* 523-524: 43-53.

Park, W.S., Y.C. Koo, B.H. Ahn and S.Y. Choi. 1994. *Process Standardization of Kimchi Manufacturing*, Korea Food Research Institute, Seoul.

Rolle, R. and M. Satin. 2002. Basic requirements for the transfer of fermentation technologies to developing countries. *International Journal of Food Microbiology* 75: 181-187.

Ryu, B.H., G.S. Sim, J.H. Lee and W.K. Ha. 2012. Plant originated *Lactobacillus plantarum* DSR CK10, DSR M2 to keep freshness and use thereof. *Korean Patent* 10-1124056.

Settanni, L. and A. Corsetti. 2007. The use of multiplex PCR to detect and differentiate food- and beverage-associated microorganisms: A review. *Journal of Microbiological Methods* 69: 1-22.

Shim, S.M., J.Y. Kim, S.M. Lee, J.B. Park, S.K. Oh and Y.S. Kim. 2012. Profiling of fermentative metabolites in kimchi: volatile and non-volatile organic acids. *Journal of the Korean Society for Applied Biological Chemistry* 55: 463-469.

Shin, D.H., M.S. Kim, J.S. Han, D.K. Limand W.S. Bak. 1996. Changes of chemical composition and microflora in commercial kimchi. *Korean Journal of Food Science and Technology* 28: 137-145.

Song, Y.O. 2004. The functional properties of kimchi for the health benefits. *Food Industry and Nutrition* 9: 27-33.

Surh, J., Y.H.L. Kim and H. Kwon. 2008. Korean fermented foods, kimchi and doenjang. Pp. 333-351. *In*: E.R. Farnworth (Eds.). *Handbook of Fermented Functional Foods*, CRC Press, Boca Raton, London.

Swain, M.R., M. Anandharaj, R.C. Ray, R.P. Rani. 2014. Fermented fruits and vegetables of Asia: A potential source of probiotics. *Biotechnology Research International* 2014: Article ID 250424, 19 pages. http://dx.doi.org/10.1155/2014/250424

Wu, A.H., D. Yang and M.C. Pike. 2000. A meta-analysis of soyfoods and risk of stomach cancer: The problem of potential confounder. *Cancer Epidemiology Biomarkers and Prevention* 9: 1051-1058.

Yoon, J.Y., K.O. Jung, S.H. Kim and K.Y. Park. 2004. Antiobesity effect of baek-kimchi (whitish baechu kimchi) in rats fed high-fat diet. *Journal of Food Science and Nutrition* 9: 259-264.

7

Cucumber Fermentation

Wendy Franco[1,2]*, Suzanne Johanningsmeier[3]*, Jean Lu[4], John Demo[5], Emily Wilson[6] and Lisa Moeller[7]

1. Introduction

Humans have consumed fermented cucumber products since before the dawn of civilization. The earliest record of fermentation dates before 6000 BC in the Fertile Crescent (Demarigny 2012). There is fossil evidence that fruits and vegetables were undergoing lactic acid bacteria (LAB) fermentation before mankind inhabited the earth (Schopf and Packer 1987). Through time, human populations have developed these processes to create different products and prolong the shelf-life of highly perishable vegetables, such as cucumbers, thereby increasing food security. Almost every culture around the globe include specific fermented foods in their dietary customs and traditions. Fermented cucumbers are turned into a product called jiang-gua in Taiwan, khalpi in Nepal and India, paocai in China, oiji in Korea and are referred to as pickles in many parts of the United States, Europe, and Canada (Das and Deka 2012, Di Cagno et al. 2013, Kumar et al. 2013, Tamang 2010, Jung 2012). Most cucumbers are fermented in a salt solution. But to make khalpi, cucumbers are cut into pieces, sun dried for two days, put into bamboo vessels and left to

[1] Chemical and Bioprocess Engineering. Pontificia Universidad Catolica de Chile. Vicuña Mackena 4860, Santiago, Chile. E-mail: wfranco@ing.puc.cl
[2] Nutrition Science, Pontificia Universidad Catolica de Chile Vicuña Mackena 4860, Santiago, Chile. E-mail: wfranco@ing.puc.cl
[3] USDA-ARS, SEA Food Science Research Unit, 322 Schaub Hall, Box 7624, North Carolina State University, Raleigh, NC, 27695 USA. E-mail: suzanne.johanningsmeier@ars.usda.gov
[4] Department of Molecular and Cellular Biology, Kennesaw State University, 370 Paulding Ave. NW, MD #1202, Kennesaw, GA 30144, USA. E-mail: jean_lu@kennesaw.edu
[5] John Demo Consulting, Inc., 230 Barton Rd, Greenfield, MA 01301, USA. E-mail: johnjdemo@gmail.com
[6] North Carolina State University, Department of Food, Bioprocessing and Nutrition Sciences, 400 Dan Allen Drive, Raleigh, NC. E-mail: emwolter2@gmail.com
[7] Fermenting Solutions International, LLC, 135 Sea Dunes, Emerald Isle, North Carolina 28594 USA. E-mail: LisaMoeller4@gmail.com
* Corresponding author: E-mail: suzanne.johanningsmeier@ars.usda.gov and wfranco@ing.puc.cl

ferment at room temperature for three to seven days. While hundreds of different commercial vegetable fermentation processes are practiced across the globe, cucumbers, cabbage and table olives are currently the most economically relevant commodities.

With an increasing world population, lactic acid fermentation is expected to play an increasingly important role in preserving fresh fruits and vegetables. According to Steinkraus (1994), fermentation plays the following five key roles in modern society: 1) preservation of substantial amounts of otherwise perishable products; 2) biological enhancement of the raw materials with protein, vitamins, essential amino acids and fatty acids; 3) enrichment of human diets through improved texture, appearance, flavor and aromas; 4) decreased energy requirements for preservation; and 5) cottage industry opportunities. Cucumbers are fermented as an economical means of storage in a well-preserved state between harvests, making them available for processing into finished products year round. Cucumber plants take only a couple of months to reach maturity and can produce many fruits. The mature fruits, which can be picked in various sizes, are composed of 95 per cent water and contain about 2.5 per cent sugar. The plants can produce high yields on different soils and complement crop rotations. In small plots, the plants can be grown on trellises to increase the yield per area. The production statistics for the top five cucumber-growing nations are given in Table 1. China has produced the largest amount of cucumbers for the past 50 years, nearly 25 times that of Turkey, the second top producer.

Table 1. Cucumber Production Trends 2008-2012

Country	Cucumber Production (millions of metric tons)				
	2008	2009	2010	2011	2012
China	42.2	44.2	45.7	47.3	48.0
Turkey	1.70	1.70	1.80	1.70	1.70
Iran	1.50	1.60	1.70	1.50	1.60
Russian Federation	1.10	1.10	1.20	1.20	1.30
Ukraine	0.80	0.90	0.90	1.00	1.00
United States	0.90	0.90	0.90	0.80	0.90

Adapted from FAO Stat. 2015

High yields of cucumber require a means of bulk storage. Because the plants grow best when there are warm days and nights, its shelf-life is very short, especially if there is no means of cooling the fruits after harvest. For this reason alone, prior to 1940, virtually all commercial cucumber products were fermented, which is still the case in the developing world today. Cucumber fermentations are energy efficient as they require no heat or refrigeration. In many areas of the world, brined cucumbers are stored in earthen pots, barrels or concrete basins where they undergo a natural lactic acid fermentation. When managed properly, the fermented products can be edible for years and are ready for direct consumption or turned into sauces and condiments to complement meals. Today, the majority of shelf-stable cucumber products in industrialized societies are acidified, and then pasteurized or refrigerated. The fermented cucumber category is now represented primarily by institutional hamburger dill chips and institutional and retail relishes. The sale of fermented cucumbers represents greater than one billion dollars in sales annually in the United States alone. Value is added throughout the pickle manufacturing and distribution system, culminating in finished products that are valued much higher than the unprocessed vegetable commodities. In this way, the fermented cucumber industry supports the livelihood of growers, packers, manufacturers, transporters and a variety of related suppliers and distributors. However, the ultimate beneficiaries remain the consumers whom have access to safe, affordable and high quality food.

This chapter highlights the current knowledge associated with cucumber fermentation. We start with a detailed description of the industrial-scale process, followed by information that explains the scientific basis of the microbial and chemical changes that occur during cucumber fermentation, and finish with trends, current innovations and future directions for cucumber fermentation.

2. Industrial Process Overview

The experience of visiting a pickling tank-yard is quite interesting. As you approach the field of well-organized tanks, the sight of large tanks frothing with foam alerts one to the scale of operations of the modern pickle plant. Hundreds or thousands of tanks are seen, each holding 15 to 20 tons of fermenting cucumbers. The sounds of air blowers and machinery to move pickles to and from tanks is continuous. The strong smell can be quite pleasing or quite disturbing, depending on the quality of fermentation and one's predisposition to pickles. In large tank-yards, teams of dedicated personnel sample, test and adjust fermentation brines and devices in an effort to control the fermentation process guided by science-based written protocols and experiences of local experts. In colder climates, one may see crews chopping ice from frozen caps. On a closer look at the brine and pickles, you can notice different stages of fermentation even within a matter of days. Newly filled tanks of brined cucumbers will have clear and salty-tasting brine devoid of any subtle flavors. After two or three days of the initial stages of fermentation, the brine tends to turn turbid and develop a pleasing aroma, a subtle acidity, and fresh cucumber flavor. At this stage, the light green chlorophyll color still remains and the cucumber itself retains the opaque white interior. A frothy foam begins to form on the tank surface as a result of the active gas purging initiated at brining to control levels of dissolved carbon dioxide (CO_2) that can destroy cucumbers and tanks alike if not properly managed.

As the fermentation progresses in the surrounding brine during the first two weeks, subtle changes take place in the cucumber. Measured levels of acidity, primarily lactic acid, begin to increase, corresponding with decreasing pH and increasing brine turbidity. Rates of CO_2 development remain high. Carbon dioxide can be introduced from carbonates in the water that makes up the brine, residual respiration of the cucumber, and from the fermentation itself, requiring nearly continuous sparging. Levels of salt may need adjustment to compensate for absorption by the cucumbers (now becoming pickles), dilution from rain-water or evaporation and loss of water from the system. The color of the cucumbers begin to change from bright green skin and white opaque flesh, to a duller olive green skin and flesh that begins to move towards translucence in appearance. Ideally, textures remain crisp at this stage and, hopefully, throughout storage. The density and buoyancy of the cucumber also changes from very buoyant to more buoyant neutral allowing the cucumbers to settle, relieving much of the upward pressure on the cover boards that keep cucumbers below the brine surface, which itself is exposed to the environment. Odors and flavors change substantially as the primary lactic acid fermentation reaches completion. The subtle fresh cucumber aromas and flavors are lost or masked at this stage. A mildly acidic and pleasing taste from lactic acid fermentation is now predominant, but is still too salty for consumption without further processing. If the purging of CO_2 gas was successful, the structure of the cucumber remains intact, aside from some minor shriveling and shrinkage due to osmotic action of the salt brine. If dissolved CO_2 is not adequately controlled, hollow pickles can result that will pop when broken, revealing poor quality and an associated loss of value.

After the completion of fermentation (conversion of the fermentable sugars to lactic acid) and a few more weeks for curing (the loss of internal opacity) for a total of about four weeks, the pickle is ready for processing. At this stage, the processor can use the pickle as a finished product or boost salt levels for long-term storage in the same tank for up to one year. Continued monitoring of brines, including assays for softening enzymes and pH, help identify secondary fermentations by opportunistic fungi and bacteria in the earliest stages, prior to significant product losses.

Not all cucumber varieties yield high-quality, firm pickles. In the US, pickling variety cucumbers are bred to yield firm pickles after brining and fermentation. The most common pickling varieties in the US and Europe have approximately a 3:1 length to diameter ratio. In other countries, pickling varieties can be 10:1 or higher. Lebanon is a good example where the local varieties are 10:1 and make excellent quality pickles. In China, most cucumbers are grown for fresh market with very few being pickled by lactic acid fermentation on a commercial scale.

While most cucumber fermentations are done on whole cucumbers, many processors will also ferment cut cucumbers that are by-products from fresh pack operations, e.g. spears, or they will pre-dice green cucumbers for relish and ferment the dices in closed tanks. These fermentations will occur very rapidly due to the large surface area of the cuts and rapid diffusion of sugars into the surrounding brine. Products treated in this way should be used quickly as storage is more complex than that with whole pickles, which have retained their protective outer skin. In tank-yards which ferment mainly relish, whole nubs and crooks (whole cucumbers, which are broken or misshapen) are often mixed with finely cut cucumber pieces to eliminate the tight wads of relish that form if only cut products are packed at a high pack-out ratio. However, it is important to monitor these fermentation tanks closely as many losses are incurred with this combination due to a higher incidence of spoilage.

What makes industrial cucumber fermentation different than that at home or in the laboratory? While the microenvironment should be very similar, the problem becomes one of scale. Materials, tank design, compressive and buoyant forces, diffusion distances and exposure to the environment — all present problems when scaling up from a few small vessels to many very large tanks. Sanitation and preventing cross contamination are challenging due to the sheer scale. While technology has advanced in other high-value fermentation industries, e.g. dairy and alcohol, cucumber fermentation remains strikingly similar to historical methods in a crock pot with a porous cover held down with a heavy stone. Part of this has to do with the challenges presented by the cucumber fruit shape and internal structure itself, a large cylindrical structure with three internal carpal sections susceptible to damage on account of rough handling, osmotic pressures or high levels of diffused CO_2. Gentle handling methods and avoidance of excessive drops are requirements for good quality pickles in addition to methods of purging CO_2 from the tank and maintaining as close to an anaerobic environment as possible given the tank design. Another reason for slow change is due to the fact that fermented pickles have had an extraordinary safety record throughout the years.

2.1. Fermentation Vats

It is speculated that the first scale-up was most likely from stone or clay crocks (4-10L) to wooden barrels (60-200L). Wooden barrels built by coopers would have made a more efficient fermentation vessel for commercialization and transport by land or by sea. Wooden barrels were still being used as late as the 1970s; however, where barrels are still currently used, plastic has become the material of choice. The majority of the first large pickle tanks were cast offs from closed breweries or those upgrading to larger tanks. Cylindrical wooden tanks (10,000-30,000L) made from beveled staves dado cut and fit to thick plank flooring were held together with thick steel hoops to withstand extreme pressures encountered during fermentation. Wooden boards laid on the surface of the leveled cucumbers and held secure with 4″ × 4″ crossboards and locked to the tank with 4″ × 6″ planks counteracted the extreme upward buoyant forces, keeping the cucumbers below the brine surface during the fermentation. Wooden tanks had to be kept full with brine or water even when emptied of pickles in order to keep the wood swelled and tight fitting. After a time, the exposed wood above brine level dries out and shrinks, allowing brine to leak through the gaps. As the tanks aged, leaks between the staves had to be stopped by pushing oakum into the gaps. Many similarities are seen between the maintenance of wooden tanks and wooden sailing ships. In the 1970s and 80s, as many of the old wooden tanks came in disrepair, there was a controversy about whether the wood harbored the essential bacteria required for the fermentation, but this myth was quickly dispelled and there seems to be no difference in the quality of fermentation in plastic or wood tanks. Over time, fiberglass reinforced plastic (FRP) with an inner layer made of an approved food contact material dominated over the polypropylene tanks, primarily due to ease of repair and structural stability over wide temperature extremes. FRP can be multi-layered and cross wound for strength and impregnated with UV light barriers to extend the tank life to more than 15 years.

The tank design has not changed much throughout the years and is typically an open top cylindrical tank with flat bottom that can rest on a concrete pad or be partially set into the ground. As the tanks are open top, slightly tapered walls allow nesting of the tanks for economical freight when shipping new tanks from the manufacturer to the factory. In the US, the standard tank diameter is limited to 11.5 ft to allow legal over the road delivery with one driver. Larger diameter tanks would be designated as wide loads and would be more costly to ship. Tank heights are typically equal to or just slightly greater than the tank diameter, providing a reasonable volume capacity to surface area ratio which optimizes FRP material. FRP tanks are often reinforced with a thick band of fiberglass near the top that maintains the structural integrity even when the tank walls are relatively thin. The top of the tank contains some hardware, such as lifting lugs for proper movement of tanks during installation and brackets for securing the 'hold down' cover which is made from wood boards as described earlier or from semicircular FRP covers with holes that allow free movement of brine. The covering system of these open top tanks holds the cucumbers at least 15 cm below the surface of the brine, which is exposed to air and sunlight. While any exposure to air (oxygen) is not ideal, today's technology relies on the strong UV light from the sun to suppress growth of aerobic yeast and molds on the surface. The open top component of the design has been a necessity due to the enormous pressures developed during the fermentation and will be discussed in more detail later. The barriers to a fully-closed tank system are primarily due to CO_2 removal, material handling of the pickles during loading and unloading and the cost of replacement of open-top tanks throughout the industry. As open-top tanks are exposed to the environment and sources of contamination and security risk, only a concerted effort by the industry could overcome the technical barriers, which include material handling and effective purging, to prepare for the eventual replacement of the open-top tank technology over time. In fact, a number of cucumber fermentation facilities around the world have adopted closed-tank fermentations. Closed tanks eliminate the threat of contamination that the current open-tank system imposes, reduce color and flavor defects associated with photo-oxidation during storage and improve microbial stability during bulk storage.

Because an active fermentation process takes place in these tanks, frequent monitoring of chemical, physical and organoleptic qualities is necessary. Samples may be pulled from the surface and from internal regions by sampling at the discharge of the side arm purger (discussed in detail later) or from sampling wells attached to the side of the tank with access from the surface near a platform or walkway. Testing can be done from tank side in some cases or in a laboratory on the premises.

In the following paragraphs and in Figure 1, the steps in commercial cucumber fermentation are described.

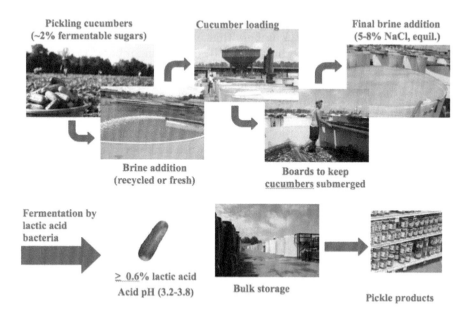

Fig. 1. Process Steps in Cucumber Fermentation. (*Adapted from Johanningsmeier* (2011)).

Prior to the cucumber harvest, open top tanks are cleaned thoroughly. In some cases, a sanitizing solution is used as well as a good potable-water rinse. Accessories are checked for functionality, especially the side-arm purgers and diffusers. An initial salt brine is prepared of proper salinity and if other chemicals are to be added, like calcium chloride, then it should be done during preparation of the final brine solution. Typically, the salt level of this initial cover brine (and cushion brine) will equilibrate to a salt concentration lower than the target of 6 per cent in order to prevent severe osmotic shriveling of the cucumbers. Dry salt adjustments after two or three days will then get the equilibrated salt levels to the target values.

The cucumbers are typically graded by size, sorted and inspected, culling out any spoiled cucumbers and foreign material. Pre-grading of cucumbers at this stage allows better control of fermentation and makes production planning much simpler when the product is ready to be processed for use in finished products. Management of sizes in the pickle industry is a big challenge and success in the business depends on balancing the supply and demand by size. Each size cucumber will have a different bulk density and impact the pack out ratio in any given tank. The pack out ratio coupled with the size and maturity of the cucumber influences the amount of sugars present and the buffering capacity of the system, affecting fermentation rates and degree of completion as measured by pH and lactic acid production. In some cases, with very large diameter cucumbers, fermentation would stop before all sugars were consumed, allowing for secondary yeast fermentation to take place during the storage phase.

Just prior to filling the tank with cucumbers, a portion of the brine is added as a cushion to the cucumbers that will be dropped into the tank. This cushion brine serves to break the fall of the cucumbers. The second function is to begin the osmotic shrinking process immediately, so that the targeted quantity of cucumbers fit into the tank. A typical target weight is calculated by taking the working volume of the tank as weight of water and multiplying by 0.65 to calculate the weight of the cucumbers within. This will vary slightly by the bulk density which is a function of the cucumber size. The correct pack out ratio in the tank is critical in order to achieve the proper final acidity of the system for preservation.

At the end of filling, the cucumbers are generally heaped in a mound that look impossible to fit; however, within a matter of hours, osmotic shrinking will begin to show a settling of the mound. This action is due to the salt brine cushion that was applied using only half of the calculated brine required. The brine rises in the tank as the cucumbers are added. Manual leveling of the cucumber mounds is typical and the head cover is applied and secured prior to final brining. If the tank is large, a person may walk on the surface of the cucumbers in order to apply and secure the cover. It is critical to avoid contamination and damage from boots or leveling equipment at this stage. Once the hold-down cover is secured, the remaining brine is added about 15 cm over the head cover, and the side-arm purger is started. It is desirable to complete all of these steps within 24 hours after the start of filling; however, in reality it may not always be possible. In some conditions, satisfactory fermentations have been observed when this process is completed within 48 hours. However, softening and spoilage of cucumbers becomes very likely when filling and brining a tank takes more than 48 hours.

2.2. Controlling the Fermentation Process

Cucumber fermentation relies on the presence of naturally-occurring LAB or in some cases inoculum from a pure starter culture. The brining process creates a competitive advantage for the LAB to flourish while inhibiting the growth of spoilage organisms. Historically, while salt (sodium chloride – NaCl) was used for preservation, there was no established consensus on how much to use. The use of very high salt concentrations hindered a rapid lactic acid fermentation and allowed growth of halophilic yeast, leading to uncertain quality and shelf-life. Too little salt allows spoilage bacteria to proliferate. Many processors today use an equilibrated salt level of about 6–7 per cent (wt/vol) to achieve the desired fermentation based on research done starting in the 1940s. After the active fermentation, it is a common practice to use the pickles immediately or add more salt to protect the product from subsequent spoilage if long-term storage is desired. In colder climates, the extra salt is intended to protect the product from freeze damage. Although the freezing point is lowered, it is generally not sufficient to completely prevent freezing of the brines.

2.3. Purging

Purging is started immediately after capping and brining and continues on an established on-off schedule for at least 14 days, which is the normal period required to complete the fermentation. From the very start of brining and during the entire fermentation process, large quantities of CO_2 (carbon dioxide) are generated due to the presence of carbonates in the water used for brine, continued respiration of the cucumbers after harvest and as a result of malic acid degradation to lactic acid and CO_2 by fermentation microbiota. Practically, purging of CO_2 is only required during the initial five to seven days of fermentation, but, as a precaution, it is usually continued for 14 days or until the sugars are depleted. When the fermentation vessels are small, dissolved CO_2 can be eliminated by various mechanical interventions. Just shaking a crock or kicking a drum can release quantities of CO_2 gas seen visually with the gas bubbles escaping from the surface. In large tanks, the volumes and distances are much larger and CO_2 cannot easily escape, generally staying in the solution until saturation is reached. If unaddressed, CO_2 can create pressures inside the cucumber and destroy the internal tissue causing pickles to become hollow to various degrees. This phenomenon is referred to as bloater damage and is a major source of lost product and inefficiency. Prior to the wide implementation of purging routines, dedicated inspection operations were needed to hand-squeeze every pickle in search of hollow centers. In some cases, pressures were so great that hold-down covers would blow off, splitting the 4 by 6 inch wooden planks. In the 1980s, all this changed when a purging device called the side-arm purger was invented (Costilow et al. 1977). Almost overnight, bloater defects and the labor-intensive sorting lines disappeared. The solution was low cost and technically simple, bubbling nitrogen gas (to prevent oxygenation) through a small pore-size diffuser in a side-arm gas lift pump which processed brine from the bottom of the tank and lifted it to the tank surface where the CO_2-laden nitrogen bubbles escaped into the atmosphere. The entire volume of brine is processed within a matter of two or three hours, depending on the liquid pumping rate. Later, cost pressures forced processors to look at lower-cost compressed air or blower air as replacement for the nitrogen gas. While effective at eliminating CO_2 and bloating, issues with soft centers and yeast growth developed. Interventions to mitigate these negative consequences, included additions of potassium sorbate, acetic acid and intermittent purging cycles (to allow time each day for the tank to go anaerobic) were used to minimize aerobic yeasts and molds from proliferating. Today, as the salt levels are being further decreased, there is the need to re-evaluate the value of pure nitrogen sparging versus air.

There are a number of factors that make side-arm purging effective. The first is production of many small bubbles that create a very large surface area for a given volume of purging gas. Exposure of the large surface area to the CO_2-laden brine allows a fast transfer of CO_2 into the gas bubble by simple diffusion. Secondly, as the bubbles rise in the purging tube, brine is pumped from the bottom of the tank on to the tank surface where the bubbles pop and are released into the atmosphere, releasing the CO_2 with them. In an efficient system, the treated brine that is spread over the tank surface falls downward in a layered flow pattern, replacing the brine that is pumped up from the bottom. Short circuiting flow patterns are not typical as long as the brine is spread fairly evenly over the surface. Thirdly, the purging gas flow rate must adequately remove dissolved CO_2 at the rate that it is generated. Today, there are purging protocols that work fairly well for standard tank sizes and the conditions experienced by individual processors, e.g. the carbonates present and fermentation rates experienced. Ideally, a purging control system based on CO_2 measurement and feedback control to the purger would optimize the use of purging gas and may be a financial incentive to return to nitrogen as the purging gas of choice.

2.4. Monitoring and Record Keeping

It is critical to monitor the condition of the tanks at a programmed frequency throughout fermentation and storage. Frequency of testing will vary, being very frequent during the 14–21 days of active fermentation and less frequent during storage up to one year. Examples here are typical of many sampling programs but should not be construed as a recommendation nor as a full quality program for fermented cucumber pickles. Complete records of each tank and daily activities in the tankyard are critical.

The first step in monitoring is to check the condition of the tank and equipment prior to filling and brining. Proper sanitation and cleanliness is critical to avoid any gross contamination that would influence the fermentation. Next is to monitor the tank preparation, filling, capping, brining and purging that initiate the fermentation. A simple status-control board is typical where workers and management can see the progress of each tank at every stage. Entire tanks of cucumbers can be lost without a reliable control system. Salt measurement is the first and most important monitoring and control function as proper equilibrated salt concentration initiates the fermentation by selecting the naturally-occurring lactic acid bacteria and inhibiting spoilage organisms present on the fresh cucumbers. Salt is monitored daily for the first week and adjusted as needed with dry salt on the surface to achieve the processor's desired equilibrated target, typically 6 per cent although many within a range of 5–7 per cent. Monitoring the performance of the side-arm purger is done daily along with a daily visual examination of the tank surface. Observation, by an experienced and trained employee of bubbles along with a good brine flow is generally sufficient to assure proper function. Brine flow can also be measured periodically and should be sufficient to process the entire tank within two to three hours (60 L per min brine flow for a 20-ton tank would be typical). However, as mentioned before, removal of CO_2 is controlled by measuring the dissolved CO_2 levels and adjusting the purging protocols accordingly. In practice this is rarely done. The pH is also monitored daily during this phase. Depending on whether the salt brine was made fresh or from recycled brine, the pH of the brine may start at neutral or acidic values and within a matter of days decrease to levels below 4.6 and continue to decrease throughout the fermentation as lactic acid is produced. Concurrent visual examination of the brine will show increasing turbidity, indicating the growth of sufficient LAB for proper fermentation. In rare cases, when the pH does not decrease and no brine turbidity develops, it means that a stalled fermentation has occurred and an immediate intervention by an expert authority must be initiated.

After the first week, when monitoring indicates that fermentation is underway, tests for titratable acidity (assumed as primarily lactic acid) and fermentable sugars is tracked at a frequency of once or twice per week. Salt and pH testing frequencies can also be reduced at this point. Monitoring the cucumber itself is also desirable if a cucumber sampling port is available. The cucumber will be sliced, looking for internal integrity and tasted for both flavor and texture. After 14 to 21 days, when lactic acid levels are more than 0.6 per cent and reducing sugars are negative, it can be presumed that primary fermentation is complete and the cucumbers are almost ready for processing, depending on the cured appearance standard of the processor. If the product is to be stored for an extended period of time in the tanks, the processor will monitor the brine chemistry and tank conditions no less than once a month, watching for any signs of off-odor or pH rise, which indicate an undesirable secondary fermentation.

After the completion of fermentation, it is ideal to test every tank for the activity of polygalacturonase (PG) softening enzymes in the brine. Knowledge of the presence of these enzymes is a useful tool for quality-improvement strategies and also to manage inventories. Although the presence of softening enzymes may not indicate poor quality, it is generally a good predictor of soft pickles in the future. Therefore, utilizing and properly processing tanks identified with PG activity early while pickles are still firm can optimize overall yield and quality. Brines intended for recycling are frequently tested prior to reuse for endo-PG activity, using a diffusion plate assay (Buescher and Burgin 1992).

2.5. Material Handling

Material handling involves moving both fresh and fermented cucumbers as well as the fermentation brine. Pickles and brine can be handled separately or sometimes together where the brine acts as the transport medium. The movement of pickles is done in a manner that avoids hard impacts, since the cucumber pickle can be damaged quite easily. Under the right conditions, a cucumber hitting a hard surface on end can develop internal damage in both the tissue and the internal structure (Marshall et al. 1972). This can happen in as little as a 3-ft drop. Damaged cucumber tissue occurring after harvest, but before brining, can trigger internal development of softening enzymes which can affect the texture weeks or months later during storage.

With very small stationary tanks, loading and unloading can be done manually by dumping buckets of cucumbers into the top of the open tank and onto the cushion brine. The fresh cucumbers contain about 5–6 per cent air by volume (Corey et al. 1983a, b), making them less dense than the brine and buoyant. Once fermented and fully cured, the cucumbers are denser than the brine and tend to settle. Unloading small tanks can be achieved manually with long handle nets.

Loading large tanks is fairly straightforward and involves dry conveyance of cucumbers elevated and directed toward the tank center. While many processors just drop them on to the cushion brine, some take precautions to avoid drops by using canvas lowerators which gently provide a cascading fall as the tank gets filled. Unloading large tanks is a bit more complicated. With large open tanks there are two basic methods in use today — one is a conveyor with buckets inserted into the tank. With some manual assistance, it will dig into the tank and convey pickles into a bin placed outside the tank. The buckets have holes to drain the brine as the pickles are being lifted out. Brine needs to be lowered as the tank becomes empty of pickles but sufficient brine needs to remain to allow the pickles to move freely. Towards the end of the process, a worker will direct the final pickles into the conveyor buckets with a long handle net. This conveyance method would not work in a closed tank. The other popular method today is a pumping system. Special large inlet centrifugal pumps with recessed impellors pump both pickles and brine out of the tank and on to a screen where the brine is separated and returned to the tank. This method is about four times faster than the bucket conveyor method, but requires about ten times the investment in equipment. With the right design and equipment, pickles of all sizes can be pumped and carried to various locations in the processing plant, using brine as the transport medium. Other methods of handling fermented pickles include the use of flumes and air-lift pumps (Demo 2002).

2.6. Processing for Finished Product

Fermented pickles in the completed state are too salty for consumption. The fermented fruits are generally over 7 per cent salt when the fermentation and storage process is complete. In order to make most pickled products, the fruits are 'desalted' by soaking in water. Pickles are typically transported to designated processing tanks, separated from the fermentation brine and flushed with potable water so that sufficient salt is removed. This diffusion process can take place within 24 hours, depending on the amount of salt. Enough salt is removed to meet the finished product specifications with a formulated cover brine. During this process, the natural flavors created by fermentation are also diluted, allowing the finished product to take on the characteristics of the other ingredients added. Sweet items are often flavored with vinegar, sugar, cinnamon and cloves; kosher dills are flavored with vinegar, garlic, onions and herbs.

3. Fermentation Microbiota

3.1. Natural Cucumber Fermentation

Cucumber fermentation is possible due to the presence of a number of microorganisms on the fresh fruits that are responsible for the chemical changes observed with time. Raw cucumbers contain a wide variety of different microorganisms mostly on the surface of the cucumbers, including aerobic bacteria, LAB, yeasts and molds. When properly handled (i.e. not washed excessively or treated with anti-microbials), raw cucumbers will contain LAB as a minor part of their natural microbiota. On average, aerobic plate counts of raw cucumbers account for 4 to 5 log CFU/g while yeast and LAB account for 1.5 and 2.5 log CFU/g, respectively. Raw cucumber also contains species from the *Enterobacteriaceae* family in orders of 1 to 4 log CFU/g. Several enterobacteria may grow at the beginning of fermentation, producing CO_2 and hydrogen, which may influence the initial development of anaerobic conditions. However, the numbers decrease drastically at the beginning of fermentation due to sensitivity to the acidic environment that develops as the fermentation proceeds (Etchells et al. 1945). Relatively low numbers of LAB naturally present at the beginning of fermentation outcompete the other natural microbiota due to the ability to survive in extreme environments, characterized by high salt and acid (Breidt 2006, Hutkins 2006).

LAB comprise a versatile group of microorganisms that are present at different stages of the fermentation process. Singh and Ramesh (2008) reported that the dominant LAB microbiota observed during cucumber fermentation were comprised of the genera *Lactobacillus*, *Pediococcus*, *Lactococcus* and *Leuconostoc*. Following a PCR approach with and without enrichment, the authors were able to observe *Lactobacillus* as the genus predominant after 72 hours of fermentation. *Leuconostoc* spp. were observed after 12 hours of fermentation in co-existence with *Lactobacillus* spp. After 36 hours of fermentation, the *Leuconostoc* population decreased and was not observed thereafter. At this point, species from the *Pediococcus* genus increased in number and remained in co-existence with *Lactobacillus* spp. until 72 hours of experimentation. *Lb. plantarum* is the predominant LAB species in cucumber fermentations. This homofermentative organism produces primarily lactic acid from glucose and fructose via the Embden-Meyerhoff-Parnas pathway (Breidt et al. 2007). Other LAB present during fermentation, such as *Pd. pentosaceus*, *Lb. brevis* and *Ln. mesenteroides* (Singh and Ramesh 2008), are in general heterofermentative and use the phosphoketolase pathway to produce lactic acid, CO_2, ethanol and acetic acid (White 2007). The ability of LAB to dominate fermentation depends on the capability to overcome the fermentation environment, rich in salt (6–7 per cent of NaCl) and high in acid. Although a simple NaCl brine has a pH near neutrality when the fermentation starts, it sharply decreases in a short period of time due to lactic acid production achieving values commonly between 3.1–3.5 (Breidt et al. 2013a). Among the LAB reported, *Lb. plantarum* and *Pd. pentosaceus* are more acid tolerant (Daeschel et al. 1987, Sandhu and Shukla 1996, Harris 1998) and therefore can be observed during the late stages of fermentation.

Yeasts, also naturally present on fresh cucumbers, may participate in the fermentation process to varying degrees. Two types of yeasts are commonly observed in cucumber fermentation: film-forming yeasts from the *Debaryomyces*, *Endomycopsis*, *Zygosaccharomyces*, and *Candida* genera that use an oxidizing metabolism to increase biomass (Etchells and Bell 1950a); and subsurface yeasts, such as *Saccharomyces cerevisiae* and *S. rosei*, which primarily carry out ethanol fermentation, converting a portion of glucose to ethanol and CO_2 (Etchells and Bell 1950b, Daeschel et al. 1988). Yeasts have been considered as contributors of flavor and growth factors during lactic acid fermentation in cucumber pickles (Etchells 1941b, Daeschel et al. 1985), and it has been hypothesized that the presence of yeasts helps establish

LAB as the dominant bacterial species (Daeschel et al. 1988). Yeasts may do this by contributing vitamins, nitrogen, amino acids and peptides to the fermentation brine which are important for the metabolic activity of LAB. During the initiation of fermentation, which lasts between two to three days, the number of LAB and yeast increases rapidly while undesirable bacteria and molds are eliminated by competition. LAB populations can reach up to 8 log CFU/mL while yeast may reach counts of ~5 log CFU/mL. This is greatly favored by the fermentation conditions, the ability of LAB and yeasts to tolerate relatively high salt conditions, and the ability of oxidizing yeasts to increase in biomass due to the presence of oxygen from air-purging routines commonly implemented during the first week of fermentation. At the end of the fermentation process, which might last up to three weeks, LAB counts average 6 log CFU/mL while yeast counts decrease to 4 log CFU/mL.

Although lactic acid is the major product of the fermentation process, some other by-products are formed. Carbon dioxide is generated from respiration of cucumbers when submerged in brine (Potts and Fleming 1979) and due to decarboxylation of malic acid by *Lb. plantarum* during fermentation (McFeeters et al. 1982a). Several LAB present in vegetable fermentations have an inducible malolactic enzyme which converts malate to lactate and CO_2 (Johanningsmeier et al. 2004). The presence of coliforms and yeasts also increases the chances of CO_2 production. Excessive CO_2 can lead to bloater pickles, an undesirable quality produced by the formation of gas pockets in the cucumber flesh (Corey et al. 1983b). To remove CO_2, Fleming et al. (1975) recommended nitrogen-purging routines and, if possible, maintaining anaerobic conditions during fermentation and bulk storage. Currently, the pickle industry commonly uses air-purging to prevent bloater damage. The change in the practice has been mainly based on costs since air displaces CO_2 from the fermentation tanks in a fashion similar to the proposed nitrogen-purging. Air-purging is commonly applied during active lactic acid fermentation (seven to 10 days in summer months and up to a month in colder temperatures); however, there are processors that follow a continuous air-purging schedule even during storage of the fermented product (personal communication, unpublished). Potts and Fleming (1979) observed that introduction of air into fermentation might lead to changes in the microbiota present in the fermenting cucumbers. Oxygen availability induces the growth and establishment of aerobic microbiota, including oxidizing yeast and undesirable aerobic spoilage bacteria, which might alter the characteristics of the fermented product. To limit the growth of aerobic microorganisms, particularly molds and yeasts, 0.1 per cent sorbic acid (0.13 per cent as potassium sorbate) or 0.9 per cent acetic acid can be used (Bell and Etchells 1952, Bell et al. 1959, Etchells et al. 1961, Binsted et al. 1962). Excessive growth of aerobic microorganisms which can cause spoilage problems is also controlled by stopping the purging for several hours each day (Breidt et al. 2007).

3.2. *Ecology of Bacteriophage*

Cucumber fermentations are driven by a variety of LAB naturally present on cucumbers. The metabolic activities of LAB determine the quality and safety of the final fermentation product (Pederson and Albury 1969). Many factors influence the metabolic activities of LAB. Bacteriophages (phages) are one such important factor because phages are natural killers of bacteria. The presence of phages against LAB in the fermentations can potentially lead to significant mortality of LAB, thereby influencing the bacterial ecology, the dynamics of the fermentation and subsequently the quality of the fermented products (Lu et al. 2012).

Phages are ubiquitous in nature. Many phages have been isolated from food environments. Phages in dairy fermentation have been extensively studied for decades. In contrast, phages in cucumber fermentation have only recently been studied. A few reports on phages from cucumber fermentation are found in the literature and all these reports focus on the phages infecting LAB. The two well-studied phages isolated from cucumber fermentations are phages ΦJL-1 and φps05. ΦJL-1 infects the starter culture *Lb. plantarum* MU45 while φps05 infects another starter culture *Pd. acidilactici* LA0281 (Lu et al. 2003a, Yoon et al. 2007). A recent pioneer study explored the phage ecology in a commercial cucumber fermentation and provided a glimpse into the diversity of LAB phages (Lu et al. 2012). The study obtained 576 LAB isolates from fermentation. Using these LAB isolates as potential hosts, 57 phage isolates were obtained from the same fermentation, indicating that about 10 per cent of LAB isolates were sensitive to phage attacks. The phage hosts included a variety of LAB such as *Lb. brevis*, *Lb. plantarum*, *Weissella paramesenteroides*, *W. cibaria* and *Pd. ethanolidurans*. However, most of the phages infected the two predominant LAB in fermentation — *Lb. brevis* and *Lb. plantarum*. The study showed that all the phages were isolated during the active period of fermentation, 3 to 30 days, and no phage active against LAB was isolated on Day One or after Day 30 in fermentation. The number of phages isolated on each sampling day correlated well with LAB counts (Fig. 2). It is known that LAB on fresh vegetables account for a very small portion (< 1 per cent) of the bacterial population (Mundt et al. 1967, Mundt and Hammer 1968, Mundt 1970) and this small LAB population is dominated by *Ln. mesenteroides*, one of the least salt-tolerant (Konisky 1989, Lund et al. 2000) and least acid-resistant (Pederson and Albury 1969, McDonald et al. 1990) LAB involved in vegetable fermentation. The high salt concentration (6 per cent NaCl, much higher than 2 per cent NaCl used in sauerkraut fermentation) and the low pH (4.0–4.4) of recycled or acidified brine used in the beginning of cucumber fermentation could greatly inhibit *Ln. mesenteroides*. Therefore, very few LAB hosts were present on Day One for phage replication, and phage activity was under the detection limit. After the fermentation started, the concentration of LAB increased rapidly, which provided more hosts for phage replication. As a result, the number of phages being isolated increased (Fig. 2). As the fermentation continued, the massive production of acids and the resulting low pH became increasingly inhibitory to both LAB and their phages. On Day 30, only one phage against an acid-resistant LAB host was isolated because the brine pH was low (3.4), which inhibited most other phages and their hosts.

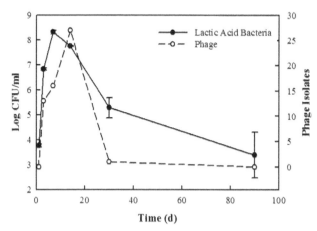

Fig. 2. Lactic Acid Bacteria (LAB) Counts and the Numbers of LAB Phage Isolates in Commercial Cucumber Fermentation (*Adapted from Lu et al. 2012*)

The phage ecology study showed that LAB phages in cucumber fermentation are highly diverse. Morphologically, most of the phages isolated from cucumber fermentation are tailed phages with icosahedral heads, but they vary in head and tail structures, belonging to different phage families, *Myoviridae* or *Siphoviridae* (Fig. 3). These phages also differ in host ranges. Some phages are species- or strain-specific, while other phages are capable of infecting multiple species. Although rare, two phages were found to be able to infect *W. cibaria*, *Lb. plantarum* and *Lb. brevis* (Lu et al. 2012). In contrast, most phages isolated from sauerkraut fermentations are species-specific and no phages were found to be able to infect LAB from different genera (Lu et al. 2003b). A variety of molecules on LAB cells can serve as receptors for phage infection, such as polysaccharides, (lipo) teichoic acids and membrane proteins (McGrath and van Sinderen 2007). Phages with a broad host range may be able to use more than one type of receptors present on different hosts, or the same receptors present on different hosts, allowing those phages to attack a wider variety of host LAB species in one genus or in different genera. These phages can also play important roles in phage ecology and genetic transfer through transduction among different LAB hosts, thereby promoting genetic diversity in microbial communities.

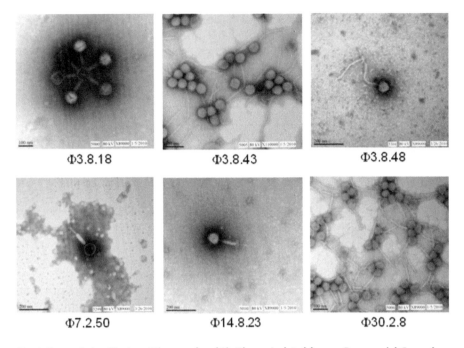

Φ3.8.18   Φ3.8.43   Φ3.8.48

Φ7.2.50   Φ14.8.23   Φ30.2.8

Fig. 3. Transmission Electron Micrographs of Six Phages Isolated from a Commercial Cucumber Fermentation (*adapted from Lu et al. 2012*).

The currently available data from SDS-PAGE analysis and restriction endonuclease digestion revealed a variety of structural protein profiles and restriction-banding patterns of the phages isolated from cucumber fermentations (Lu et al. 2003a, Yoon et al. 2007, Lu et al. 2012). However, the details and extent of the genetic diversity of phages in cucumber fermentation are largely unknown due to the scanty genome sequence database. So far, only one phage, ΦJL-1, from cucumber fermentations has been sequenced (Lu et al. 2005). Sequence analysis showed that the genome of ΦJL-1 is made of a linear double-stranded DNA (36,674 bp) containing 46 possible open reading frames (ORFs). Proven or putative functions were assigned to 17 ORFs, including five structural protein genes. Similar to several other LAB phage genomes, ΦJL-1 genome had a modular organization with functionally-related genes clustered together (Lu et al. 2005). The genome did not contain a lysogeny module, indicating that ΦJL-1 is not a temperate phage. To better understand the genetic diversity of phages in cucumber fermentation, a larger phage sequence database is needed for comparative sequence analysis.

The currently available data provide only a glimpse into the abundance and diversity of the phages active against LAB in cucumber fermentation that could cause significant mortality to LAB populations. As the disposal of high-salt waste from cucumber fermentation causes increasing concern, technologies for low-salt cucumber fermentation are under development. These technologies may require the use of LAB starter cultures to ensure normal fermentation. The naturally-present phages could lead to starter culture failure. Therefore, phage-control strategies may be essential in those types of fermentation. Further study is greatly needed in order to get the whole picture of phage ecology in cucumber fermentations. This may include, but is not limited to: 1) studying phage ecology in cucumber fermentation in various geographic locations; 2) evaluating the impact of phages on these fermentations; and 3) sequencing more phages from cucumber fermentations. In addition, the potential of starter culture failure caused by phage infection should also be investigated if the fermentations require the use of starter cultures.

4. Secondary Fermentation and Spoilage

Over the years, there have been sporadic reports of spoilage in commercial tanks of brined cucumbers after apparently normal fermentation. The spoilage is characterized by a gradual increase in pH and decrease in titratable acidity followed by a very rapid increase in pH above 4.6, gas and odor production, and the potential for germination and outgrowth of clostridium spores. More detailed information on how this spoilage proceeds and the organisms involved has been limited in part to the sporadic occurrence of the event and the inability to predict the conditions that lead to secondary fermentation. The diversity in microbiota that is present during and after the lactic acid fermentation makes the isolation and identification of potential causative agents very challenging. The first documented post-fermentation cucumber spoilage was reported by Fleming et al. (1989). As part of experiments to reduce salt levels during fermentation, cucumbers that fermented normally in a closed 1000 litres tank with low NaCl concentration (2.3 per cent) spoiled several months after completion of the fermentation. The spoilage was characterized by complete depletion of lactic acid and production of butyric and propionic acids. Butyric acid production was attributed to *Clostridium tertium*, which was isolated from the spoiled brine. However, the authors concluded that this bacterium was not able to initiate the spoilage process since it was only able to convert lactic acid into butyric acid at pH 5 and above.

Later studies demonstrated that pH and NaCl content are important factors to modulate the spoilage process under anaerobic conditions (Kim and Breidt 2007, Johanningsmeier et al. 2012). Spoilage has been reproduced at pH 3.8 and above and NaCl concentrations of 4 per cent and below under anaerobic conditions. Organisms isolated from early laboratory-spoilage experiments were identified as *Propionibacterium*, *Clostridium* and *Lactobacillus* spp., but these microorganisms were not responsible for the initiation of secondary fermentation at the conditions prevailing once the primary fermentation is completed. More recently, *Lb. buchneri*, an aciduric heterofermentative LAB, was isolated from both laboratory-reproduced spoilage (Johanningsmeier et al. 2012) and commercial fermentations (Franco and Pérez-Díaz 2012a) and shown capable of initiating spoilage by metabolizing lactic acid into acetic acid and 1,2-propanediol in fermented cucumber media under both aerobic and anaerobic conditions (Johanningsmeier and McFeeters 2013).

The natural microbiota present on and in cucumber fruits includes LAB, yeasts, enterobacteria and *Clostridium* spp. If oxygen availability is considered, it is reasonable to postulate that organisms other than those reported under anaerobic conditions may have a role in the post-fermentation spoilage of fermented cucumber pickles. For instance, under aerobic conditions, *Lb. plantarum* is able to grow faster and reach higher cell densities as compared to anaerobic conditions (Bobillo and Marshall 1991). This bacterium produces mainly lactate when glucose is present. However, once the sugar is exhausted, an oxygen-dependent pathway promotes the formation of acetic acid at the expense of lactic acid (Murphy and Condon 1984, Murphy et al. 1985, Bobillo and Marshall 1991). Other microorganisms with similar 'lactate oxidizing' systems are *Pd. pentosaceus*, *Lb. casei*, *Lb. sakei*, *Streptococcus faecium* and *Str. faecalis* (Thomas et al. 1985, Malleret et al. 1998).

During bulk storage in open tanks, yeasts may grow on the surface of the brine and oxidize the organic acids produced during primary fermentation (Etchells and Bell 1950a, b, Bell and Etchells 1952). It has been observed in the laboratory that common pickling-spoilage yeasts, such as *Zygosaccharomyces globiformis* (Belland Etchells 1952) may grow even in the presence of extremely low concentrations of oxygen (personal communication/unpublished). Additionally, various yeasts have been related to spoilage problems in the table olive industry (Vaughan et al. 1969, Durán Quintana et al. 1979). Under aerobic conditions, species from the genera *Candida*, *Pichia* and *Saccharomyces* are capable of utilizing lactic and/or acetic acids (Dakin and Day 1958, Ruiz-Cruz and Gonzalez-Cancho 1969).

Other microorganisms of interest belong to the *Enterobacteriaceae* family. These bacteria, commonly present in fresh produce, are usually inhibited by the acidic conditions and low pH that develop as the primary fermentation proceeds (Etchells et al. 1945). However, a recent study reported that *Enterobacter cloacae* might be a vector of contamination in fermented green olives (Bevilacqua et al. 2009). *Enterobacter* sp. were also identified as the causative agents in an unusual gaseous spoilage in high-salt fermentation that resulted in the production of hydrogen gas (Etchells 1941a, Etchells et al. 1945). A number of anaerobic organisms have been isolated that relate to fermented cucumber spoilage (Fleming et al. 1989, Kim and Breidt 2007). Among those, *Clostridia* spp. are of interest due to their ability to sporulate under stress conditions and germinate once environmental conditions are favorable.

Investigations into the development of secondary cucumber fermentations have been limited by the inability to predict the event on the commercial scale and by its sporadic occurrence in a small number of fermentation tanks. In the 2010-cucumber-brining season, a considerable number of commercial tanks spoiled due to secondary fermentation, which resulted in the loss of lactic acid and increased pH to an extent that led to discarding of fermented cucumbers (Franco et al. 2012). Microbiological analysis of the brines showed a diverse microbiota composed of yeasts and LAB (Fig. 4). Oxidative yeasts, identified as *Pichia manshurica* and *Issatchenkia occidentalis*, were observed only in spoiled samples, while *Candida etchelsii* was observed in stable commercial samples. The study of this outbreak also confirmed that selected LAB, different from those that carry out primary fermentation, are capable of proliferating during secondary fermentation. These spoilage LAB occurred simultaneously with the spoilage yeasts. Among these bacteria, the LAB *Lb. buchneri* and *Pd. ethanolidurans* were frequently observed in spoiled samples. Figure 4 shows the different colony morphologies of these spoilage microorganisms. Although *Pd. ethanolidurans* persists in these spoilage conditions, it is not capable of metabolizing lactic acid or other spoilage intermediates (Johanningsmeier et al. 2012, Johanningsmeier and McFeeters 2013). Under aerobic conditions, *P. manshurica* and *I. occidentalis* were able to initiate lactic acid utilization at pH values as low as 3.5, resulting in an increase in pH and decrease in redox potential. These changes favored the establishment of bacteria, such as *Lb. buchneri*, *Pd. ethanolidurans*, *Enterobacter* and *Clostridium* spp. whose metabolic activity resulted in acetic, propionic and butyric acid production (Franco and Pérez-Díaz, 2012 a, b, 2013).

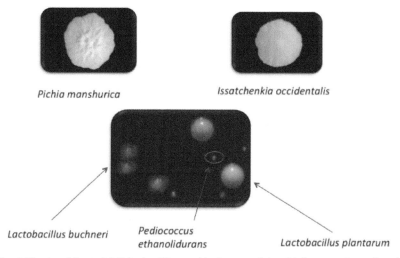

Pichia manshurica   *Issatchenkia occidentalis*

Lactobacillus buchneri   *Pediococcus ethanolidurans*   *Lactobacillus plantarum*

Fig. 4. Yeast and Bacterial Colonies Observed in Commercial and Laboratory Reproduced Spoiled Fermented Cucumber. (*Adapted from Franco and Pérez-Díaz (2012a)*)

Although, lower NaCl concentrations were considered as important for the development of spoilage in fermented cucumber pickles under anaerobic conditions (Kim and Breidt 2007, Johanningsmeier et al. 2012), data gathered from these commercial spoilage samples indicated that fermentations carried out at 0–6 per cent NaCl and 0.2–1.1 per cent $CaCl_2$ are susceptible to spoilage with the typical air purging routines used in commercial production. Certain spoilage yeasts and LAB are known for being halotolerant (Deák 2008), which, when combined with the ability to utilize organic acids as a source of carbon, gives these organisms a unique competitive advantage in an environment characterized by high NaCl and lactic acid concentrations.

A number of LAB species have been isolated from commercial and laboratory reproduced spoilage samples (Johanningsmeier et al. 2012). Among these bacteria, *Lb. buchneri* and *Lb. parafarraginis* were able to utilize lactic acid in fermented cucumber while converting it to acetic acid and 1,2-propanediol. Subsequent to such conversion, 1,2 propanediol can be converted into propionic acid and propanol by organisms like *Lb. rapi*, also isolated from spoiled fermented cucumbers (Johanningsmeier et al. 2012, Johanningsmeier and McFeeters 2013). A secondary fermentation initiated by *Lb. buchneri* in the absence of the oxidative spoilage yeasts would be preferable on the commercial scale over spoilage induced by oxidative yeasts. This is due to the possibility of having slower lactic acid utilization and the conversion of lactic acid into acetic and propionic acids, which would maintain a more acidic pH with time. The presence of oxidative yeasts and aerobic conditions accelerates the removal of lactic acid, leading to a rapid rise in pH and gas formation that could cause bloating of whole cucumbers. Additionally, the presence of oxidative yeasts accelerates the spoilage associated with the lactic acid bacterium *Lb. buchneri* (Franco et al. 2012). Recently, Breidt et al. (2013b) reported that gram-positive *Firmicutes* dominated the early stages of spoilage fermentation. As pH values approach 5, gram-negative anaerobes mostly representative of the *Bacteriodetes/Chlorobi* group dominated, along with the presence of butyric acid in the spoiled brine samples. Metabolic activity of *Propionibacterium* and *Pectinatus* species isolated from more advanced spoilage samples was demonstrated in brine or media with a pH above 4.9, resulting in lactic acid conversion into propionic acid with a further increase in pH.

Based on current scientific evidence, it is logical to conclude that the spoilage process is a complex microbiological process in which biochemical changes are often related to the establishment of specific and diverse microorganisms in a successive fashion. During long-term storage, environmental conditions could favor the establishment of oxidative yeasts, such as *P. manshurica* and *I. occidentalis* when sufficient oxygen is available (Franco and Pérez-Díaz 2012a, b, Franco et al. 2012). While under anaerobic conditions, fermented cucumbers are susceptible to spoilage by *Lb. buchneri* and *Lb. parafarraginis* (Johanningsmeier and McFeeters 2013) unless the pH of 3.2 and 6 per cent NaCl are maintained (Johanningsmeier et al. 2012).

These microorganisms can utilize the organic acids produced during primary fermentation, and thus initiate secondary fermentation with increases in brine pH and chemical reduction in the environment. Under these new conditions, opportunistic bacteria, such as *Cl. tertium* (Fleming et al. 1989), *Cl. bifermentans*, *Enterobacter cloacae* (Franco and Pérez-Díaz, 2012a, b, Franco et al. 2012), *Propionibacterium acidipropionici* and/or *Pectinatus sottacetonis* (Breidt et al. 2013, Caldwell et al. 2013) are able to metabolize lactic acid and other carbon sources, resulting in an increase in propionic acid and production of butyric acid (Fig. 5).

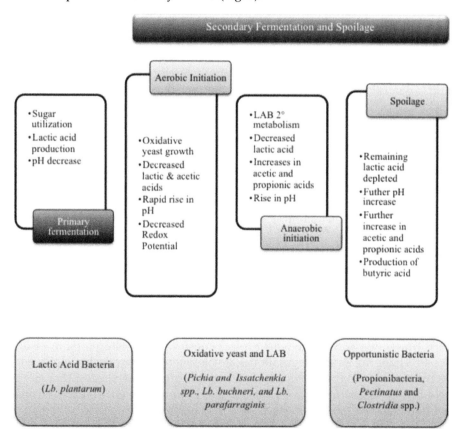

Fig. 5: Succession of Microbiota and Biochemical Changes Associated with Primary and Secondary Fermentations during Storage of Fermented Cucumber Pickles

5. Microbial Safety of Fermented Cucumbers

Acids and acidified foods are defined in the United States Code of Federal Regulations (21 CFR part 114) as having a pH value equal to or lower than 4.6. This pH is the upper limit that prevents *Cl. botulinum* spore outgrowth and neurotoxin production (Ito et al. 1976). In addition, the regulation specifies that acid and acidified foods, such as fermented cucumbers, should be processed "to the extend that is sufficient" to destroy vegetative cells of microorganisms of public health concern.

Fermented cucumbers, characterized for having low pH values and the presence of organic acids, are considered safe because these conditions inhibit pathogen growth and eventually lead to bacterial death. So far, the safety record for fermented cucumbers has been excellent. During the last five years of reported data (2009-2013), no single food-borne outbreak related to fermented cucumbers was reported (CDC, FOOD Database). Proper control of pH below 4.6 at all times and studies confirming the 5-log reduction of vegetative pathogens of concern are reasons for this good record.

Food-borne pathogens, such as *Escherichia coli*, *Listeria monocytogenes* and *Salmonella enterica* may be present on the surface of vegetables, including pickling cucumbers (Beuchat 1996, 2002, Brackett 1999, Taormina and Beuchat 1999). Of these, *E. coli* strains, in particular *E. coli* O157:H7, have been reported as the most acid-resistant bacteria in acidified, non-heated vegetables (Breidt et al. 2007). Breidt and Caldwell (2011) reported that a 5-log reduction of *E. coli* O157:H7 in fermented cucumber brine was possible in four days when brine pH was kept below 3.3, regardless of the holding temperature (10°C and above). These results were observed for both active fermentation and stable brines. When lower temperatures and higher pH were used, the authors reported that the 5-log reduction required longer periods of time, but was still achieved within the time frame of most commercial-scale fermentations (i.e. 23 days for pH 3.9 and 23°C). In contrast, a 5-log reduction in *L. monocytogenes* strains in home-prepared, fermented and refrigerated dill pickles required about 50 days, regardless of salt concentration (1.3–6.7 per cent), and, therefore, did not accomplish the pathogen kill required by the Code of Federal Regulations in a timeframe consistent with typical consumption patterns (Kim et al. 2004). Changing safety regulations, development of new technologies for fermentation, and the evolution of pathogenic organisms suggest that more research is needed in this area to ensure a continued excellent safety record for fermented cucumber pickles.

6. Chemical and Physical Changes During Fermentation, Bulk Storage and Processing

6.1. Quality of Fermented Cucumbers

The quality of fermented cucumber pickles is influenced by the presence or absence of physical defects, such as hollow cavities formed due to CO_2 buildup during brining, development of a cured appearance, texture properties and flavor. The greatest body of scientific literature is focused on the prevention of hollow centers and softening of cucumber tissue during fermentation and bulk storage because these defects, when severe, can result in devastating economic losses for processors.

6.2. Physical Defects

The major physical defect that causes economic losses for processors of fermented cucumbers is the development of hollow cavities in the interior of whole fruits, commonly known as 'bloater formation'. Different patterns of hollow cavities, referred to as balloon, lens and honeycomb, and varying levels of severity of bloating can occur, depending on the conditions of brining (Etchells et al. 1974, Fleming 1979). Severe bloating is related to trapped gas volumes of 10–12 per cent and expansion volumes greater than 200 mL in 4 litres fermentation jars (Fleming et al. 1973a). Early studies showed that increased incidence of severely bloated cucumbers was related to high salt concentrations used for natural cucumber fermentation and was associated with higher gas content in the brines (Veldhuis and Etchells 1939, Jones et al. 1941, Jones and Etchells 1943). The excess gas accumulation occurred between 12–50 days after brining and was composed of both CO_2 and hydrogen between Day 12 and Day 23 (Veldhuis and Etchells 1939). It was later discovered that the unusual production of hydrogen gas occurs in brinestock fermented in 10.6–15.8 per cent NaCl as a result of an aberrant secondary fermentation by *Aerobacter* species, now known as *Enterobacter* species (Etchells 1941a, Etchells et al. 1945). In most cases of bloater formation, the gas inside the cucumber is primarily CO_2 (Etchells and Jones 1941) and excessive gas production during fermentation coincides with an increase in yeast count (Etchells 1941b). Addition of acetic or lactic acids and sugars to fermentation brines significantly increased the incidence of bloating (Jones et al. 1941, Veldhuis et al. 1941), due to shifting of the fermentation environment to one that was more favorable for yeasts. However, even in well-controlled and pure culture lactic acid fermentations, there are a number of sources of CO_2 which peak in concentration between two to seven days of fermentation and is sufficient to cause bloater defects (Etchells et al. 1975). The cucumber fruits themselves contain CO_2 and produce additional CO_2 through respiration for several days after brining (Fleming et al. 1973b). Thus, storage conditions of cucumbers prior to brining that result in higher accumulation of CO_2 or higher respiration rates may also contribute to bloater formation. LAB are another source of CO_2, which in combination with CO_2 from the cucumbers is enough to cause substantial bloating defects. *Lb. brevis*, a heterofermentative LAB that produces one mole of CO_2 for every mole of glucose or fructose consumed, produced severe bloating that was equivalent to that of yeasts, *Saccharomyces rosei* and *S. cerevisiae* (Etchells et al. 1968, Fleming et al. 1978). Interestingly, the decarboxylation of malic acid by homofermentative *Lb. plantarum* during the active part of fermentation is also a source of CO_2 (McFeeters et al. 1982a) that may contribute significantly to bloater formation (Fleming et al. 1973b). All these sources contribute to the accumulation of CO_2 in the brine and fermenting cucumbers, and a strong relationship exists between the percentage CO_2 saturation of brine and bloater index as higher levels of saturation encourage diffusion of CO_2 into the cucumber fruits. Generally, a saturation of less than 50 per cent is associated with substantially reduced risk of bloater formation (Fleming et al. 1978).

The ability of the accumulated CO_2 to cause bloating of cucumbers depends on several factors. Cucumber cultivar (Wehner and Fleming 1984) and maturity influence the susceptibility of pickling cucumbers to bloating, with larger sized cucumbers exhibiting a higher incidence of bloating under similar fermentation conditions (Jones et al. 1941, Fleming et al. 1973a, Fleming et al. 1977). High NaCl concentration increases the susceptibility of cucumbers to bloating independently of the effects on microbiota discussed previously (Fleming et al. 1978). These effects are believed to be due to differences in the internal structure of cucumbers that influence the absorption of brine. High buoyancy forces increase bloater index as a result of the physical internal damage to the fruits, but an increase in hydrostatic pressure (influenced primarily by tank depth) decreases the cucumber susceptibility to bloating (Fleming et al. 1977).

Control of bloater defects is therefore a balance between the microbiological, chemical and physical conditions of the system. NaCl concentration, temperature and availability of oxygen significantly influence the composition of microbiota, which are substantial sources of CO_2. The sugar and malic acid content of cucumbers may vary due to cultivar, maturity and growing conditions (McFeeters et al. 1982b). The impact of these variations in substrates for CO_2 production will depend on the composition of the microbial community at any given time during cucumber fermentation. Natural fermentations in reduced NaCl brines (2.7 per cent) accumulated more CO_2 than those in 5.4 or 7.0 per cent of NaCl (Fleming et al. 1973b), perhaps due to the increased solubility of CO_2 at lower NaCl concentrations (Fleming et al. 1975) or possibly due to differences in microbiota. The solubility of CO_2, a governing factor in bloater defect formation (Corey et al. 1983b), decreases with increasing NaCl and temperature in fermented cucumber brines at pH 3.3 (Fleming et al. 1975), which influences the movement of gases in the cucumber (Corey et al. 1983b). It has been observed that less bloating occurs at lower fermentation temperatures (Etchells et al. 1975). Though pH does not substantially influence the solubility of CO_2, it does influence the proportion that exists as the gas vs. bicarbonate ion, which is relevant for efficient removal of CO_2 using nitrogen or air-purging techniques (Fleming et al. 1975). The microbiological and chemical factors that lead to CO_2 accumulation in the brine challenge the physical limitations of cucumber to resist damage caused by CO_2 trapped inside the cucumber tissues. It was determined that the critical period for susceptibility to bloating was between one to 32 days (Fleming et al. 1978) and that physical pre-treatments that increase the ability for brine to move into the cucumber, such as piercing (Etchells and Moore 1971, Fleming et al. 1973a), peeling (Fleming et al. 1973a) and oxygen-exchange (Fleming et al. 1980), are effective in eliminating bloater formation. Since bloater defects occur frequently, even in controlled fermentations, purging systems were developed to remove CO_2 from brines, effectively eliminating the occurrence of bloaters (reviewed by Fleming 1979). Nitrogen purging is effective in reducing CO_2, eliminating bloating (Fleming et al. 1973a, 1977) and in reducing the incidence of bloaters from 60 per cent to less than 10 per cent for size 3B cucumbers in a commercial system employing a small-pore sized, gas diffuser (Costilow et al. 1977).

6.3. Color Changes and Cured Appearance Development

The appearance of fermented cucumber pickles is largely influenced by the exterior skin color and the change in mesocarp from opaque white in raw cucumbers to a translucent appearance in the completely fermented fruit (Figure 6). The intense green color of raw cucumber fruit changes to a dark olive-green during fermentation due to the conversion of chlorophylls to pheophorbides and to a lesser extent pheophytins (White et al. 1963). Differences in the skin color may occur due to cultivar selection and growing conditions (Shetty and Wehner 2002, Jasso-Chaverria et al. 2005, Gómez-López et al. 2006, Aghili et al. 2009) and/or undesirable changes associated with photooxidation during brining and storage that result in a lightening and yellowing of the surface color (Buescher and Hamilton 2000). Cucumbers and the salts used for brining contain trace amounts of metals, such as iron, zinc, and copper, which can promote oxidation of pigments (Eisenstat and Fabian 1953). Protection from light or addition of 100–200 ppm $CaNa_2EDTA$, a stable compound with metal-chelating capabilities, to cucumber fermentations inhibits the bleaching of color associated with photoxidation (Buescher and Hamilton 2000).

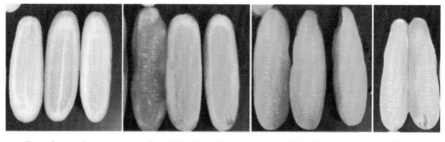

| Raw Cucumbers | Partially Cured | Fully Cured | Photo-oxidation |

Fig. 6. Changes in Cucumber Fruit Appearance during Fermentation and Bulk Storage

In finished products, a red color defect can occur due to the metabolism of yellow 5 (tartrazine), a common additive in pickle products, by *Lb. casei* or *Lb. paracasei* during shelf storage of unpasteurized or improperly pasteurized pickles (Pérez-Díaz et al. 2007, Pérez-Díaz and McFeeters 2009). Other commonly observed, but not scientifically documented color defects can occur during storage and processing of fermented cucumber pickles, including bright green and blue-green surface pigmentation due to higher than normal levels of contaminant metals and browning/rust, color development that may be related to oxidation or polymerization reactions.

The development of a cured appearance is considered a desirable characteristic of fermented cucumber pickles and the cucumbers are not processed further until they are fully cured. The cured appearance is due to the absorption of brine into intercellular spaces that are filled with ~60–94 mL/kg gas in the raw cucumber (Veldhuis and Etchells 1939, Corey et al. 1983a, c). The rate of curing varies significantly with fermentation. These authors have observed fully cured cucumbers as early as eight days after brining as well as only partially cured cucumbers 30 days after brining under current commercial production conditions that employ air purging during fermentation (Johanningsmeier, unpublished data). Several factors may influence the rate of curing, including internal cucumber pressure, temperature, redox potential of the system and proteolytic activity from the cucumber tissue or fermentation microbiota. Application of vacuum or heat speeds the curing process (Veldhuis and Etchells 1939). Air-purged fermentation achieve a cured appearance more rapidly than unpurged or nitrogen-purged fermentation; higher fermentation temperatures favor faster curing rates; and oxygen-exchanged cucumbers rapidly absorb brine and appear fully cured within 24 hours of brining (Fleming et al. 1980). Additional research on development of cured appearance has been conducted in refrigerated and pasteurized cucumber pickles where the cured appearance is considered a quality defect. Pasteurized cucumber pickle products are recommended to have no more than 10 ins. Hg vacuum to prevent rapid curing of the product upon opening (Etchells and Jones 1944). In acidified cucumbers, curing is related to proteolytic changes in cell wall constituents during shelf storage (Howard and Buescher 1993, Mok and Buescher 2012, Buescher et al. 2013). Oxidizing agents slow the curing and reducing agents hasten it (Howard and Buescher 1993). Therefore, changes in redox potential during fermentation (Olsen and Pérez-Díaz 2009) and potential proteolytic activity of fermentation microbiota (Liu et al. 2010) may also influence the rate of curing. In most commercial facilities today, long-term bulk storage is desired, so the variations in the rate of curing have little impact. However, implementation of new fermentation technologies and a growing demand for fermented vegetables by consumers may drive a faster turnover of the product and stimulate more research in this area to better predict and control cure rates.

6.4. Texture Properties

Texture quality of fermented cucumber pickles is primarily associated with sensory attributes of crunchiness, crispness, firmness and fracturability (Breene et al. 1972, Ennis and O'Sullivan 1979, Rosenberg 2013). Because of the high degree of cucumber to cucumber variability and the costs associated with maintaining trained descriptive sensory panels, mechanical texture measurements have been developed for monitoring softening of fermented cucumbers at various stages of processing. The first such method that was widely adopted by the pickling industry was the use of a penetrometer with a 5/16" tip (USDA Fruit Pressure Tester) to measure the entire fruit firmness of various cultivars after brining and storage (Jones and Etchells 1950). This method is quick, inexpensive, and a good indicator of whether the fermented fruits are firm enough to be processed into finished products. Since this method measures the force required to puncture through the exocarp (cucumber skin) and the mesocarp tissue below the surface (flesh), it does not differentiate between the small changes in the mesocarp that may relate to the sensory quality of sliced products. Texture profile analysis of fresh and brined cucumbers with and without skin showed that the sensory perception of crispness is more closely correlated with instrumental brittleness measurements for the cucumber tissue with skins removed (Jeon et al. 1973, 1975). Other methods that use a puncture test to measure mesocarp firmness also correlate well with fermented cucumber pickle sensory attributes of firmness (Thompson et al. 1982), crunchiness (Rosenberg 2013) and crispness (Yoshioka et al. 2009, 2010, Buescher et al. 2011, Rosenberg 2013). Significant differences in peak puncture force of hamburger dill chips were also related to consumer perception of texture quality with higher liking of texture clearly associated with higher peak puncture forces of pickle mesocarp (Wilson et al. 2015).

Many factors are known to influence the texture quality of fermented cucumbers. It has been shown that NaCl concentration, calcium addition, temperature, pH, bulk storage time, cucumber cultivar and maturity and the presence or absence of softening enzymes all influence the ability to maintain firm, fermented cucumber fruits for processing into pickle products (Table 2). These factors exert their effects in an interdependent fashion (Thompson et al. 1979, McFeeters et al. 1995), which makes it challenging to extrapolate the results of any one study to a commercial environment. That said, there are consistent patterns that have emerged from the many research studies performed to date. The concentration of NaCl used for brining and bulk storage is a key factor. Softening of cucumbers is consistently observed if NaCl concentrations of 1.8 per cent or less (equilibrated) are used for fermentation and bulk storage (Hudson and Buescher 1985, Fleming et al. 1987, 1996), and variable softening occurs with fermentation and storage in 2–2.8 per cent NaCl. However, the addition of 0.2–0.44 per cent $CaCl_2$ to fermentation is known to improve the retention

Table 2: Factors that Influence the Texture Quality of Fermented Cucumber Pickles

Brine Ingredient or Intrinsic Factor	Function	Optimal Range	Legal Limit
Sodium chloride concentration	Selects for proper fermentation microbiota; influences microbial stability and texture quality during bulk storage	3–7 per cent for natural fermentation; 4–10 per cent for bulk storage	GRAS status; Limit is for chlorides in wastewater (230 ppm)
Calcium (usually added as $CaCl_2$)	Firming agent; protects against pH-induced softening and inhibits softening due to pectinolytic or cellulytic enzyme action	25–72 mM (0.28-0.8 per cent as $CaCl_2$) depending on NaCl concentration, cucumber size and downstream processing	36 mM (0.4 per cent) $CaCl_2$ in finished pickle products
Potassium sorbate	Anti-fungal to prevent spoilage yeasts in closed, anaerobic systems and inhibit molds in air-purged fermentations	6–20 mM (0.1–0.3 per cent) at pH 4.5 or below	
Acetic acid (as vinegar)	Reduces initial brine pH to give a selective advantage to lactic acid bacteria; fungal inhibitor; inhibits Enterobacteriaceae	27–50 mM (0.16–0.3 per cent) depending on purging routines and other brine ingredients	GRAS status
Alum	Firming/crisping agent used in desalting waters or finished products	Varies	Considered a processing aid
pH	Influences initiation of fermentation; induces softening at pH's below 3.5 and at neutral pH during extended hold times	3.5–4.3	None. Food safety standard: pH should reach less than 4.6 as quickly as possible and be maintained below this value throughout storage and processing
Temperature	Impacts initiation, rate and extent of fermentation; softening rates during storage increase with increasing temperatures	Storage conditions: As low as is feasible without freezing	Not Applicable
Pectin methylesterase (PME)	Cell wall enzyme that demethylates pectin	Unknown	Not Applicable

Contd...

Table 2: (*Contd.*)

Brine Ingredient or Intrinsic Factor	Function	Optimal Range	Legal Limit
Polygalacturonase	Cell wall enzyme that cleaves polygalacturonic acid (demethylated pectin), resulting in texture quality loss	None	Not Applicable
Cellulase	Cell wall enzyme that cleaves cellulose, resulting in texture quality loss	None	Not Applicable

of firmness in cucumbers fermented and stored in low salt (1.8–2.6 per cent NaCl) brines (Buescher et al. 1979, Hudson and Buescher 1985, Fleming et al. 1987, McFeeters et al. 1995), including the prevention of soft center defects in size 4 (5.1–5.7 cm diameter) cucumber fruits when brined with $CaCl_2$ and NaCl to equilibrate at 0.44 and 1.8 per cent, respectively (Hudson and Buescher 1980). Even in fermentation brines containing higher NaCl concentrations for fermentation (5–7 per cent wt/vol) and storage (5–11 per cent wt/vol), the addition of 0.10–0.44 per cent $CaCl_2$ was found to improve firmness retention during bulk storage (Thompson et al. 1979, Tang and McFeeters 1983, Fleming et al. 1987, Buescher and Burgin 1988, Guillou et al. 1992, Guillou and Floros 1993) and in finished pickle products (Buescher and Burgin 1988). Conversely, simple addition of 0.2 per cent $CaCl_2$ to fermentation brines in the absence of NaCl was not able to maintain the cucumber firmness during bulk storage, despite implementation of controlled-fermentation procedures (Fleming et al. 1995). These results are consistent with the multiresponse optimization model which predicted that the optimal $CaCl_2$ concentration for brining and storage of cucumbers relying on a natural fermentation is 0.28 per cent in combination with 0.3–0.32 per cent potassium sorbate and 3–7 per cent NaCl (Guillou and Floros 1993). Increasing calcium concentrations from 0 to 25 mM (equivalent to 0.28 per cent $CaCl_2$) dramatically reduced the softening rate of fermented cucumber tissue during storage in 2 per cent NaCl with varying pH and temperature combinations (McFeeters et al. 1995). Thus, it is currently a common practice in the U.S.A. to add $CaCl_2$ (0.1–0.3 per cent, equilibrated) to brines for commercial production of fermented cucumbers (Pérez-Díaz et al. 2014).

Calcium ions added to fermentation brines help maintain cucumber firmness regardless of the form that is used for addition. Calcium acetate or calcium hydroxide used in buffered systems for controlled cucumber fermentation, $CaCl_2$, and sea salts that naturally contain calcium and magnesium (Yoo et al. 2006) have all been used to improve firmness retention during fermentation and storage. The natural level of calcium in pickling cucumbers ranges from 1.8–8.3 mM (Hudson and Buescher 1985, McFeeters and Fleming 1989, McFeeters et al. 1995) and has been observed to significantly influence the softening rates of cucumbers stored in acid brines (McFeeters and Fleming 1989). To prevent softening during fermentation in low NaCl-brined cucumbers, calcium needs to be added at brining (Hudson and Buescher 1985, McFeeters et al. 1995). However, the softening rates during subsequent storage are mostly dependent on the concentration of calcium during storage (McFeeters et al. 1995). An empirical model was developed for the effects of pH (2.6–3.8), temperature (25–65°C), and calcium concentration (0–72 mM) on fermented cucumber softening rates during storage (McFeeters et al. 1995). This model suggests that a high concentration of calcium ions may be used to override the softening effects of low pH at temperatures up to 40°C (104°F). Although calcium chloride has been shown to inhibit PG-induced softening (Buescher et al. 1979, 1981a), it appears that it frequently aids in firmness retention that is independent of detectable PG activity (Buescher et al. 1981b). It was shown that increasing the $CaCl_2$ to 0.8 per cent for fermentation and bulk storage in NaCl brines resulted in a higher residual calcium concentration in the finished products (~31 mM) that helped maintain sensory crispness during shelf storage for one year without the use of alum in the desalting water (Buescher et al. 2011).

Alum, the common name for a variety of aluminum containing additives, such as potassium aluminum sulfate, sodium aluminum sulfate, ammonium aluminum sulfate or aluminum sulfate, is used during the desalting process or in finished pickle products as an ingredient to increase or maintain the pickle firmness and crispness. Despite the historical use of alum in pickling, a number of studies demonstrate that calcium is a more effective firming agent during storage of finished pickle products (Buescher and Burgin 1988) and that 31 mM calcium in finished products will achieve crispness retention better than the typical practice that includes 0.35 per cent wt/wt alum during desalting (Buescher et al. 2011). However, alum is still commonly added to desalting waters as insurance against texture quality losses as reflected by the range in aluminum content in fermented cucumber pickle products from 0–200 ppm (Pérez-Díaz et al. 2014).

In addition to the important role that salts play in texture quality, pH and temperature can have a significant impact on the ability to maintain high-quality texture attributes during storage in bulk or as finished products. Softening rates in fermented cucumbers increase with decreasing pH from 4.3 to 2.6 (McFeeters et al. 1995, Fleming et al. 1996). This relationship is also apparent in blanched cucumber tissue from pH 2–5 (McFeeters and Fleming 1991) and for *in vitro* cucumber pectin hydrolysis rates from pH 2–3.5 (Krall and McFeeters 1998), suggesting that the effect of pH in this region is independent of enzymatic or microbial mechanisms. The lowest softening rates were observed with the addition of 20 mM $CaCl_2$ to blanched cucumber mesocarp tissues with pHs of

3.5–4.5 (McFeeters and Fleming 1991). Addition of $CaCl_2$ to brined whole cucumber dramatically improved firmness retention for fruits stored at pH 3.3 as compared to those at pH 3.8 (Thompson et al. 1979), and the optimal pH for assuring both texture quality and microbial stability of fermented cucumbers in bulk storage was a pH of 3.5 (Fleming et al. 1996). Interestingly, *in vitro* acid hydrolysis of cucumber pectin was unaffected by 20 mM $CaCl_2$, indicating that some other mechanisms may be responsible for the softening of fermented cucumber pickles at low pH (Krall and McFeeters 1998). The effect of temperature appears to be additive and linear in most cases. Higher temperatures for storage result in greater softening rates (McFeeters et al. 1995). Temperature and pH can also influence enzymatic softening rates. However, this relationship is more complex as there are many isozymes of common softening enzymes, each with different temperature and pH optima.

Much less is known about the influence of cucumber cultivar and maturity on the texture after fermentation and bulk storage. However, a few studies have found differences in texture properties among cucumber cultivars using the raw fruits for evaluation (Breene et al. 1972, Yoshioka et al. 2009) and evaluation of cultivars after brining and storage (Jones and Etchells 1950, Jones et al. 1954). A study of 20 pickling cucumber cultivars showed that cultivar differences in crispness and firmness were only apparent after brining (Ennis and O'Sullivan 1979), indicating the challenges in evaluating the cucumber fruit suitability for brining. Therefore, brining trials are often conducted among seed companies, farmers and briners to select the best varieties for cultivation and pickling. Extended storage of pickling cucumbers prior to brining can also lead to changes in the microstructure, causing textural defects in the finished products (Walter et al. 1990), and the ideal conditions for storage were deemed to be 10°C and high relative humidity (Etchells et al. 1973).

Efforts to understand the root causes of softening of fermented cucumber tissues have focused on changes in the pectic substances of the middle lamella and the role of softening enzymes of native and fungal origin. Microscopic examination of soft and firm fermented cucumbers indicated that mesocarp tissue softening was related to swollen cell walls and dissolution of the middle lamella, resulting in decreased strength of the cell-to-cell junctions (Walter et al. 1985). Cucumber pectin isolated from cell walls was found to be ~74,200 Daltons with an average degree of polymerization of 402 (Tang and McFeeters 1983) and 34–64 per cent degree of methylation (Tang and McFeeters 1983, Hudson and Buescher 1985, McFeeters and Lovdal 1987, McFeeters et al. 1995, Krall and McFeeters 1998). Extensive demethylation occurs during fermentation, resulting in no (Tang and McFeeters 1983) to 14–15 per cent (Hudson and Buescher 1985, McFeeters et al. 1995) methylated residues in fermented cucumber pectic fractions. Changes in the degree of methylation of pectin were reflected in a decrease in acid soluble and an increase in EDTA soluble pectic fractions

(Hudson and Buescher 1985). However, the degree of pectin demethylation did not consistently relate to calcium concentration or cucumber tissue firmness among these studies. Cell-wall neutral sugar content is relatively low compared to other fruits and vegetables: 13–16 per cent of cell-wall by weight, comprised of primarily galactose (50 per cent) followed by xylose, glucose, mannose and arabinose (Tang and McFeeters 1983), which changes in composition to a small degree during maturation (McFeeters and Lovdal 1987). Softening of fermented cucumber tissue was related to losses in galacturonic acid, galactose, arabinose and rhamnose content using an alternative cell-wall extraction procedure that allowed detection of small changes in solubility of cell-wall polysaccharides (McFeeters 1992). Changes in the solubility of the pectin appear to be related to changes in cell-wall polysaccharides that influence texture quality, but the exact mechanisms of softening have not been fully elucidated.

Polygalacturonase, an enzyme that catalyzes the hydrolytic cleavage of pectic acid, is known to cause varying degrees of softening during storage of fermented cucumbers (Bell et al. 1950, Buescher et al. 1981a). Pectinase, PG and cell-free mold extracts caused softening of cucumbers and the extent of softening was dependent on NaCl concentration (Bell and Etchells 1961). Pectolytic enzyme activity can come from the cucumbers themselves, especially the seeds and ripe fruits (Bell 1951, Pressey and Vants 1975, McFeeters et al. 1980, Saltveit and McFeeters 1980, Cho and Buescher 2012), pollinated flowers that are not removed before brining (Bell 1951, Bell et al. 1958) and other fungal sources (Etchells et al. 1958a). Certain species of yeasts, such as *S. fragilis*, are unique in their ability to cause softening of cucumber brinestock (Bell and Etchells 1956), whereas an abundance of molds isolated from cucumbers and blossoms possess both PG and cellulase activity (Etchells et al. 1958a). Cellulases can also induce softening in fermented cucumbers during long-term storage (Buescher and Hudson 1984). The leaves of certain plants, including leaves from Scuppernong grapes, contain a natural inhibitor against softening enzymes (Bell et al. 1960, 1962, 1965a, b, Bell and Etchells 1961, Porter et al. 1961, Etchells et al. 1958b), which may be why many traditional home recipes suggest adding a couple of grape leaves to the ferment. In commercial fermentations that employ air-purging to reduce bloater defects, acetic acid can be added to reduce mold-induced softening of cucumbers (Potts and Fleming 1982).

6.5. *Volatile Flavor Compounds in Fresh and Fermented Cucumbers*

The characteristic aroma of fresh cucumbers is attributed to (E,Z)-2,6-nonadienal and E-2-nonenal, which are enzymatically synthesized from linolenic and linoleic acids by the action of lipoxygenase during tissue disruption (Fleming et al. 1968, Buescher and Buescher 2001). E-2-nonenal has been shown to have approximately 2 per cent of the odor impact of (E,Z)-2,6-nonadienal (Xu et al. 2012). Interestingly, significantly less (E,Z)-2,6-nonadienal is produced in the exocarp as compared to the mesocarp and endocarp tissues, which explains why smaller-diameter fruits are associated with less (E,Z)-2,6-nonadienal production (Buescher and Buescher 2001). Exocarp tissues have higher levels of lipoxygenase, hydroperoxide lyase and unsaturated fatty acids; thus, the low (E,Z)-2,6-nonadienal production by exocarp tissues is attributed to limited substrate, enzyme inhibitors, or production with concomitant degradation (Pederson et al. 1964, Wardale et al. 1978, Wardale and Lambert 1980, Buescher and Buescher 2001). Palma-Harris et al. (2002) demonstrated that an increase in (E,Z)-2,6-nonadienal resulted in an increase in the intensity of 'fresh cucumber flavor' in refrigerated cucumber pickles as assessed by 24 trained panelists. These fresh cucumber flavors are not typically observed in fermented cucumbers due to the inactivation of lipoxygenase under acidic pH conditions (Wardale and Lambert 1980). Similarly, other conditions that lead to inactivation of lipoxygenase, such as freezing or pasteurization, also result in the observed loss of (E,Z)-2,6-nonadienal production (Buescher and Buescher 2001).

Fresh cucumber juice also contains ethanol, propanal, (E)-2-pentenal, hexanal, (E)-2-hexenal and (Z)-6-nonenal (Cho and Buescher 2011). Linoleic acid is believed to be the precursor of hexanal, (E)-2-heptenal, (E)-2-octenal, and E-2-nonenal; while propanal, (E)-2-hexenal, (E)-2-pentenal, ethanal and (E,Z)-2,6-nonadienal are formed from linolenic acid (Grosch and Schwarz 1971, Zhou et al. 2000). Cucumbers have a very high fatty acid α-oxidation activity, in which fatty acids are enzymatically broken down into $C_{(n-1)}$ long-chain fatty aldehydes and CO_2 (Borge et al. 1998). For example, pentadecanal is identified as the product of α-oxidation of palmitic acid in cucumber homogenate (Borge et al. 1998). Due to the high water activity of cucumber pickles, oxidation of these products involves enzymatic or microbial mechanisms; direct, non-enzymatic oxidation tends to be outcompeted by other modes of deterioration at high water activities (St. Angelo 1992).

The production of these flavor-active aldehydes is hypothesized to include breakage of double bonds in the unsaturated fatty acids by a dioxygenase-like reaction and the formation of hydroperoxide intermediates (Grosch and Schwarz 1971, Zhou et al. 2000). Among these compounds, (E,Z)-2,6-nonadienal and hexanal were identified in fermented cucumber brines with characteristic aromas of 'fresh cucumber' and 'green', respectively (Marsili and Miller 2000). Additionally, (E)-2-nonenal and (E)-2-octenal were identified by Cordero et al. (2010) in roasted hazelnuts with fatty/green and green aromas, respectively.

Although more often associated with raw fruits or refrigerated pickles, 'Green' is a term sometimes used to describe the aroma of freshly-fermented cucumbers. Eight volatile compounds are characteristic of the green odor emitted by plant leaves: (Z)-3-hexenol, (E)-3-hexenol, (E)-2-hexenol, (Z)-3-hexenal, (E)-3-hexenal, (E)-2-hexenal, hexanol and hexanal (Hatanaka 1996). Forss et al. (1962) found that (E,Z)-2,6-decadienal was characterized by the flavor notes of 'green', 'plant-like', and 'like cucumbers'. After further review of literature, Forss et al. (1962) hypothesized that the cis-non-conjugated unsaturation is characteristic of a compound that produces a 'green' or 'plant-like' flavor. Interestingly, 2-nonenal and (Z,Z)-2,6-nonadienal were described as 'oily' and 'tallow', and (Z,E)-2,6-nonadienal as 'green' or 'like cucumbers' when evaluated near their threshold concentrations; but, as the concentration of the compounds increased, panelists gave the description of 'like cucumbers' to all three compounds (Forss et al. 1962).

A number of secondary reactions occur during fermentation of cucumber pickles, contributing to the flavor profile of finished products. Fermented cucumbers evaluated directly from the tankyard have been described as silage-like, sour, slightly sweet and green (Marsili and Miller 2000). These authors identified trans- and cis-4-hexenoic acid, low volatility compounds, as key flavor compounds with aromas characteristic of fermented cucumber brine by gas chromatography-olfactometry. Additional compounds with high odor impact values in fermented cucumber brines included 2-heptanol, cis-2,4-hexadienoic acid (tentative identification), phenyl ethyl alcohol, 2,6-nonadienal and 2-dodecen-1-al (tentative identification). While pure solutions of trans-4-hexenoic acid were characterized as being similar to authentic brine samples, addition of phenyl ethyl alcohol (a rose/floral note) to the pure solution resulted in a closer match. Phenylacetaldehyde was also present in brine samples and is produced by oxidation of phenyl ethyl alcohol (Marsili and Miller 2000). Murray and Whitefield (1975) reported the presence of 3-isopropyl-2-methoxypyrazine in fresh cucumbers. Since the compound does not change significantly during fermentation and has an odor threshold of only 2 ppt, it possibly plays a role in the overall aroma profile of fermented cucumbers (Zhou and McFeeters 1998, Whitefield and Last 1991).

Previous work by Zhou and McFeeters (1998) identified high levels of linalool in 2 per cent salt fermentation brines, but Marsili and Miller (2000) observed linalool in only 5 per cent of commercial brine samples with 8–10 per cent salt concentration, indicating linalool is really not a major aroma impact compound in industrial-scale fermented cucumbers. However, varying salt levels in fermentations could alter the microbiota, which could impact the production of by-products that contribute to flavor and aroma (Marsili and Miller 2000). Zhou and McFeeters (1998) compared volatile compounds found in fresh cucumbers to those in fermented cucumbers: ethyl benzene, o-xylene and benzaldehyde were the only compounds identified in fermented cucumbers that were not present in fresh cucumbers. Four compounds of interest were found in lower concentrations in fermented cucumbers as compared to fresh cucumbers: hexanal, (E)-3,7-dimethyl-1,3,6-octatriene, (E,Z)-2,6-nonadienal and 2-undecanone (Zhou and McFeeters 1998). More recently, Johanningsmeier and McFeeters (2011) identified 314 volatile compounds in the brine of cucumbers fermented in 6 per cent NaCl, including hydrocarbons, aldehydes, alcohols, ketones, acids, esters, ethers, furans, pyrans, phenols, nitrogenous compounds and sulfur-containing compounds.

Oxidized flavor is an off-flavor of particular concern for pickle products. By definition, oxidation is simply the loss of electrons from a compound (Smith 2008). The oxidation of polyunsaturated fatty acids is of particular concern in flavor chemistry (Andreou and Fuessner 2009). Oxidation of unsaturated lipids is the process in which carbon-centered alkyl radicals and peroxy radicals are formed due to the presence of initiators, such as enzymes, light, heat, metals, metalloproteins and microorganisms. After initiation, propagation in the presence of oxygen leads to the formation of hydroperoxides as the primary products (St. Angelo 1992). In the presence of light, a non-free radical process known as photooxidation may form hydroperoxides due to the reaction of unsaturated fatty acids with singlet oxygen resulting from the excitation of oxygen by a photosensitizer, such as chlorophyll (Frankel 1991). For plant tissues, the polyunsaturated fatty acids found in greatest abundance are linoleic and linolenic acid (Baysal and Demirdoven 2007). After completion of cucumber fermentation, Pederson et al. (1964) observed a five-fold increase in free fatty acids and a reduction in phospholipids greater than 90 per cent of that found in the raw cucumber. Linoleic acid was found in the flesh, skin and seed of raw cucumbers at concentrations of 11, 38, and 26 mg/100 g. After fermentation, linoleic acid in the flesh, skin and seed increased to 215, 444, and 908 mg/100 g. Similarly, linolenic acid increased in the flesh, skin and seed from 10, 255, and 18 mg/100 g to 450, 545, and 1067 mg/100 g, respectively. The lipid content of cucumber flesh is much lower than that of the seed or skin, which suggests that some aromatic compounds might partition to the endo- or exocarp (Pederson et al. 1964). However, Zhou and McFeeters (1998) found that volatile compounds in fermented cucumber slurry and brine differed less than twofold in relative peak areas for the 37 identified compounds. Consistent with expectations, hexanal and ethyl-benzene were found in higher concentrations in the brine, while α-caryophyllene, 2-undecanone and (E,Z)-2,6-nonadienal were found in higher concentrations in fermented cucumber.

During fermentation and bulk storage, cucumbers are susceptible to oxidation due to the use of open tanks exposed to sunlight and the common practice of purging the tanks with air to mix the contents and remove CO_2 (Buescher and Hamilton 2000). Additionally, cucumbers and the salt used for brining contain trace amounts of metals, such as iron, zinc, and copper, which can play a role in promoting oxidation of pigments and flavor compounds (Eisenstat and Fabian 1953). Fermented cucumber homogenates incubated in the presence of oxygen exhibited significant increases in hexanal, (E)-2-heptenal, (E)-2-pentenal, (E)-2-hexenal and (E)-2-octenal that were highly correlated to oxidized odor intensity, as assessed by 20 trained sensory panelists (Zhou et al. 2000). Furthermore, Zhou et al. (2000) determined that the aldehydes were not formed due to lipoxygenase activity since heat treatment to inactivate the enzyme prior to oxygen exposure did not result in a significant reduction in aldehyde production. However, recent studies show that several of these aldehydes decrease during long-term bulk storage of fermented cucumbers, giving rise to ketones and other compounds highly correlated with oxidized off-flavor in finished hamburger dill chips (Wolter 2013). Venkateshwarlu et al. (2004) modeled oxidative off-flavors commonly found in fish oil by mixing pure volatile flavor compounds in milk and evaluating the intensity of fishy and metallic off-flavors in these samples. A trained 16-member descriptive analysis panel determined that (E,Z)-2,6-nonadienal and 1-penten-3-one were the two main compounds characteristic of the off-flavors described as metallic and fishy. Of note, the fishy and metallic flavors were not present when the compounds were added to milk alone, confirming the importance of compound interaction in flavor production. In addition, the metallic odor of penten-3-one was enhanced in the presence of heptenal, while a synergistic relationship between (E,Z)-2,6-nonadienal and (Z)-4-heptenal contributed to fishy off-flavors (Venkateshwarlu et al. 2004). It is possible that in the earlier studies of oxidized off-flavor in fermented cucumber (Zhou et al. 2000), the unsaturated aldehydes were acting as potentiators for undetected compounds responsible for the oxidized flavor perceived by the sensory panel. More research is needed to fully understand the relationship between the changes in chemical composition and off-flavor development.

7. Sodium Chloride Reduction Strategies

7.1. *Environmental Considerations*

One of the biggest drivers of process change in cucumber fermentation has been the regulatory pressures to reduce salt discharge in the environment. In the US, there are many different regions with varying regulations depending on the state and location, with proximity to large municipal systems being more favorable. Nonetheless, waste treatment is required in one form or another, including removal of vegetable solids, adjustment of pH and control of hydraulic flow, BOD and chloride content.

7.2. *Brine Recycling*

Brine recycling is one way for the cucumber fermentation industry to reduce its salty effluent waste stream. The primary barriers to reusing the salty tank brine in the past were the carryover of softening enzymes and off-flavors. A diffusion plate assay for testing recycled brines for PG activity (Buescher and Burgin 1992) is used routinely for management of recycled brines. Various strategies were developed to inactivate or remove the softening enzymes from brines prior to recycling, including coagulation and settling, heat inactivation, ultrafiltration, activated carbon and clay treatment (Geisman and Henne 1973, Little et al. 1974, Palnitkar and McFeeters 1975, Mercer et al. 1971, Buescher and Hamilton 2007). Today, because of its low investment and operating cost, clay treatment is the most commonly used process in commercial production. Pure-Flo B80 from Oil Dry-Corp has been a widely used bentonite based clay for the removal of softening enzymes in spent pickle brines. Even with active brine recycling programs in place, the disposal of large volumes of effluent waste water from the desalting process remains a challenge for large processing operations.

7.3. Salt Reduction in Cucumber Fermentation

Simple reduction of NaCl below the 6-7 per cent wt/vol concentration used in most commercial fermentations significantly increases the susceptibility to secondary fermentations and/or tissue softening, thereby increasing the risk of economic loss. However, research to develop new technologies that use less NaCl or even no NaCl have been developed to provide sustainable alternatives to the current process. A 'process-ready' bag-in-box fermentation technology was developed (Fleming et al. 2002a, b) that resulted in high-quality finished products (Johanningsmeier et al. 2002) and the ability to filter the brine for use in finished products (Fasina et al. 2002, 2003), such that the entire contents of the fermentation could be consumed with very little waste. This technology has not yet been widely adopted primarily because the investment in current infrastructure in tankyards was not conducive to an immediate shift to the new technology and investments would need to be made in new types of equipment. Guillou and Floros (1993) demonstrated that natural fermentations and bulk storage of cucumbers for 6 mo could be accomplished in 2.5-3.0 per cent NaCl in combination with 0.3 per cent $CaCl_2$ and 0.32 per cent potassium sorbate, but there is no public record of commercial trials with this process. Most recently, a fermentation process that relies entirely on calcium chloride as the only salt promises to eliminate sodium salt waste from the tankyard and reduce chloride levels by 60-80 per cent (McFeeters and Pérez-Díaz 2010). This process employs 1.1 per cent of $CaCl_2$ to aid in firmness retention, a *Lb. plantarum* starter culture, potassium sorbate to inhibit yeasts, and fumaric acid or sodium benzoate to stabilize the tanks for long-term bulk storage (Pérez-Díaz et al. 2015) and was capable of producing acceptable quality-hamburger dill chips on the commercial scale (Wilson et al. 2015). Commercial trials (2011-2014) have illuminated the challenges associated with scale-up to a production environment and active research in this area includes: 1) selection of appropriate starter cultures; 2) optimization of texture quality of finished products; 3) investigation of the suitability of this process in cold climates; and

4) adapting process-ready technologies for the current commercial environment to eliminate the desalting step.

8. Conclusions and Future Developments

Although cucumber fermentation remains largely a traditional process, it has proven to be a consistently safe process by which raw cucumbers are transformed into high-quality pickles that have a long shelf-life at ambient temperatures. Lactic acid bacteria, especially *Lb. plantarum*, which drives the fermentation, in community with yeast, both aerobic and anaerobic, confer the flavor and aroma characteristics of the end-product. Associated with these changes, the production of organic acids and decrease in pH results in a safe and innocuous final product. Conditions for natural fermentation that consistently result in high quality products that are microbially stable for many months include: 1) brining with NaCl to equilibrate with the cucumbers at 6–7 per cent (wt/vol); 2) purging the fermentation brines during active fermentation to remove CO_2 which can cause hollow cavities inside the whole cucumber fruits; and 3) excluding oxygen from the process as much as possible during fermentation and storage. However, to create more sustainable industrial processes, continued efforts are needed to further reduce the salt used for brining while delivering safe, high-quality fermented cucumber products to consumers.

Tremendous increases in demand for 'fresh-like' products containing natural ingredients and changing food consumption patterns have increased the desirability of products produced using LAB fermentations. As such, the process of LAB fermentations is being rediscovered in exciting ways all around the globe and the indigenous fermented foods that have played a vital role in the history of humankind's development are finding their way back into modern society as artisanal products. New molecular approaches to study the composition of microbiota and to select starter cultures targeted for fruits and vegetables, including those with health-promoting properties (probiotic and bioactive compound-generating), are being pursued to allow controlled fermentation processes and develop novel products as done for other fermented foods (cheese, yogurt, sausage, etc.). Lactic acid bacteria tailored to various environmental conditions and desired finished products will provide new potential for further development of cucumber fermentation.

Acknowledgements

The authors acknowledge Dr. Roger F. McFeeters for his helpful discussions and review of the chapter. We also thank Sandra Parker for her excellent editorial assistance.

References

Aghili, F., A.H. Khoshgoftarmanesh, M. Afyuni and M. Mobli. 2009. Relationships between fruit mineral nutrients concentrations and some fruit quality attributes in greenhouse cucumber. *Journal of Plant Nutrition* 32(12): 1994-2007.

Andreou, A. and I. Feussner. 2009. Lipoxygenases – Structure and reaction mechanism. *Phytochemistry* 70: 1504-1510.

Baysal, T. and A. Demirdoven. 2007. Lipoxygenase in fruits and vegetables: A review. *Enzyme and Microbial Technology* 40: 491-496.

Bell, T.A. and J.L. Etchells. 1956. Pectin hydrolysis by certain salt-tolerant yeasts. *Applied Microbiology* 4(4): 196-201.

Bell T.A., J.L. Etchells and R.N. Costilow. 1958. Softening enzyme activity of cucumber flowers from northern production areas. *Food Research* 23(2): 198-204

Bell, T.A. and J.L. Etchells. 1952. Sugar and acid tolerance of spoilage yeasts from sweet-cucumber pickles. *Food Technology* 6(12): 468-472.

Bell, T.A. and J.L. Etchells. 1961. Influence of salt (NaCl) on pectinolytic softening of cucumbers. *Journal of Food Science* 26(1): 84-90.

Bell, T.A. 1951. Pectolytic enzyme activity in various parts of the cucumber plant and fruit. *Botanical Gazette* 113(2): 216-221.

Bell, T.A., J.L. Etchells and J.D. Jones. 1950. Softening of commercial cucumber salt-stock in relation to polygalacturonase activity. *Food Technology* 4(4): 157-163.

Bell, T.A., J.L. Etchells, C.F. Williams and W.L. Porter. 1962. Inhibition of pectinase and cellulase by certain plants. *Botanical Gazette* 123(3): 220-223.

Bell, T.A., J.L. Etchells and W.W.G. Smart Jr. 1965b. Pectinase and cellulase enzyme inhibitor from sericea and certain other plants. *Botanical Gazette* 126: 40-45.

Bell, T.A., J.L. Etchells and A.F. Borg. 1959. The influence of sorbic acid on the growth of certain species of bacteria, yeasts and filamentous fungi. *Journal of Bacteriology* 7(5): 573-580.

Bell, T.A., J.L. Etchells, J.A. Singleton and W.W.G. Smart Jr. 1965a. Inhibition of pectinolytic and cellulolytic enzymes in cucumber fermentations by sericea. *Journal of Food Science* 30(2): 223-239.

Bell, T.A., L.W. Aurand and J.L. Etchells. 1960. Cellulase inhibitor in grape leaves. *Botanical Gazette* 122(2): 143-148.

Beuchat L.R. 1996. Pathogenic microorganisms associated with fresh produce. *Journal of Food Protection* 59: 204-16.

Beuchat, L.R. 2002. Ecological factors influencing survival and growth of human pathogens on raw fruits and vegetables. *Microbes and Infection* 4: 413-423.

Bevilacqua, A., M. Cannarsi, M. Gallo, M. Sinigaglia and M.R Corbo. 2009. Characterization and implications of *Enterobacter cloacae* strains, isolated from Italian table olives Bella di Cerignola. *Journal of Food Science* 75(1): M53-M60.

Binsted, R., J.D. Devey and J.C. Dakin. 1962. *Pickle and Sauce Making*, second editon. Food Trade Press, London.

Bobillo, M. and V.M. Marshall. 1991. Effect of salt and culture aeration on lactate and acetate production by *Lactobacillus plantarum*. *Food Microbiology* 8(2): 153-160.

Borge, G.I.A., E. Slinde and A. Nilsson. 1998. Fatty acid alpha-oxidation of tetradecylthioacetic acid and tetradecylthiopropionic acid in cucumber (*Cucumis sativus*). *Biochimica et Biophysica Acta, Lipids and Lipid Metabolism* 1394(2): 158-168.

Brackett, R.E. 1999. Incidence, contributing factors and control of bacterial pathogens in produce. *Post Harvest Biology and Technology* 15: 305-11.

Breene, W.M., D.W. Vis and H. Chou. 1972. Texture profile analysis of cucumbers. *Journal of Food Science*. 37(1): 113-117.

Breidt, F. 2006. Safety of minimally processed, acidified, and fermented vegetable products. Pp. 314-327 *In*: G.M. Sapers, J.R. Gorny and A.E. Yousef (Eds.). *Microbiology of Fruits and Vegetables*. CRC Press, Inc., Boca Raton, Florida.

Breidt, F. and J.M. Caldwell. 2011. Survival of *Escherichia coli* O157:H7 in cucumber fermentation brines. *Journal of Food Science* 76(3): M198-M203.

Breidt, F., E. Medina-Pradas, D. Wafa, I.M. Pérez-Díaz, W. Franco, H. Huang, S.D. Johanningsmeier and J. Kim. 2013b. Characterization of cucumber fermentation spoilage bacteria by enrichment culture and 16S rDNA cloning. *Journal of Food Science* 78(3): M470-M476.

Breidt, F., R.F. McFeeters and I. Díaz-Muñíz. 2007. Fermented vegetables. Pp. 783-793. *In*: M.P. Doyle and L.R. Beuchat (Eds.). *Food Microbiology: Fundamentals and Frontiers*, fourth edition. ASM Press, Washington, D.C.

Breidt, F., R.F. McFeeters, I. Pérez-Díaz and C.-H. Lee. 2013a. Fermented Vegetables. Chapter 33.*In*: M.P. Doyle and R.L. Buchanan (Eds.). *Food Microbiology: Fundamentals and Frontiers*, fourth editon. ASM Press, Washington, D.C.

Buescher, R. and C. Hamilton. 2000. Protection of cucumber pickle quality by CaNa$_2$EDTA. *Journal of Food Quality* 23(4): 429-441.

Buescher, R. and C. Hamilton. 2007. Adsorption of polygalacturonase from recycled cucumber pickle brines by pure-flo B80 CLAY1. *Journal of Food Biochemistry* 26(2): 153-165.

Buescher, R. and M. Hudson. 1984. Softening of cucumber pickles by cx-cellulase and its inhibition by calcium. *Journal of Food Science* 49(3): 954-955.

Buescher, R., M.J. Cho and C. Hamilton. 2013. Heat conditioning of cucumbers improves retention of sliced refrigerated pickle texture and appearance. *Journal of Food Biochemistry* 37(5): 564-570.

Buescher, R.H. and R.W. Buescher. 2001. Production and stability of (E,Z)-2,6-nonadienal, the major flavor volatile of cucumbers. *Journal of Food Science* 66(2): 357-361.

Buescher, R.W. and C. Burgin. 1988. Effect of calcium chloride and alum on fermentation, desalting, and firmness retention of cucumber pickles. *Journal of Food Science* 53(1): 296-297.

Buescher, R.W. and C. Burgin. 1992. Diffusion plate assay for measurement of polygalacturonase activities in pickle brines. *Journal of Food Biochemistry* 16: 59-68.

Buescher, R.W., C. Hamilton, J. Thorne and J.C. Mi. 2011. Elevated calcium chloride in cucumber fermentation brine prolongs pickle product crispness. *Journal of Food Quality* 34(2): 93-99.

Buescher, R.W., J.M. Hudson and J.R. Adams. 1981a. Utilization of calcium to reduce pectinolytic softening of cucumber pickles in low-salt conditions. *Lebensmittel Wissenschaft und Technologie* 14(2): 65-69.

Buescher, R.W., J.M. Hudson and J.R. Adams. 1979. Inhibition of polygalacturonase softening of cucumber pickles by calcium chloride. *Journal of Food Science* 44(6): 1786-1787.

Buescher, R.W., J.M. Hudson, J.R. Adams and D.H. Wallace. 1981b. Calcium makes it possible to store cucumber pickles in low salt brine. *Arkansas Farm Research* 30(4): 2.

Caldwell, J.M., R. Juvonen, J. Brown and F. Breidt. 2013. *Pectinatus sottacetonis* sp. nov., isolated from commercial pickle spoilage tank. *International Journal of Systematic and Evolutionary Microbiology* 63(10): 3609-3616.

Cho, M.J. and R. Buescher. 2011. Degradation of cucumber flavor aldehydes in juice. *Food Research International* 44(9): 2975-2977.

Cho, M.J. and R.W. Buescher. 2012. Potential role of native pickling cucumber polygalacturonase in softening of fresh pack pickles. *Journal of Food Science* 77(4): C443-C447.

Code of Federal Regulations (CFR). 2010. *Acidified Foods*. 21CFR114.

Cordero, C., E. Liberto, C. Bicchi, P. Rubiolo, P. Schieberle, S.E. Reichenbach and Q. Tao. 2010. Profiling food volatiles by comprehensive two-dimensional gaschromatography coupled with mass spectrometry: Advanced fingerprinting approaches for comparative analysis of the volatile fraction of roasted hazelnuts (*Corylus avellana* L.) from different origins. *Journal of Chromatography* A 1217(37): 5848-5858.

Corey, K.A., D.M. Pharr and H.P. Fleming. 1983a. Pressure changes in oxygen-exchanged, brined cucumbers. *Journal of the American Society of Horticultural Science* 108(1): 61-65.

Corey, K.A., D.M. Pharr and H.P. Fleming.1983c. Role of the osmoticum in bloater formation of pickling cucumbers. *Journal of Food Science* 48(1): 197-201.

Corey, K.A., D.M. Pharr and H.P. Fleming. 1983b. Role of gas diffusion in bloater formation of brined cucumbers. *Journal of Food Science* 48(2): 389-393, 399.

Costilow, R.N., C.L. Bedford, D. Mingus and D. Black. 1977. Purging of natural salt-stock pickle fermentations to reduce bloater damage. *Journal of Food Science* 42: 234-240.

Daeschel, M.A., R.E. Anderson and H.P. Fleming. 1987. Microbial ecology of fermenting plant materials. *FEMS Microbiology Reviews* 46: 357-367.

Daeschel, M.A., H.P. Fleming and E.A. Potts. 1985. Compartmentalization of lactic acid bacteria and yeasts in the fermentation of brined cucumbers. *Food Microbiology* 2(1): 77-84.

Daeschel, M.A., H.P. Fleming, and R.F. McFeeters. 1988. Mixed culture fermentation of cucumber juice with *Lactobacillus plantarum* and yeasts. *Journal of Food Science* 53(3): 862-864, 888.

Dakin, J.C. and M.P. Day. 1958. Yeast causing spoilage in acetic acid preserves. *Journal of Applied Bacteriology* 21(1): 94-96.

Das, A. and S. Deka. 2012. Fermented foods and beverages of the north-east India. *International Food Research Journal* 19: 377-392.

Deák T. 2008. *Handbook of Food Spoilage Yeast,* second edition, Boca Raton, Miami: CRC Press.

Demarigny, Y. 2012. Fermented food products made with vegetable materials from tropical and warm countries: microbial and technological considerations. *International Journal of Food Science and Technology* 47: 2469-2476.

Demo, J. 2002. Bulk tank technology: A pickle pumping system using an air-lift pump. *Pickle Pak Science* VIII (1): 44-46.

Di Cagno, R., R. Coda, M. De Angelis and M. Gobbetti. 2013. Review: Exploitation of vegetables and fruits through lactic acid fermentation. *Food Microbiology* 33: 1-10.

Durán Quintana, M.C., F. González and A. Garrido. 1979. Aceitunas negras al natural en salmuera. Ensayos de producción de alambrado. Inoculación de diversos microorganismos aislados de salmueras de fermentación. *Grasas y Aceites* 30361-30367.

Eisenstat, L. and F.W. Fabian. 1953. Factors which produce bleaching of pickles-I. *The Glass Packer.* April 29-5178.

Ennis, D.M. and J. O'Sullivan. 1979. Sensory quality of cucumbers before and after brining. *Journal of Food Science* 44(3): 847-849.

Etchells J.L. and I.D. Jones. 1944. Procedure for pasteurizing pickle products. *Glass Packer* 23(7): 519-523, 546.

Etchells, J.L. and I.D. Jones. 1941. An occurrence of bloaters during the finishing of sweet pickles. *Fruit Products Journal* 20(12): 370, 381.

Etchells, J.L. and T.A. Bell. 1950a. Film yeasts on commercial cucumber brines. *Food Technology* 4(3): 77-83.

Etchells, J.L. and T.A. Bell. 1950b. Classification of yeasts from the fermentation of commercially brined cucumbers. *Farlowia* 4(1): 87-112.

Etchells, J.L. and W.R. Jr. Moore. 1971. Factors influencing the brining of pickling cucumbers — Questions and answers. *Pickle Pak Science* 1(1): 1-17

Etchells, J.L. 1941a. A new type of gaseous fermentation occurring during the salting of cucumbers. *Univ Microfilms Abstr* III (2): 7-8.

Etchells, J.L. 1941b. Incidence of yeasts in cucumber fermentations. *Food Research* 6(1): 95-104.

Etchells, J.L., A.F. Borg and T.A. Bell. 1961. Influence of sorbic acid on populations and species of yeasts occurring in cucumber fermentations. *Applied Microbiology* 9(2): 139-144.

Etchells, J.L., A.F. Borg and T.A. Bell. 1968. Bloater formation by gas-forming lactic acid bacteria in cucumber fermentations. *Applied Microbiology* 16(7): 1029-1035.

Etchells, J.L., F.W. Fabian and I.D. Jones. 1945. The *aerobacter* fermentation of cucumbers during salting. *Michigan Agricultural Experiment Station Technical Bulletin,* No. 200. P. 56.

Etchells, J.L., T.A. Bell, H.P. Fleming, R.E. Kelling and R.L. Thompson. 1974. Q-BAT *Instruction Sheet with Bloater Chart.* Advisory statement published and distributed by Pickle Packers International, Inc. St. Charles, IL.

Etchells, J.L., H.P. Fleming, L.H. Hontz, T.A. Bell and R.J. Monroe R.J. 1975. Factors influencing bloater formation in brined cucumbers during controlled fermentation. *Journal of Food Science* 40(3): 569-575.

Etchells, J.L., T.A. Bell and C.F. Williams. 1958b. Inhibition of pectinolytic and cellulolytic enzymes in cucumber fermentations by Scuppernong grape leaves. *Food Technology* 12(5): 204-208.

Etchells, J.L., T.A. Bell, R.N. Costilow, C.E. Hood and T.E. Anderson. 1973. Influence of temperature and humidity on microbial, enzymatic and physical changes of stored, pickling cucumbers. *Applied Microbiology* 26(6): 943-950.

Etchells, J.L., T.A. Bell, R.J. Monroe, P.M. Masley and A.L. Demain AL. 1958a. Populations and softening enzyme activity of filamentous fungi on flowers, ovaries and fruit of pickling cucumbers. *Applied Microbiology* 6(6): 427-440.

Fasina, O.O., H.P. Fleming and L.D. Reina. 2002. BAG-IN-BOX TECHNOLOGY: Membrane filtration of cucumber fermentation brine. *Pickle Pak Science* VIII (1): 19-25.

Fasina O.O., H.P. Fleming, E.G. Humphries, R.L. Thompson, and L.D. Reina. 2003. Crossflow filtration of brine from cucumber fermentation. *Applied Engineering in Agriculture*. 19(1): 107-113.

Fleming, H.P., W.Y. Cobb, J.L. Etchells, T.A. Bell. 1968. The formation of carbonyl compounds in cucumbers. *Journal of Food Science* 33(6): 572-576.

Fleming, H.P. 1979. Purging carbon dioxide from cucumber brines to prevent bloater damage — A review. *Pickle Pak Science* 6(1): 8-22.

Fleming, H.P., E.G. Humphries, O.O. Fasina, R.F. McFeeters, R.L. Thompson and F. Breidt Jr. 2002a. BAG-IN-BOX TECHNOLOGY: Pilot system for process-ready, fermented cucumbers. *Pickle Pak Science* VIII (1): 1-8.

Fleming, H.P., E.G. Humphries, R.L. Thompson and R.F. McFeeters. 2002b. Bag-in-Box Technology: Storage stability of process-ready, fermented cucumbers. *Pickle Pak Science* VIII (1): 14-18.

Fleming, H.P., J.L. Etchells, R.L. Thompson and T.A. Bell. 1975. Purging of CO_2 from cucumber brines to reduce bloater damage. *Journal of Food Science* 40(6): 1304-1310.

Fleming, H.P., L.C. McDonald, R.F. McFeeters, R.L. Thompson and E.G. Humphries. 1995. Fermentation of cucumbers without sodium chloride. *Journal of Food Science* 60(2): 312-319.

Fleming, H.P., M.A. Daeschel, R.F. McFeeters and M.D. Pierson. 1989. Butyric acid spoilage of fermented cucumbers. *Journal of Food Science* 54(3): 636-639.

Fleming, H.P., R.F. McFeeters and R.L. Thompson. 1987. Effects of sodium chloride concentration on firmness retention of cucumbers fermented and stored with calcium chloride. *Journal of Food Science* 52(3): 653-657.

Fleming, H.P., R.F. McFeeters and R.L. Thompson. 1996. Assuring microbial and textural stability of fermented cucumbers by pH adjustment and sodium benzoate addition. *Journal of Food Science* 61(4): 832-6.

Fleming, H.P., R.L. Thompson and D.M. Pharr. 1980. Brining properties of cucumbers exposed to pure oxygen before brining. *Journal of Food Science* 45(6): 1579-1582.

Fleming, H.P., R.L. Thompson and R.J. Monroe. 1978. Susceptibility of pickling cucumbers to bloater damage by carbonation. *Journal of Food Science* 43(3): 892-896.

Fleming, H.P., R.L. Thompson, J.L. Etchells, R.E. Kelling and T.A. Bell. 1973a. Bloater formation in brined cucumbers fermented by *Lactobacillus plantarum*. *Journal of Food Science* 38(3): 499-503.

Fleming, H.P., R.L. Thompson, J.L. Etchells, R.E. Kelling and T.A, Bell. 1973b. Carbon dioxide production in the fermentation of brined cucumbers. *Journal of Food Science* 38(3): 504-506.

Fleming, H.P., R.L. Thompson, T.A. Bell and R.J. Monroe. 1977. Effect of brine depth on physical properties of brine-stock cucumbers. *Journal of Food Science* 42(6): 1464-1470.

Food and Agriculture Organization of the United Nations.FAO. 2015. FAO Crop Production Data Base. Economic and Social Development Department. Available at http://faostat3.fao.org/home/E. Accessed April 2015.

Forss, D.A., E.A. Dunstone, E.H. Ramshaw and W. Stark. 1962. The flavor of cucumbers. *Journal of Food Science* 27(1): 90-93.

Franco, W. and I. Pérez-Díaz. 2012b. Role of selected oxidative yeasts and bacteria in cucumber secondary fermentation associated with spoilage of the fermented fruit. *Food Microbiology* 32: 338-344.

Franco, W. and I.M. Pérez-Díaz. 2012a. Development of a model system for the study of spoilage associated secondary cucumber fermentation during long term storage. *Journal of Food Science* 77(10): M586-M592.

Franco, W. and I.M. Pérez-Díaz. 2013. Microbial interactions associated with secondary cucumber fermentation. *Journal of Applied Microbiology* 114: 161-172.

Franco, W., I.M. Pérez-Díaz, S.D. Johanningsmeier and R.F. McFeeters. 2012. Characteristics of spoilage-associated secondary cucumber fermentation. *Applied and Environmental Microbiology* 78(4): 1273-1284.

Frankel, E.N. 1991. Recent advances in lipid oxidation. *Journal of the Science of Food and Agriculture* 54: 495-511.

Geisman, J.R. and R.E. Henne. 1973. *Recycling Brine from Pickling*. Ohio Report 58 (4): 76-77.

Gómez-López, M.D., J.P. Fernández-Trujillo and A. Baille. 2006. Cucumber fruit quality at harvest affected by soil-less system, crop age and preharvest climatic conditions during two consecutive seasons. *Scientia Horticulturae* 110(1): 68-78.

Grosch, W. and J.M. Schwarz. 1971. Linoleic and linolenic acid as precursors of the cucumber flavor. *Lipids* 6(5): 351-352.

Guillou, A.A. and J.D. Floros. 1993. Multiresponse optimization minimizes salt in natural cucumber fermentation and storage. *Journal of Food Science* 58(6): 1381-1389.

Guillou, A.A., J.D. Floros and M.A. Cousin MA. 1992. Calcium chloride and potassium sorbate reduce sodium chloride used during natural cucumber fermentation and storage. *Journal of Food Science* 57(6): 1364-1368.

Harris, L.J. 1998. The microbiology of vegetable fermentations. Pp. 45-72. *In*: B.J.E. Wood (Ed.). *Microbiology of Fermented Foods*, vol. 1, second edition. Blackie Academic &Professional, London.

Hatanaka, A. 1996. The fresh green odor emitted by plants. *Food Reviews International* 12(3): 303-350.

Howard L.R. and R.W. Buescher. 1993. Effect of redox agents on visual cure in refrigerated pickles. *Journal of Food Biochemistry* 17(2): 125-134.

Hudson, J.M. and R.W. Buescher. 1980. Prevention of soft center development in large whole cucumber pickles by calcium. *Journal of Food Science* 45(5): 1450-1451.

Hudson, J.M. and R.W. Buescher. 1985. Pectic substances and firmness of cucumber pickles as influenced by $CaCl_2$, NaCl and brine storage. *Journal of Food Biochemistry* 9(3): 211-229.

Hutkins, R.W. 2006. *Microbiology and Technology of Fermented Foods*. Blackwell Publishing, Ames, Iowa.

Ito, K.A., J.K. Chen, P.A. Lerke, M.L. Seeger and J.A. Unverferth. 1976. Effect of acid and salt concentration in fresh-pack pickles on the growth of *Clostridium botulinum* spores. *Applied and Environmental Microbiology* 32(1): 121-124.

Jasso-Chaverria, C., G.J. Hochmuth, R.C. Hochmuth and S.A. Sargent. 2005. Fruit yield, size, and color responses of two greenhouse cucumber types to nitrogen fertilization in perlite soil-less culture. *HortTechnology* 15(3): 565-571.

Jeon, I.J., W.M. Breene and S.T. Munson. 1973. Texture of cucumbers: Correlation of instrumental and sensory measurements. *Journal of Food Science* 38(2): 334-337.

Jeon, I.J., W.M. Breene and S.T. Munson. 1975. Texture of salt stock whole cucumber pickles: Correlation of instrumental and sensory measurements. *Journal of Texture Studies* 5(4): 411-423.

Johanningsmeier, S.D. 2011. Biochemical characterization of fermented cucumber spoilage using non-targeted, comprehensive two-dimensional gas chromatography-time-of-flight mass spectrometry: Anaerobic lactic acid utilization by lactic acid bacteria. Ph.D. Dissertation, North Carolina State University, Raleigh, NC.

Johanningsmeier, S.D. and R.F. McFeeters. 2011. Detection of volatile spoilage metabolites in fermented cucumbers using nontargeted, comprehensive two-dimensional gas chromatography-time-of-flight mass spectrometry (GCxGC-TOFMS). *Journal of Food Science* 76(1): C168-C177.

Johanningsmeier, S.D. and R.F. McFeeters. 2013. Metabolism of lactic acid in fermented cucumbers by *Lactobacillus buchneri* and related species, potential spoilage organisms in reduced salt fermentations. *Food Microbiology* 35(2): 129-135.

Johanningsmeier, S.D., H.P. Fleming and F. Breidt FJ. 2004. Malolactic activity of lactic acid bacteria during sauerkraut fermentation. *Journal of Food Science* 69(8): M222-M227.

Johanningsmeier, S.D., R.L. Thompson and H.P. Fleming. 2002. Bag-in-Box Technology: Sensory quality of pickles produced from process-ready, fermented cucumbers. *Pickle Pak Science* VIII (1): 26-33.

Johanningsmeier, S.D., W. Franco, I.M. Pérez-Díaz and R.F. McFeeters. 2012. Influence of sodium chloride, pH and lactic acid bacteria on anaerobic lactic acid utilization during fermented cucumber spoilage. *Journal of Food Science* 77(7): M397-M404.

Jones I.D. and J.L Etchells. 1950. *Cucumber Varieties in Pickle Manufacture*. Canner 110(1): 34, 36, 38, 40.

Jones I.D. and J.L. Etchells JL. 1943. Physical and chemical changes in cucumber fermentations. *Food Ind* 15(1):62-64.

Jones, I.D., J.L. Etchells, O. Veerhoff and M.K. Veldhuis. 1941. Observations on bloater formation in cucumber fermentation. *Fruit Products Journal* 20(7): 202-206, 219-220.

Jones I.D., J.L. Etchells, and R.J. Monroe. 1954. Varietal differences in cucumbers for pickling. *Food Technology* 8(9): 415-418.

Jung, D.H. 2012. *Great Dictionary of Fermented Food*. Yuhanmunwhasa, Inc., Kayang-dong, Seoul. ISBN: 978-89-9722-570-1.

Kim, J. and F. Breidt Jr. 2007.Development of preservation prediction chart for long-term storage. *Korean Journal of Life Sciences* 17(12):1616-1621.

Kim, J.K., E.M. D'Sa, M.A. Harrison, J.A. Harrison and E.L. Andress. 2004. *Listeria monocytogenes* survival in refrigerator dill pickles. Paper 33C-1. Presented at the Institute of Food Technologists Annual Meeting, Las Vegas, NV, July 14, 2004.

Konisky, I. 1989. Colicins and other bacteriocins with established modes of action. *Annual Reviews of Microbiology* 36: 125-144.

Krall, S.M. and R.F. McFeeters. 1998. Pectin hydrolysis: Effect of temperature, degree of methylation, pH and calcium on hydrolysis rates. *Journal of Agricultural and Food Chemistry* 46(4): 1311-1315.

Kumar, R. Satish, P. Kanmani, N. Yuvaraj, A. Paari, V. Pattukamar and V. Arul. 2013. Traditional Indian fermented foods: a rich source of lactic acid bacteria. *International Journal of Food Science and Nutrition* 64(4): 415-428.

Little, L.W., J.C. Lamb, and L.F. Horney. 1974. Characterization and treatment of brine wastewaters from the cucumber pickle industry. Thesis. University of North Carolina at Chapel Hill.

Liu, M., J.R. Bayjanov, B. Renckens, Nauta A, and Siezen R.J. 2010. The proteolytic system of lactic acid bacteria revisited: A genomic comparison. *BMC Genomics* 11(36).doi: 10.1186/1471-2164-11-36

Lu, Z., E. Altermann, F. Breidt, P. Predki, H.P. Fleming and T.R. Klaenhammer. 2005. Sequence analysis of the *Lactobacillus plantarum* bacteriophage ΦJL-1. Gene 348: 45-54.

Lu, Z., F. Breidt, H.P. Fleming, E. Altermann and T.R. Klaenhammer. 2003a. Isolation and characterization of a *Lactobacillus plantarum* bacteriophage ΦJL-1 from a cucumber fermentation. *International Journal of Food Microbiology* 84: 225-235.

Lu, Z., F. Breidt, V. Plengvidhya and H.P. Fleming. 2003b. Bacteriophage ecology in commercial sauerkraut fermentations. *Applied and Environmental Microbiology* 69(6): 3192-3202.

Lu, Z., I.M. Pérez-Díaz, J.S. Hayes and F. Breidt. 2012. Bacteriophage ecology in an industrial cucumber fermentation. *Applied and Environmental Microbiology* 78(24): 8571-8578.

Lund, B.M., T.C. Baird-Parker and G.W. Gould. 2000. *The Microbiological Safety and Quality of Food*. pp. 160-161. Aspen Publishers, Inc. Gaithersburg, MD.

Malleret, C., R. Lauret, S.D. Ehrlich, F. Morel-Deville, and M. Zagorec. 1998. Disruption of the sole ldhL gene in *Lactobacillus sakei* prevents the production of both L- and D-lactate. *Microbiology* 144: 3327-3333.

Marshall, D.E., B.F. Cargill, and J.H. Levin. 1972. Physical and quality factors of pickling cucumbers as affected by mechanical harvesting. *Transactions of the ASAE*-1972: 604-608, 612.

Marsili, R.T. and N. Miller. 2000. Determination of major aroma impact compounds in fermented cucumbers by solid-phase microextraction-gas chromatography-mass spectrometry-olfactometry detection. *Journal of Chromatographic Science* 38: 307-314.

McDonald, L.C., H.P. Fleming and H.M. Hassan. 1990. Acid tolerance of *Leuconostoc mesenteroides* and *Lactobacillus plantarum*. *Applied and Environmental Microbiology* 56: 2120-2124.

McFeeters, R.F., and I.M. Perez-Diaz. 2010. Fermentation of cucumbers brined with calcium chloride instead of sodium chloride. *Journal of Food Science* 75(3): C291-C296.

McFeeters, R.F. 1992. Cell wall monosaccharide changes during softening of brined cucumber mesocarp tissue. *Journal of Food Science* 57(4): 937-940.

McFeeters, R.F. and H.P. Fleming. 1989. Inhibition of cucumber tissue softening in acid brines by multivalent cations: Inadequacy of the pectin 'egg box' model to explain textural effects. *Journal of Agricultural and Food Chemistry* 37(4): 1053-1059.

McFeeters, R.F. and H.P. Fleming. 1991. pH effect on calcium inhibition of softening of cucumber mesocarp tissue. *Journal of Food Science* 56(3): 730-732, 735.

McFeeters, R.F. and L.A. Lovdal. 1987. Sugar composition of cucumber cell walls during fruit development. *Journal of Food Science* 52(4): 996-1001.

McFeeters, R.F., H.P. Fleming and M. Brenes Balbuena. 1995. Softening rates of fermented cucumber tissue: Effects of pH, calcium and temperature. *Journal of Food Science* 60(4): 786-788, 793.

McFeeters, R.F., H.P. Fleming and R.L. Thompson. 1982a. Malic acid as a source of carbon dioxide in cucumber fermentations. *Journal of Food Science* 47(6): 1862-1865.

McFeeters, R.F., H.P. Fleming and R.L. Thompson. 1982b. Malic and citric acids in pickling cucumbers. *Journal of Food Science* 47(6): 1859-1861, 1865.

McFeeters, R.F., T.A. Bell and H.P. Fleming. 1980. An endo-polygalacturonase in cucumber fruit. *Journal of Food Biochemistry* 4(1): 1-16.

McGrath, S. and D. van Sinderen. 2007. *Bacteriophage: Genetics and Molecular Biology*. Caister Academic Press, Northfolk, VA.

Mercer, W.A., J.W. Ralls, and H.J. Maagdenberg. 1971. *Reconditioning Food Processing Brines with Activated Carbon*. Chemical Engineering Progress Symposium Series 67: 435-438.

Mok, Y.N. and R. Buescher. 2012. Cure appearance development in pasteurized cucumber pickle mesocarp tissue associated with proteolysis. *Journal of Food Biochemistry* 36(5): 513-519.

Mundt, J.O. 1970. Lactic acid bacteria associated with raw plant food material. *Journal of Milk and Food Technology* 33: 550-553.

Mundt, J.O. and J.L. Hammer. 1968. Lactobacilli on plants. *Applied Microbiology* 16: 1326-1330.

Mundt, J.O., W.F. Graham and I.E. McCarty. 1967. Spherical lactic acid-producing bacteria of southern-grown raw and processed vegetables. *Applied Microbiology* 15: 1303-1308.

Murphy, M.G. and S. Condon. 1984. Comparison of aerobic and anaerobic growth of *Lactobacillus plantarum* in a glucose medium. *Archives of Microbiology* 138: 49-53.

Murphy, M.G., L. O'Connor, D. Walsh, and S. Condon. 1985. Oxygen-dependent lactate utilization by *Lactobacillus plantarum*. *Archives of Microbiology* 141:75-79.

Murray, K.E. and F.B. Whitefield. 1975. The occurrence of 3-alkyl-2-methoxypyrazines in raw vegetables. *Journal of the Science of Food and Agriculture* 26: 973-986.

Olsen, M.J. and I.M. Pérez-Díaz. 2009. Influence of microbial growth on the redox potential of fermented cucumbers. *Journal of Food Science* 74(4): M149-M153.

Palma-Harris, C., R.F. McFeeters and H.P. Fleming. 2002. Fresh cucumber flavor in refrigerated pickles: Comparison of sensory and instrumental analysis. *Journal of Agricultural and Food Chemistry* 50(17): 4875-4877.

Palnitkar, M.P., and R.F. McFeeters. 1975. Recycling spent brines in cucumber fermentati ons. *Journal of Food Science* 40(6): 1311-1315.

Pederson, C.S. and M.N. Albury. 1969. The sauerkraut fermentation. *New York State Agricultural Experiment Station Technical Bulletin* 824, Geneva, New York.

Pederson, C.S., L.R. Mattick, F.A. Lee and R.M. Butts. 1964. Lipid alterations during the fermentation of dill pickles. *Applied Microbiology* 12(6): 513-516.

Pérez-Díaz, I., R.E. Kelling, S. Hale, F. Breidt and R.F. McFeeters. 2007. *Lactobacilli* and tartrazine as causative agents of red-color spoilage in cucumber pickle products. *Journal of Food Science* 72(7): M240-M245.

Pérez-Díaz, I.M., and R.F. McFeeters. 2009. Modification of azo dyes by lactic acid bacteria. *Journal of Applied Microbiology* 107: 584-589.

Pérez-Díaz, I.M., F. Breidt, R.W. Buescher, F.N. Arroyo-Lopez, R. Jimenez-Diaz, J. Bautista-Gallego, A. Garrido-Fernandez, S. Yoon and S.D. Johanningsmeier. 2014. Chapter 51: Fermented and acidified vegetables. *In*: F. Pouch Downes and K.A. Ito (Eds.). *Compendium of Methods for the Microbiological Examination of Foods*, fifth edition. American Public Health Association, Washington, DC.

Pérez-Díaz, I.M., R.F. McFeeters, L. Moeller, S.D. Johanningsmeier, J. Hayes, D.S. Fornea, L. Rosenberg, C. Gilbert, N. Custis, K. Beene and D. Bass. 2015. Commercial scale cucumber fermentations brined with calcium chloride instead of sodium chloride. *Journal of Food Science* 80(12): M2827-M2836.

Porter, W.L., J.H. Schwartz, T.A. Bell and J.L. Etchells. 1961. Probable identity of the pectinase inhibitor in grape leaves. *Journal of Food Science* 26(6): 600-605.

Potts, E.A. and H.P. Fleming. 1979. Changes in dissolved oxygen and microflora during fermentation of aerated, brined cucumbers. *Journal of Food Science* 44(2): 429-434.

Potts, E.A. and H.P. Fleming. 1982. Prevention of mold-induced softening in air-purged, brined cucumbers by acidification. *Journal of Food Science* 47(5): 1723-1727.

Pressey, R. and J.K.A. Vants. 1975. Cucumber Polygalacturonase. *Journal of Food Science* 40(5): 937-939.

Rosenberg, L.B. 2013. Texture retention in pickles produced from commercial scale cucumber fermentation using calcium chloride instead of sodium chloride. Thesis. North Carolina State University, Raleigh, North Carolina.

Ruiz-Cruz, J. and F. Gonzalez-Cancho. 1969. Metabolismo de levaduras aisladas de salmuera de aceitunas aderezadas 'estilo espanol'. I. *Asimilacion de los acidos lactico, acetico y citrico*. Grasas y Aceites 20(1): 6-11.

Saltveit, M.E. and R.F. McFeeters. 1980. Polygalacturonase activity and ethylene synthesis during cucumber fruit development and maturation. *Plant Physiology* 66(6): 1019-1023.

Sandhu, K.S., and F.C. Shukla. 1996. Methods for pickling of cucumber (*Cucumis sativus*) a critical appraisal. *Journal of Food Technology* 33: 455-473.

Schopf, J.W. and B.M. Packer. 1987. Early Archean (3.3-billion to 3.5-billion-year-old) microfossils from Warrawoona Group, Australia. *Aust. Sci.* 237(4810): 70-73.

Shetty, N.V. and T.C. Wehner. 2002. Screening the cucumber germplasm collection for fruit yield and quality. *Crop Science* 42(6): 2174-2183.

Singh, A.K., and A. Ramesh. 2008. Succession of dominant and antagonistic lactic acid bacteria in fermented cucumber: Insights from a PCR-based approach. *Food Microbiology* 25(2): 278-287.

Smith, J.G. 2008. *Organic Chemistry*, second edition. McGraw-Hill, New York.

St. Angelo, A.J. 1992. *Lipid Oxidation in Food*. American Chemical Society, Washington D.C.

Steinkraus, K.H. 1994. Nutritional significance of fermented foods. *Food Research International* 27: 259-261.

Tamang, J.P. 2010. Fermented Foods and Beverages. Pp. 41-84. *In*: J.P. Tamang and K. Kailasapathy (Eds.). *Diversity of fermented foods*. CRC Press, Boca Raton, Florida.

Tang, H.C.L. and R.F. McFeeters. 1983. Relationships among cell wall constituents, calcium and texture during cucumber fermentation and storage. *Journal of Food Science* 48(1): 66-70.

Taormina, P.J. and L.R. Beuchat. 1999. Behavior of enterohemorrhagic *Escherichia coli* O157:H7 on alfalfa sprouts during the sprouting process as influenced by treatments with various chemicals. *Journal of Food Protection* 62: 850-6.

Thomas, T.R., L.L. McKay and H.A. Morris. 1985. Lactate metabolism by Pediococci isolated from cheese. *Applied and Environmental Microbiology* 49(4): 908-913.

Thompson, R.L., H.P. Fleming, D.D. Hamann and R.J. Monroe. 1982. Method for determination of firmness in cucumber slices. *Journal of Texture Studies* 13(3): 311-324.

Thompson, R.L., R.J. Monroe and H.P. Fleming. 1979. Effects of storage conditions on firmness of brined cucumbers. *Journal of Food Science* 44(3): 843-846.

Vaughan, R.H., K.T. Jakubczyk, J.D. McMillan, E. Higinio Thomas, B.A. Dave and V.M. Crampton. 1969. Some pink yeast associated with softening of olives. *Applied Microbiology* 18(5): 771-775.

Veldhuis, M.K. and J.L. Etchells. 1939. Gaseous products of cucumber pickle fermentations. *Food Research* 4(6): 621-630.

Veldhuis, M.K., J.L. Etchells, I.D. Jones and O. Veerhoff. 1941. Influence of sugar addition to brines in pickle fermentation. *Food Ind.* 13(10): 54-56; 13(11): 48-50.

Venkateshwarlu, G., M.B. Let, A.S. Meyer and C. Jacobsen. 2004. Modeling the sensory impact of defined combinations of volatile lipid oxidation products on fishy and metallic off-flavors. *Journal of Agricultural and Food Chemistry* 52(6): 1635-1641.

Walter, W.M. Jr., D.G. Epley and R.F. McFeeters. 1990. Effect of water stress on stored pickling cucumbers. *Journal of Agricultural and Food Chemistry* 38(12): 2185-2191.

Walter, W.M. Jr., H.P. Fleming and R.N. Trigiano. 1985. Comparison of the microstructure of firm and stem- end softened cucumber pickles preserved by brine fermentation. *Food Microstructure* 4(1): 165-172.

Wardale, D.A. and E.A. Lambert. 1980. Lipoxygenase from cucumber fruit: Localization and properties. *Phytochemistry* 19(6): 1013-1016.

Wardale, D.A., E.A. Lambert and T. Galliard. 1978. Localization of fatty acid hydroperoxide cleavage activity in membranes of cucumber fruit. *Phytochemistry* 17(2): 205-012.

Wehner, T.C. and H.P. Fleming. 1984. Evaluation of bloater resistance in pickling cucumbers using a brine carbonation method. *Journal of the American Society of Horticultural Science* 109(2): 261-265.

White, D. 2007. *The Physiology and Biochemistry of Prokaryotes*. Oxford University Press, Inc., New York, NY.

White, R.C., I.D. Jones and E. Gibbs. 1963. Determination of chlorophylls, chlorophyllides, pheophytins, and pheophorbides in plant material. *Journal of Food Science* 28(4): 431-436.

Whitefield, F.B. and J.H. Last. 1991. Volatile compounds in foods and beverages. *In*: H. Maarse (Ed). *Vegetables*. Dekker, New York.

Wilson, E.M., S.D. Johanningsmeier and J.A. Osborne. 2015. Consumer acceptability of cucumber pickles produced by fermentation in calcium chloride brine for reduced environmental impact. *Journal of Food Science* 80(6): S1360-S1367.

Wolter, E.M. 2013. Consumer Acceptability and Flavor Characteristics of Cucumber Pickles Produced Using an Environmentally-friendly Calcium Chloride Fermentation. Thesis. North Carolina State University, Raleigh, North Carolina.

Xu, Q., Y.L. Geng, X.H. Qi, X.H. Chen. 2012. Genetic analysis of the five major aromatic substances in cucumber (*Cucumis sativus* L.). *The Journal of Horticultural Science and Biotechnology* 87(2): 113-116.

Yoo, K.M., I.K. Hwang, G.E. Ji and B.K. Moon. 2006. Effects of salts and preheating temperature of brine on the texture of pickled cucumbers. *Journal of Food Science* 71(2): C97-C101.

Yoon, S.S., R. Barrangou-Poueys, F. Breidt and H.P. Fleming. 2007. Detection and characterization of a lytic *Pediococcus* bacteriophage from the fermenting cucumber brine. *Journal of Microbiology and Biotechnology* 17(2): 262-270.

Yoshioka, Y., H. Horie, M. Sugiyama and Y. Sakata. 2009. Quantifying cucumber fruit crispness by mechanical measurement. *Breeding Science* 59(2): 139-147.

Yoshioka, Y., H. Horie, M. Sugiyama, Y. Sakata and Y. Tamaki. 2010. Search for quantitative indicators of fruit texture for breeding in cucumber. *Acta Horticulturae* (871): 171-178.

Zhou, A. and R.F. McFeeters. 1998. Volatile compounds in cucumbers fermented in low-salt conditions. *Journal of Agricultural and Food Chemistry* 46(6): 2117-2122.

Zhou, A., R.F. McFeeters and H.P. Fleming. 2000. Development of oxidized odor and volatile aldehydes in fermented cucumber tissue exposed to oxygen. *Journal of Agricultural and Food Chemistry* 48(2): 193-197.

8

Olives Fermentation

Athena Grounta and Efstathios Z. Panagou[1],*

1. Introduction

The olive tree, *Olea europaea* L. ssp. *europaea*, is among the oldest crops of the Mediterranean basin and one of the best biological indicators of the Mediterranean climate (Blondel et al. 2010). The exact geographic origins, the time and the reasons of its early exploitation are still open to arguments (Breton et al. 2009, Kaniewski et al. 2012, Besnard et al. 2015). The domestication of the olive tree was characterized by selection and propagation of the most valuable trees and establishment of olive orchards (Breton et al. 2009). Based on archaeological remains, it is believed that the olive tree was first domesticated during the early Neolithic period in the Near East (Galili et al. 1989) and it was later introduced into the West of the Mediterranean region via human migrants. Cultivation of the olive tree has been practiced from antiquity to modern times for its wood (utensils, furnishing, manufacturing solid fuel) and for its fruit which is used in olive oil and table olive production (Kaniewski et al. 2012, Riley 2002). The fruit of the olive tree is considered as one of the most extensively cultivated fruit crop in the world. The annual world consumption of table olives has steadily increased during the last two decades, reaching approximately 2,398,300 tons from 2008-9 to 2013-14 crop seasons (IOC 2014). The world's crop season of table olives production of 2014-15 is estimated at 2,554,500 tonnes, the majority of which comes from countries in the European Union (IOC 2014). Spain is the leader among producer countries followed by Greece and Italy while Turkey, Egypt, Algeria, Argentina, Syria, Morocco, USA and Peru are major non-European producers.

Olives are botanically classified as drupes and anatomically consist of three component tissues, namely, the epicarp or skin, the mesocarp or flesh

[1] Laboratory of Microbiology and Biotechnology of Foods, Department of Food Science and Human Nutrition, Agricultural University of Athens, Iera Odos 75, 11855 Athens, Greece.
* Corresponding author: E-mail: stathispanagou@aua.gr

and the endocarp or pit which encloses one or, rarely, two seeds (Garrido-Fernández et al. 1997). The epicarp comprises a layer of epidermal cells rich in chloroplasts and is covered by a thin cuticle. The mesocarp, rich in protoplasm, surrounds the endocarp which progressively sclerifies during fruit development (Connor and Fereres 2005).

Fruit development, which starts approximately 30 days after fertilization and fruit set, is completed within four to five months and generally involves cell division, cell expansion and storage of metabolites. The primary stage of fruit development is characterized by intense cell division, resulting in rapid growth of the endocarp with little mesocarp. During the middle stage, the drupes are covered by an epicuticular wax layer, the mesocarp cells develop vacuole cells while the endocarp gets completely sclerified with a stop to its enlargement. Then a period of marked fruit growth follows due to expansion of the pre-existing mesocarp cells. At the same time, intense oil synthesis and accumulation in the mesocarp is observed which continues at a slower rate until the maturation/ripening phase. Upon maturation, the color of the drupes changes from lime green to purple-black and the texture of the flesh becomes softer and easier to squash until some juice is released. (Connor and Fereres 2005, Conde et al. 2008). Depending on the color of the fruit upon harvest, they are categorized as (i) green olives, harvested at the early stage of maturity prior to coloring and on having obtained the appropriate size, (ii) black olives, harvested at the full stage of maturity or slightly earlier on having attained deep violet black color, and (iii) turning color olives, harvested between the two stages presenting a wine-rose color (Garrido-Fernández et al. 1997, IOC 2004).

Nutrient components are present at the highest percentage in the mesocarp. They are represented by a high level of water and lipids and a low level of sugars and protein. The value of each nutrient may significantly vary depending on the cultivar, degree of maturation and post-harvest treatment (Wodner et al. 1988, Nergiz and Engez 2000, Marsilio et al. 2001a, b, Sakouhi et al. 2008, López-López et al. 2009, Lanza et al. 2013). Upon harvest, the lipid content is dominated by oleic acid followed by palmitic acid, linoleic acid and stearic acid. Carbohydrates in olive drupes are represented mainly by soluble-reducing sugars, such as glucose and fructose and non-reducing sugars, such as mannitol. Their concentration is lower in comparison to any other drupes since they act as precursors for fatty acids synthesis during fruit growth (Wodner et al. 1988). The sugar content that remains serves as a carbon source for development of the desired microorganisms during table olive production (Garrido-Fernández et al. 1997). Complex sugars, such as lignin, hemicellulose, cellulose and pectin are distributed in the olive fruit. Lignin is present in the stone while the rest of the polymers are present in the mesocarp, playing a substantial role in the structural characteristics of the olive flesh (Kailis and Harris 2007, Lanza et al. 2010). The protein content is low but of high quality due to the presence of essential amino acids for adults (threonine, valine, leucine, isoleucine, phenylalanine and lysine), and (arginine, histidine and tyrosine) for children (Lanza et al. 2010, 2013). Phenolics are also present with oleuropein, a secoiridoid glucoside, being the representative phenolic compound responsible for the bitter taste of the fruit. Its concentration decreases during ripening (Amiot et al. 1986), giving rise primarly to hydroxytyrosol and other simple phenolics (Bianchi 2003), like tyrosol, homovanillic alcohol, caffeic acid, coumaric acid, phloretic acid and vanillic acid (Boskou et al. 2006) contributing also to sensory and aromatic characteristics of the olive while imparting pharmaceutical and physiological benefits (Tassou 1993, Kountouri et al. 2007, Omar 2010, Ghanbari et al. 2012).

2. Table Olive Processing

2.1. Commercial Preparation of Table Olives

Olive fruits when freshly picked are bitter due to the high phenolic fraction they contain and thus unsuitable for direct consumption. In order to make them edible, they need to undergo a series of processes involving complete or partial removal of the bitter content and eventually the development of a final product with enhanced sensory and preservation characteristics. The final product, 'table olive', is defined by the International Olive Council as 'the product prepared from the sound fruit of the varieties of cultivated olive trees (*Olea europea* L.) that are chosen for the production of olives whose volume, shape, flesh-to-stone ratio, fine flesh, taste, firmness and ease of detachment from the stone make them particularly suitable for processing. Table olives are treated to remove their bitterness and preserved by natural fermentation, or by heat treatment, with or without the addition of preservatives; packed with or without covering liquid' (IOC 2004).

Depending on the method used for debittering, table olives are broadly categorized as treated and untreated/natural olives. The difference lies in the use of lye/alkali treatment in the case of treated olives, resulting in a quick and complete removal of bitterness. The drupes are immersed in a weak solution of sodium hydroxide which penetrates the fruit's epidermis and hydrolyzes the ester bond of the phenolic hydroxytyrosol with the rest of the oleuropein molecule (Brenes and de Castro 1998). After lye treatment, the drupes are washed to remove the excess of alkali and further stored in brine, where complete or partial lactic acid fermentation takes place. The treatment of green olives in alkaline solution is commonly applied in Spain, particularly in Seville. This type of processing, also known as 'Spanish style' method represents the first most important commercial preparation in the international table olive market (Garrido-Fernández et al. 1997, Sánchez Gómez et al. 2006).

Untreated olives, on the other hand, are naturally debittered without any prior alkali treatment. The most common practice to obtain natural olives is by directly placing the drupes in brine where the bitter phenolic compounds are diffused from the olive's mesocarp in the surrounding brine. In parallel, lactic acid fermentation develops which may further assist the debittering process due to the reported ability of certain lactic acid bacteria (LAB) species to hydrolyze oleuropein (Ciafardini et al. 1994, Servili et al. 2006, Landete et al. 2008, Kaltsa et al. 2015). Green, black and turning colour olives can be prepared thus. Natural green olives in brine are commercially known as 'Siciliano style' olives (Colmargo et al. 2001). Natural black olives in brine are the main trade preparation in Greece, also known as 'Greek style' in the international market, representing a widespread commercial preparation (Balatsouras 1990). Another special type of untreated olives is obtained by placing the raw drupes between layers of coarse salt, resulting in a product completely different from the other preparations. This is mainly applied to fruits harvested at the stage of full ripeness and the final product is known as 'naturally black dry-salted olives'. Due to high osmosis induced by the coarse salt, the water content leaks out of the fruit while the salt is taken up by the olives (Panagou 2006). It has been hypothesized that along with water, other solutes, including oleuropein, flow out, resulting in progressive debitterness of the fruit. Recently it has been reported that olive debittering during the dry-salting process is due to enzymatic oxidation, in particular due to polyphenol oxidase (PPO) activity. The use of salt causes rupture of the tissue, thereby oleuropein component comes into contact with PPO (Ramírez et al. 2013). Olives are finally shrivelled and characterized by a salty, bitter and sweet taste. The physicochemical characteristics of the product are 4.5-5.5 pH, 0.75-0.85 water activity, 30-39g/100g moisture content, 35-39g/100g oil content, 2.0-2.5g/100g reducing sugars and 10g/100g NaCl (Sodium chloride) content of the flesh (Balatsouras 1995) Due to their high salt content, their share in the international market is relatively small and limited to countries of the Mediterranean region (Panagou 2006).

Another commercially important process is the black ripe olive-processing technology which involves artificial darkening (oxidation) of the drupes in an alkaline solution. This technology was initiated in California at the beginning of the 20th century; thus olives processed in this way are known as 'Californian' olives. The fruits are subjected to three successive treatments with sodium hydroxide solution on three consecutive days, penetrating skin, 1–2 mm into the flesh and to the pit, respectively. Between treatments, the drupes are immersed in water or dilute brine through which air is bubbled. Olives darken progressively, both in the flesh and on the surface and, once the colour is obtained, more water is added. Aeration continues until the pH is around 7–8 units. Then iron salts (ferrous gluconate or lactate) are added to stabilize the black color. After one day at equilibrium, the product is canned and sterilized (Brenes-Balbuena et al. 1992, Marsilio et al. 2001b).

The processing procedures of the most broadly known commercial preparations, namely, 'Spanish-style', 'Greek-style', 'Californian-style' and 'dry-salted' olives are schematically illustrated in Figures 1–4.

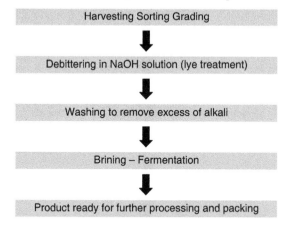

Fig. 1: **Flow diagram for the processing of 'Spanish' green olives in brine**

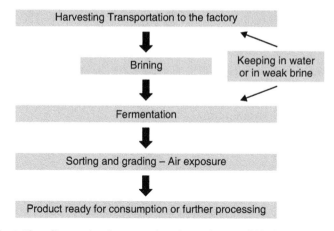

Fig. 2: **Flow diagram for the processing of 'Greek' natural black olives in brine**

The above-mentioned processes reflect the fact that olive preparation may significantly vary, depending on regional and national practices. All the different commercial preparations under which table olives are distributed in the market have been previously elucidated in detail (Sánchez Gómez et al. 2006, Heperkan 2013), while the name of each one of them includes information on the ripeness stage of the harvested drupe (green, black or turning colour), the procedure used in the debittering step (treated or natural) and the preservation method (brining or dry salting) (Garrido-Fernández et al. 1997). In terms of international trade, the International Olive Council has adopted a unified qualitative standard for table olives wherein five commercial/trade preparations are described, namely treated

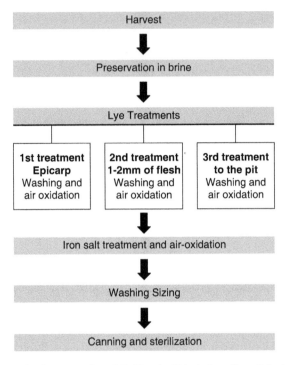

Fig. 3: Flow diagram for the processing of 'Californian' black ripe olives (*Adapted from Marsilio et al. 2001b*)

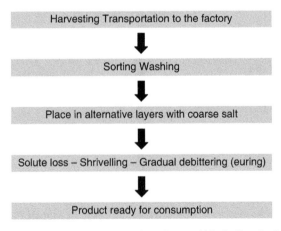

Fig. 4: Flow diagram for the processing of natural black olives in dry salt

olives, natural olives, dehydrated and/or shrivelled olives, olives darkened by oxidation and specialities (IOC 2004). Based on this classification, treated and natural olives undergo a fermentation process, the aspects of which will be further discussed.

2.2. Fermentation Process

Fermentation is one of the oldest and most widespread methods of food preservation that appeared with the dawn of civilization. There are numerous examples of meat, milk, grain and vegetable foods and beverages consumed in fermented form, such as wine, beer, cheese, yogurt, kefir, sauerkraut, pickles, olives, sausages and many others. The action of microorganisms, mainly LAB and yeast, results in the production of organic acids, alcohol and carbon dioxide through which preservation is achieved while the other metabolites that are produced enhance the sensory attributes of the product (Campbell-Platt 1994, Caplice and Fitzgerald 1999, Ross et al. 2002, Farnworth 2005, Bourdichon et al. 2012).

Table olives are one of the most important fermented vegetables. The fermentation process starts spontaneously once the drupes, previously treated with alkali or not, are placed in brine. The whole process is divided into three stages, namely; primary stage, lasting up to 15 days; middle stage, lasting up to three weeks and last stage, until the end of fermentation (Tassou 1993, Balatsouras 1995, Garrido-Fernández et al. 1997).

During the primary fermentation stage, the brine is conditioned by nutrients diffused from the mesocarp serving as substrate for the growth of microorganisms coming from the olive's indigenous biota that is highly heterogeneous, constituted by gram negative and gram positive bacteria as well as yeast and fungi (Tassou 1993, Balatsouras 1990). Species belonging to *Pseudomonas*, *Aeromonas* and *Flavobacterium* genera as well as entero-bacteria form the group of gram negatives that are present during the initial stage of olive fermentation. Their presence has been associated with the formation of 'gas pockets' resulting in softening of the fruit. These microorganisms, while metabolizing sugars, release carbon dioxide which is accumulated as pockets of gas either hypocuticularly or intramesocarpically (Lanza 2013). Macroscopically, hypocuticular gas appears as bubbles on the olive surface. Intramesocarpical gas results in the formation of gas fissures in the mesocarp and a narrow belt on the skin in a condition known as 'fish eye' or 'alambrado' (Sánchez Gómez et al. 2006, Lanza 2013). Other important gram positive bacteria that may prevail during this stage and cause malodorous fermentation are *Clostridium* and *Bacillus* species. Butyric fermentation is caused by the action of butyric acid bacteria, such as *Clostridium butyricum*, *Cl. beijerinckii*, *Cl. multifermentans* and *Cl. fallax*. Apart from the rancid butter-like smell, they are also responsible for gas fissures in the olive's mesocarp (Gililland and Vaughn 1943). The presence of *Bacillus* species, such as *Bacillus subtilis* and *B. pumilus* is associated with pectinolytic activity resulting in softening of the drupes (Nortje and Vaughn 1953). These microorganisms are present during the initial days of fermentation and progressively decrease. By the end of the primary stage, LAB, such as *Lactococcus*, *Pediococcus* and *Leuconostoc* prevail. Heterofermentative *Leuconostoc* and homofermentative *Pediococcus* cocci dominate the middle stage of fermentation, producing lactic acid which increases the acidity and lowers the pH in the brine. The developed acidity further favors the growth of *Lactobacilli*. The dominance and fast growth of *Lactobacilli* coupled with the production of lactic acid in high amounts characterize the last stage of fermentation and guarantee the preservation properties of the final product. The process is considered complete once the fermentable substrates are exhausted. Meanwhile, the pH in the brine is maintained at around 4.0 (Hurtado et al. 2012). The *Lactobacilli* group is represented mainly by *Lactobacillus plantarum* and *Lb. pentosus* and to a lesser extent by *Lb. paracasei*, *Lb. casei* and *Lb. brevis*. Non-*Lactobacilli* species, such as *Enterococcus faecium*, *Ec. casseliflavus* and *Leuconostoc mesenteroides* have also been reported (Hurtado et al. 2012).

Once the fermentation process is over, the drupes are stored in the brine. Brines with properly fermented olives contain mainly lactic and acetic acids whereas the presence of succinic, citric, malic and tartaric acids has been reported in lower concentrations (Bleve et al. 2015). During the storage period, the development of 'zapatera' spoilage is of great concern. Metabolites, such as formic, propionic, butyric, valeric, caproic and caprylic acids in the brine are associated with 'zapatera' spoilage. The first off-odors to appear are described as 'cheesy' and as the spoilage progresses a characteristic fecal odor develops. *Propionibacterium* species via propionic and acetic acid production (Plastourgos and Vaughn 1957) and *Clostridium* species (Kawatomari and Vaughn 1956) are the microorganisms associated with this condition. To control development of these bacteria, the pH should be kept <4.0 and the NaCl in the brine should be raised to >8 per cent after the end of fermentation (Lanza 2013).

Apart from LAB, yeasts are present throughout the process. *Candida*, *Pichia*, *Debaryomyces*, *Wickerhamomyces* and *Saccharomyces* are the main genera associated with table olive processing while their role can be either beneficial or detrimental (Arroyo-López et al. 2012b). For a proper fermentation, which demands high levels of free acidity and low pH (pH<4.5) in the brine, LAB should predominate over yeasts. This is generally achieved when using low salt brines. Under high salt brining (> 8 (w/v) per cent NaCl) the yeasts outnumber the LAB, resulting in a final product with milder taste and less self-preservation characteristics due to lower acid production (Tassou et al. 2002, Hurtado et al. 2009, Aponte et al. 2010). Excessive growth of fermentative yeasts has been linked with gas pockets formation due to vigorous carbon dioxide production and fruit softening (Vaughn et al. 1972). Some strains of *Rhodotorula minuta*, *W. anomalus* and *D. hansenii* are reported to produce enzymes such as proteases, xylanases and pectinases, causing softening of the fruits (Hernández et al. 2007, Bautista-Gallego et al. 2011). Furthermore, their presence in packaged olives may cause clouding of the brines (Arroyo-López et al. 2005). On the beneficial side, yeasts may produce metabolites, such as higher alcohols, esters and other volatile compounds, contributing in aroma and flavor development (Garrido-Fernández et al. 1997). Other technological properties, such as degradation of polyphenols, production of killer toxins and enhancement of LAB growth have been reported (Psani and Kotzekidou 2006, Hernández et al. 2007, Aponte et al. 2010, Bautista-Gallego et al. 2011, Silva et al. 2011, Bonatsou et al. 2015) while the scientific interest in their probiotic potential and use as starter cultures is also increasing (Psani and Kotzekidou 2006, Etienne-Mesmin et al. 2011, Silva et al. 2011, Bevilacqua et al. 2013).

2.3. Control of the Fermentation Process

Olive fermentation is a spontaneous process influenced by factors, such as pH, water activity, availability of nutrients, the level of salt concentration in the brine and phenol content (Spyropoulou et al. 2001, Tassou et al. 2002, Hurtado et al. 2009). During an uncontrolled fermentation, development of spoilage microorganisms and deterioration of the product are likely to occur, leading to economic losses for the table olive industry.

Constant adjustment of salt concentration and acidification of brine are the main industrial practices to control the process and avoid abnormal fermentation (Balatsouras 1990, Garrido-Fernández et al. 1997). During brining, the initial salt concentration is progressively taken up by the olive mesocarp and its level in the brine decreases until an equilibrium between fruit and brine is reached. Under low-salt conditions, the survival of gram negatives may be prolonged. In order to avoid spoilage, the salt level should be progressively raised to 8.5–9.5 per cent in equilibrium. Acidification of brine at pH 4.0–4.5 at the primary stage of fermentation is another common industrial practice in order to restrain the action of gram negatives that are abundant during this stage. As a result, the duration of the primary stage decreases, giving rise to the desired LAB growth. Lactic acid and acetic acid are widely used as acidifying agents for this purpose.

The use of starter cultures has also been proposed as a way to better control olive fermentation since it decreases the risk of spoilage and accelerates the course of fermentation and acidification of the brine (Montaño et al. 1993, Garrido-Fernández et al. 1997, Spyropoulou et al. 2001, Panagou and Tassou 2006, Corsetti et al. 2012). Different cultures were studied for different cultivars and preparation methods with the focus mainly on *Lb. plantarum* and *Lb. pentosus* as the main species associated with fermented olives (Montaño et al. 1993, Panagou and Tassou 2006, Peres et al. 2008, Sabatini et al. 2008, Panagou et al. 2008, Hurtado et al. 2010, Perricone et al. 2010, Caballero-Guerrero et al. 2013, Pistarino et al. 2013, Randazzo et al. 2014, Rodríguez-Gómez et al. 2014). Their use was employed as single or co-inoculated cultures with their effectiveness depending on the type of cultivar and processing method (Panagou and Tassou 2006).

The selection of potential starter cultures is generally based on the following criteria: (i) fast and predominant growth, (ii) homofermentative metabolism, leading to lactic acid production and consequently to high acidification rate, (iii) tolerance to the harsh environment of the brine, such as salt, organic acids and phenolic compounds, (iv) minimum nutritional requirements, (v) ability to grow at low temperatures, and (vi) ability to resist freezing or lyophilization if required for commercial purposes (Hurtado et al. 2012). Apart from their use as protective cultures and their effectiveness in accelerating fermentation, other functional properties, such as bacteriocin production, are desired. The bacteriocin producer strain *Lb. plantarum* LPCO10 was first isolated from green olive fermentation (Jiménez-Díaz et al. 1993) and reported to produce bacteriocins S and T with the former being active against *Propionibacterium* and *Clostridium* spoilage bacteria. Its later use as a starter culture in Spanish-style green olive fermentation (Ruiz-Barba et al. 1994) resulted in bacteriocin production in brine and predomination over other *Lactobacillus* strains. Further, the final concentration of lactic acid was higher in brines inoculated with *Lb. plantarum* LPCO10 strain than in uninoculated brines or brines inoculated with another non-bacteriocin-producing strain. Other bacteriocin producer strains were studied and it seems that factors such as temperature, salt concentration in the brine and the presence of other bacteria affect bacteriocin production (Delgado et al. 2005, 2007, Ruiz-Barba et al. 2010, Hurtado et al. 2011). Recently, research on table olives was shifted to the probiotic potential of starter cultures, enhancing thus the significance of table olives from a traditional agricultural product to a high added-value functional food that provides new perspectives for the table olive sector (Peres et al. 2012).

3. Table Olives as Functional Food

'Let food by thy medicine and medicine be thy food' — this philosophy, first adopted by Hippocrates, the Father of Medicine, was again embraced by the modern man in the 20th century. Modern nutrition lays stress on consumption of a diet rich in vegetables, fruits, grains, legumes and low saturated fat, in order to reduce the risk of chronic diseases. As a result, there is an increasing interest in production and consumption of health-promoting products, prompting the food industry to develop a diverse range of food products with enhanced characteristics and known as 'functional foods' (Hasler 2002, Betoret et al. 2011).

Functional foods were first introduced in Japan in the 1980s (Arai 1996) and today the Functional Food Science in Europe (FUFOSE) proposes a working definition that says, "Foods that beneficially affect one or more target functions in the body beyond adequate nutritional effects in a way that is relevant to either an improved state of health and well-being and/ or reduction of risk of disease. They are consumed as part of a normal food pattern and they are not a pill, a capsule or any form of dietary supplement". (EC 2010). With the aim of developing functional foods, probiotics have received increasing attention over the years. The term 'probiotic' means 'for life' and the concept is generally attributed to Nobel Prize-winner, Éli Metchnikoff, who first suggested that "the dependence of the intestinal microbes on the food makes it possible to adopt measures to modify the flora in our bodies and to replace the harmful microbes by useful microbes" (Metchnikoff, 1907). Today, probiotics can be defined as 'live microorganisms, which, when consumed in adequate amounts, confer a health benefit on the host' (FAO/WHO 2001). *Lactobacillus* and *Bifidobacteria* are the main genera including bacterial strains with health-promoting properties. Intake of probiotics contributes to the maintenance of the intestinal microbial balance of the host by inhibiting pathogens and lowering the risk of gastrointestinal diseases. They are considered to possess anti-carcinogenic, anti-hypertensive and immunomodulatory poperties while alleviating certain intolerances, such as lactose intolerance and prevent food allergies (Salminen et al. 1998, Wollowski et al. 2001, Cross 2002, Commane et al. 2005, Tang et al. 2010). So far, the majority of probiotic foods found in the market are milk-based products, such as yogurts, cheese and fermented milk (*kefir*). Due to the increasing number of individuals being lactose intolerant and/or suffering from milk protein allergy, the need for manufacturing non-dairy probiotic products arises (Gupta and Abu-Ghannam 2012).

In recent years, table olives have attracted the attention of the scientific community for their use as biological carriers of microorganisms with health-promoting properties, transforming thus a traditional product into a novel functional food. The first attempt to incorporate probiotic microorganisms on to table olives was published by Italian researchers (Lavermicocca et al. 2005). In this work, different types of ready-to-eat table olives were inoculated with seven probiotic strains of *Lactobacillus* and *Bifidobacteria*. The strains tested were able to adhere and colonize the olive surface and most of them exhibited high survival rates (above 10^6 CFU/g) after 90 days of storage. Furthermore, one of the strains, *Lb. paracasei* IMPC2.1, was recovered from four out of five human fecal samples from healthy volunteers after daily consumption of olives containing about 10^9 to 10^{10} viable cells. In another work (De Bellis et al. 2010), the same strain was used both as a probiotic and as a starter culture in the fermentation process of *cv* Bella di Cerignola green olives, leading to a final product with functional appeal.

The main source of probiotic isolates is the human gastrointestinal tract, meaning that the selected starter cultures are allochthonous to olives. The limitations of allochthonous starters have been previously mentioned (Di Cagno et al. 2013) while the exploitation and use of autochthonous strains with functional properties was proposed (Vitali et al. 2012, Di Cagno et al. 2013). In this sense, a number of LAB originating from table-olive environment has been screened for functional properties. Recently, a total of 144 LAB isolates obtained from naturally-fermented Aloreña green olives were studied for their potential as probiotic starter cultures (Abriouel et al. 2012). Fifteen *Lb. pentosus* strains and one *Ln. pseudomesenteroides* strain from this study exhibited anti-microbial properties and tolerance to low pH and high bile salt concentration. In another work, *Lactobacilli* isolates from different spontaneous industrial green olive fermentations were selected by Italian and Spanish researchers (Bautista-Gallego et al. 2013) by *in vitro* phenotypic tests related to probiotic potential. In the same line, four *Lb. pentosus* strains, previously isolated from diverse table olive processing and selected according to *in vitro* phenotypic tests related to probiotic potential were used as starters to better control Spanish-style green olive fermentation (Rodríguez-Gómez et al. 2013). Moreover, the probiotic potential of LAB isolated from naturally-fermented olives by *in vitro* probiotic tests was reported (Argyri et al. 2013). In this work, four strains of *Lb. pentosus*, three strains of *Lb. plantarum* and two strains of *Lb. paracasei* were found to possess desirable *in vitro* probiotic properties similar or superior to the reference probiotic strains *Lb. rhamnosus* GG and *Lb. casei* Shirota. Two of them, namely *Lb. pentosus* B281 and *Lb. plantarum* B282 were selected and later used as starters during Spanish style fermentation of green olives (Blana et al. 2014) and heat-shocked green olives (Argyri et al. 2014) in low and high salt brines. In both studies, the strain B281 was found to be more effective in terms of survival ability under different salt levels in the brine while colonizing the olive surface in higher populations than the strain B282. Changes in the population dynamics of LAB, yeasts and enterobacteria reported in the study of Blana et al. (2014) are illustrated in Fig. 5. As it can be observed, the recovered LAB population ranged from 6.0 to 7.0 log CFU/g which is particularly important since a threshold of 10^6-10^7 CFU/g is a prerequisite to deliver the health benefits (FAO/WHO 2001).

4. Biofilm Development in Table Olive Fermentation: Role and Significance

Biofilm formation in food processing environments has been the focus of extensive scientific research, especially in the context of food hygiene, as many outbreaks are associated with the presence of biofilms through food-borne parthogens in food industries (Zottola and Sasahara 1994, Srey et al. 2013). On the other hand, biofilm formation by non-pathogenic microorganisms may have positive effects. Their role is vital for the health of the ecosystem by decontaminating polluted sites and breaking down sewage (Hunter 2008). Biofilms by certain rhizobacteria species on plant roots were found to induce plant growth and protect plants from phytopathogens (Bogino et al. 2013). The exploitation of industrial *Bacillus subtilis* and related species biofilms as biocontrol and bioremediation agents has been previously proposed (Morikawa 2006), while biofilms

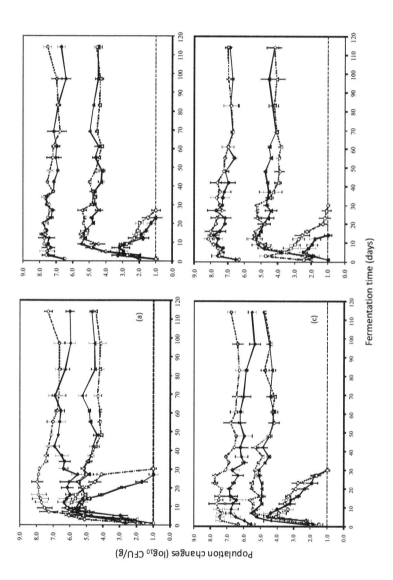

Fig. 5: Changes in the population of LAB (◆, ◇), yeasts (▲, △) and enterobacteria (●, ○) on olive drupes during fermentation in 10 per cent (solid line) and 8 per cent (dotted line) initial salt concentration: (a) spontaneous process, (b) inoculated process with *Lb. pentosus* B281, (c) inoculated process with *Lb. plantarum* B282, and (d) inoculated process with a co-culture of the two strains. Data points are average values of duplicate fermentations ± standard deviation. Horizontal dotted line indicates the detection limit of the method (1.0 log CFU/g). (Blana et al. 2014, data used with permission)

by bacteriocin-producing LAB have and reported to exhibit antilisterial activities (Guerrieri et al. 2009).

Biofilms are defined as functional consortia of microorganisms attached to a surface and embedded in the extracellular polymeric substances (EPS) produced by the microorganisms (Monds and O'Toole 2009). It has long been established that microorganisms in natural environments tend to adhere to any solid surface, either biotic or abiotic, when immersed in a liquid medium and assemble themselves in a complex multispecies consortium often embedded in a extracellular polymeric matrix (EPS) produced by the microorganisms (Sutherland 2001, Donlan 2002, Hall-Stoodley et al. 2004, Elhariry 2011). One of the prerequisites for biofilm development is a process known as 'conditioning film' which involves the accumulation of nutrients on the solid surface immersed in the liquid medium (Donlan, 2002). After film conditioning, microorganisms may adhere to the contact surface, gradually forming microcolonies and finally assembling themselves in biofilms that exhibit high microbial diversity in terms of genera, species and strain level (Rickard et al. 2003, Burmølle et al. 2006).

In the case of table olives, the fermentation process is described as a biofilm phenomenon. During the process, the brine solution is progressively enriched with fermentable sugars outflowing from the mesocarp. This may serve as a means of conditioning the film on the olive's epicarp and on the surface of the container where fermentation takes place. Thus, in accordance with the biofilm formation theory, the enriched brine supports the role of the liquid medium while the drupes and the fermentation container support the role of biotic and abiotic sites of adherence and biofilm development, respectively.

The first evidence supporting that olive drupes serve as a site of attachment and biofilm formation was published by Nychas et al. (2002). The authors, using Scanning Electron Microscope (SEM), observed aggregates of bacteria and yeasts in the stomal openings of naturally black *cv* Conservolea olives, with bacteria predominating in the intercellular spaces of the sub-stomal cells. More recently, Spanish researchers observed biofilm formation of LAB and yeast communities on the olive surface of Spanish-style green olives of Manzanilla (Arroyo- López et al. 2012a) and Gordal cultvivars (Domínguez-Manzano et al. 2012). In another study (Lavermicocca et al. 2005), the ability of different *Lactobacillus* and *Bifidobacterium* probiotic species to adhere and colonize the olive's epicarp was investigated. As observed by SEM, all strains tested were able to adhere to the olive surface retaining high populations during storage. One of them, *Lb. paracasei* IMPC2.1, was selected and successfully employed during debittered green olives *cv* Bella di Cerignola fermentation (De Bellis et al. 2010) and later proposed as a suitable strain for the development of probiotic table olives (Sisto and Lavermicocca 2012). In a recent work (Grounta and Panagou, 2014), mono and dual species biofilm formation on black oxidized olives *cv* Halkidiki, inoculated with the LAB *Lb. pentosus* B281 and yeast *P. membranifaciens* M3A under different brining treatments was demonstrated. Regardless of the brining treatment, the adhering microorganisms embedded in EPS matrix were observed inside stomatal apertures and on discontinuations of olive's epidermis (Fig. 6). The specific LAB strain was used due to its previously reported *in vitro* probiotic potential (Argyri et al. 2013) while the technological properties of the yeast strain have been previously reported (Bonatsou et al. 2015). In a follow-up work, the same strains were used as starters in the fermentation of naturally black *cv* Conservolea olives and at the end of the process a mixed association of LAB and yeasts in stomatal openings was observed under SEM (Grounta et al. 2014). The spontaneous process resulted in a fermentation dominated by yeasts and low acid production while the use of starters reassured LAB development in high populations with concurrent high acid production in brine (Fig. 7a, b).

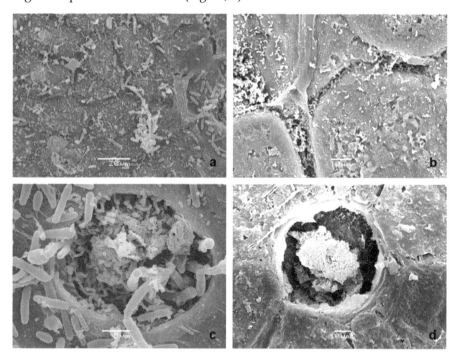

Fig. 6: Dual species biofilm on olives in low salt brines (6 per cent NaCl) supplemented with glucose and lactic acid. (a, b) Micrographs showing LAB and yeast cells on epidermis and discontinuations of the olive skin, (c) micrographs showing LAB and yeast cells in stomal aperture, (d) EPS matrix produced by the microorgansims (Grounta and Panagou 2014).

The formation of biofilms on the abiotic surface of fermentation vessels, on the other hand, was pond scant attention in table olive research. To our knowledge, only two studies have been recently published focusing on this aspect. In the first study, the establishment of polymicrobial communities

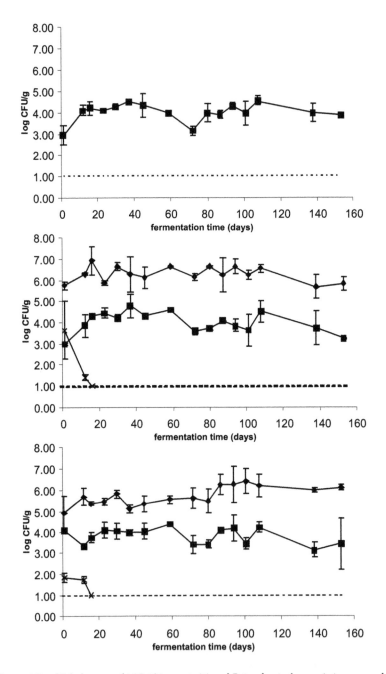

Fig. 7a: Microbial changes of LAB (♦), yeasts (■) and Enterobacteriaceae (×) expressed as log CFU/g on olive drupes during (i) spontaneous fermentation, (ii) inoculated fermentation with *Lb. pentosus* B281, and (iii) inoculated fermentation with *Lb. pentosus* B281 and *P. membranifaciens* M3A. Data points are average of duplicate fermentations ± standard deviation. Dashed line indicates the detection limit of the method (1.0 log CFU/g) (Grounta et al. 2014).

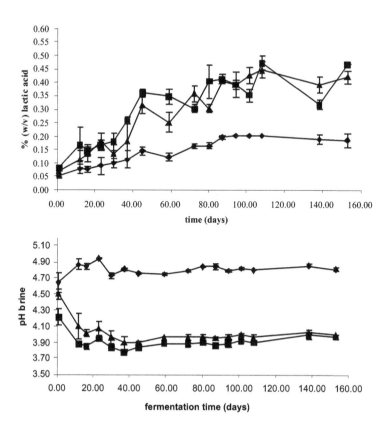

Fig. 7b: Evolution of total acidity in the brines expressed as per cent (w/v) lactic acid (a) and changes of pH values in the brines (b) during spontaneous fermentation (◆), inoculated fermentation with *Lb. pentosus* B281 (■), and inoculated fermentation with *Lb. pentosus* B281 and *P. membranifaciens* M3A (▲). Data points are average of duplicate fermentations ± standard deviation (Grounta et al. 2014).

on glass slides that came into contact with the brine during Spanish style table olive fermentation was investigated (Domínguez- Manzano et al. 2012). The authors confirmed the ability of microorganisms to adhere and produce biofilms on the abiotic surfaces. In the second study, the quantification and characterization of the biofilm community formed on the surface of plastic containers used in the fermentation of green table olives at different sampling locations and different cleaning treatments of the vessel was investigated (Grounta et al. 2015). It was seen that multi-species biofilm communities were formed on the fermentation vessels, exhibiting persistence during the applied cleaning treatments. Among LAB species, *Lb. pentosus* was the most abundant species recovered from the biofilm. The most frequently detected yeasts were characterized as *Wickerhamomyces anomalus*, *Debaryomyces hansenii* and *Pichia guilliermondii* which are common members of the yeast-fermenting microbiota of table olives. The results of these studies highlight the fact that the presence of multispecies microbial community could have a beneficial impact on table-olive processing, since biofilm detachment from the surface of the vessels into the brine could contribute to a certain extent in brine inoculation with the necessary technological microbiota supporting fermentation.

References

Abriouel, H., N. Benomar, A. Cobo, N. Caballero, M.A. Fernández Fuentes, R. Pérez-Pulido and A. Gálvez. 2012. Characterization of lactic acid bacteria from naturally fermented Manzanilla Aloreña green table olives. *Food Microbiology* 32: 308-316.

Amiot, M.J., A. Fleuriet and J.J. Macheix. 1986. Importance and evolution of phenolic compounds during growth and maturation. *Journal of Agriculture and Food Chemistry* 34: 823-826.

Aponte, M., V. Ventorino, G. Blaiotta, G. Volpe, V. Farina, G. Avellone, C.M. Lanza and G. Moschetti. 2010. Study of green Sicilian table olive fermentations through microbiological, chemical and sensory analyses. *Food Microbiology* 27: 162-170.

Arai, S. 1996. Studies on functional foods in Japan – State of the art. *Bioscience of Biotechnology and Biochemistry* 60: 9-15.

Argyri, A.A., G. Zoumpopoulou, K.A.G. Karatzas, E. Tsakalidou, G.J.E. Nychas, E.Z. Panagou and C.C. Tassou. 2013. Selection of potential probiotic lactic acid bacteria from fermented olives by *in vitro* tests. *Food Microbiology* 33: 282-291.

Argyri, A.A., A.A. Nisiotou, A. Mallouchos, E.Z. Panagou and C.C. Tassou. 2014. Performance of two potential probiotic *Lactobacillus* strains from the olive microbiota as starters in the fermentation of heat shocked green olives. *International Journal of Food Microbiology* 171: 68-76.

Arroyo López, F.N., C. Romero, M. Durán Quintana, A. López López, P. García García and A. Garrido Fernández. 2005. Kinetic study of the physicochemical and microbiological changes in seasoned olives during the shelf period. *Journal of Agricultural and Food Chemistry* 53: 5285-5292.

Arroyo-López, F.N., J. Bautista-Gallego, J. Domínguez-Manzano, V. Romero-Gil, F. Rodriguez-Gómez, P. García-García, A. Garrido-Fernández and R. Jiménez-Díaz. 2012a. Formation of lactic acid bacteria-yeasts communities on the olive surface during Spanish-style Manzanilla fermentations. *Food Microbiology* 32: 295-301.

Arroyo-López, F.N., V. Romero-Gil, J. Bautista-Gallego, F. Rodríguez-Gómez, R. Jiménez-Díaz, P. García-García, A. Querol and A. Garrido-Fernández. 2012b. Yeasts in table olive processing: Desirable or spoilage microorganisms? *International Journal of Food Microbiology* 160: 42-49.

Balatsouras, G.D. 1990. Edible olive cultivars, chemical composition of fruit, harvesting, transportation, processing, sorting and packaging, styles of black olives, deterioration, quality standards, chemical analysis, nutritional and biological value of the end product. Instituto Sperimentale per la Elaiotecnica, Pescara, Italy.

Balatsouras, G.D. 1995. Table olives: cultivars, chemical composition, commercial preparations, quality characteristics, packaging, marketing. Agricultural University of Athens (in Greek).

Bautista-Gallego, J., F. Rodríguez-Gómez, E. Barrio, A. Querol, A. Garrido-Fernández and F.N. Arroyo-López. 2011. Exploring the yeast biodiversity of green table olive industrial fermentations for technological applications. *International Journal of Food Microbiology* 147: 89-96.

Bautista-Gallego, J., F.N. Arroyo-López, K. Rantsiou, R. Jiménez-Díaz, A. Garrido-Fernández and L. Cocolin. 2013. Screening of lactic acid bacteria isolated from fermented table olives with probiotic potential. *Food Research International* 50: 135-142.

Besnard, G., B. Khadari, M. Navascués, M. Fernández-Mazuecos, A. El Bakkali, N. Arrigo, D. Baali-Cherif, V. Brunini-Bronzini de Caraffa, S. Santoni, P. Vargas and V. Savolainen.

2015. The complex history of the olive tree: From late quaternary diversification of Mediterranean lineages to primary domestication in the northern Levant. *Proceedings of the Royal Society* B 280: 1-7.

Betoret, E., N. Betoret, D. Vidal and P. Fito. 2011. Functional foods development: trends and technologies. *Trends in Food Science and Nutrition* 22: 498-508.

Bevilacqua, A., L. Beneduce, M. Sinigaglia and M.R. Corbo. 2013. Selection of yeasts as starter cultures for table olives. *Journal of Food Science* 78: M742-M751.

Bianchi, G. 2003. Lipids and phenols in table olives. *European Journal of Lipid Science and Technology* 105: 229-242.

Blana, V.A., A. Grounta, C.C. Tassou, G-J.E. Nychas and E.Z. Panagou. 2014. Inoculated fermentation of green olives with potential probiotic *Lactobacillus pentosus* and *Lactobacillus plantarum* starter cultures isolated from industrially fermented olives. *Food Microbiology* 38: 208-218.

Bleve, G., M. Tufariello, M. Durante, F. Grieco, F.A. Ramires, G. Mita, M. Tasioula-Margari and A.F. Logrieco. 2015. Physico-chemical characterization of natural fermentation process of Conservolea and Kalamata table olives and development of a protocol for the pre-selection of fermentation starters. *Food Microbiology* 46: 368-382.

Blondel, J., J. Aronson, J., J-Y. Bodiou and G. Bœuf. 2010. The Mediterranean Region: *Biological Diversity in Space and Time*. Oxford University Press, Oxford, UK.

Bogino, P.C., M. de las Mercedes Oliva, F.G. Sorroche and W. Giordano. 2013. The role of bacterial biofilms and surface components in plant-bacterial associations. *International Journal of Molecular Sciences* 14: 15838-15859.

Bonatsou, S., A. Benítez, F. Rodríguez-Gómez, E.Z. Panagou and F.N. Arroyo-López. 2015. Selection of yeasts with multifunctional features for application as starters in natural black table olive processing. *Food Microbiology* 46: 66-73.

Boskou, G., F.N. Salta, S. Chrysostomou, A. Mylona, A. Chiou and N.K. Andrikopoulos. 2006. Antioxidant capacity and phenolic profile of table olives from the Greek market. *Food Chemistry* 94: 558-564.

Bourdichon, F., S. Casaregola, C. Farrokh, J.C. Frisvad, M.L. Gerds, W.P. Hammes, J. Harnett, G. Huys, S. Laulund, A. Ouwehand, I.B. Powell, J.B. Prajapati, Y. Seto, E.T. Schure, A.V. Boven, V. Vankerckhoven, A. Zgoda, S. Tuijtelaars and E.B. Hansen. 2012. Food fermentations: Microorganisms with technological beneficial use. *International Journal of Food Microbiology* 154: 87-97.

Brenes-Balbuena, M., P. García- García and A. Garrido-Fernandez. 1992. Phenolic compounds related to the black colour formed during the processing of ripe olives. *Journal of Agricultural and Food Chemistry* 40: 1192-1196.

Brenes, M. and A. de Castro. 1998. Transformation of oleuropein and its hydrolysis products during Spanish-style green olive processing. *Journal of the Science of Food and Agriculture* 77: 53-358.

Breton, C., J-F. Terral, C. Pinatel, F. Médail, F. Bonhomme and A. Bervillé. 2009. The origins of the domestication of the olive tree. *C.R. Biologies* 332: 1059-1064.

Burmølle, M., J.S. Webb, D. Rao, L.H. Hansen, S.J. Sørensen and S. Kjelleberg. 2006. Enhanced biofilm formation and increased resistance to anti-microbial agents and bacterial invasion are caused by synergistic interactions in multi-species biofilms. *Applied and Environmental Microbiology* 72: 3916-3923.

Caballero-Guerrero, B., H. Lucena-Padrós, A. Maldonado-Barragán and J.L. Ruiz-Barba. 2013. High-salt brines compromise autoinducer-mediated bacteriocinogenic *Lactobacillus plantarum* survival in Spanish-style green olive fermentations. *Food Microbiology* 33: 90-96.

Campbell-Platt, G. 1994. Fermented foods – A world perspective. *Food Research International* 27: 253-257.

Caplice, E. and G.F. Fitzgerald. 1999. Food fermentations: role of microorganisms in food production and preservation. *International Journal of Food Microbiology* 50: 131-149.

Ciafardini, G., V. Marsilio, B. Lanza and N. Pozzi. 1994. Hydrolysis of oleuropein by *Lactobacillus plantarum* strains associated with olive fermentation. *Applied and Environmental Microbiology* 60: 4142-4147.

Colmargo, S., G. Collins and M. Sedgley. 2001. Processing Technology of the Table Olive. *Horticultural Reviews*, vol. 25: 235-242.

Commane, D., R. Hughes, C. Shortt and I. Rowland. 2005. The potential mechanisms involved in the anti-carcinogenic action of probiotics. *Mechanistic Approaches to Chemoprevention of Mutation and Cancer* 591: 276-289.

Conde, C., S. Derlot and H. Gerós. 2008. Physiological, biochemical and molecular changes occurring during olive development and ripening. *Journal of Plant Physiology* 165: 1545-1562.

Connor, D.J. and E. Fereres. 2005. The physiology of adaptation and yield expression in olive. *Horticultural Reviews* 34: 45-54.

Corsetti, A., G. Perpetuini, M. Schirone, R. Tofalo and G. Suzzi. 2012. Application of starter cultures to table olive fermentation: An overview on the experimental studies. *Frontiers in Microbiology* 3: 1-6.

Cross, M.L. 2002. Microbes versus microbes: Immune signals generated by probiotic *Lactobacilli* and their role in protection against microbial pathogens. *FEMS Immunology and Medical Microbiology* 34: 245-253.

De Bellis, P., F. Valerio, A. Sisto, S.L. Lonigro and P. Lavermicocca. 2010. Probiotic table olives: Microbial populations adhering on olive surface in fermentation sets inoculated with the probiotic strain *Lactobacillus paracasei* IMPC2.1 in an industrial plant. *International Journal of Food Microbiology* 140: 6-13.

Delgado, A., D. Brito, C. Peres, F. Noé-Arroyo,and A. Garrido-Fernández. 2005. Bacteriocin production by *Lactobacillus pentosus* B96 can be expressed as a function of temperature and NaCl concentration. *Food Microbiology* 22: 521-528.

Delgado, A., F. Noé-Arroyo, D. Brito, C. Peres, P. Fevereiro and A. Garrido-Fernández. 2007. Optimum bacteriocin production by *Lactobacillus plantarum* 17.2b requires absence of NaCl and apparently follows a mixed metabolite kinetics. *Journal of Biotechnology* 130: 193-201.

Di Cagno, R., R. Coda, M. De Angelis and M. Gobbetti. 2013. Exploitation of vegetables and fruits through lactic acid fermentation. *Food Microbiology* 33: 1-10.

Domínguez-Manzano, J., C. Olmo-Ruiz, J. Bautista-Gallego, F.N. Arroyo-López, A. Garrido-Fernández and R. Jiménez-Díaz. 2012. Biofilm formation on abiotic and biotic surfaces during Spanish style green table olive fermentation. *International Journal of Food Microbiology* 157: 230-238.

Donlan, R.M. 2002. Biofilms: Microbial life on surfaces. *Emerging Infectious Diseases* 8: 881-890.

Elhariry, H.M. 2011. Attachment strength and biofilm forming ability of *Bacillus cereus* on green-leafy vegetables: cabbage and lettuce. *Food Microbiology* 28: 1266-1274.

Etienne-Mesmin, L., V. Livrelli, M. Privat, S. Denis, J.M. Cardot, M. Alric and S. Blanquet-Diot. 2011. Effect of a new probiotic *Saccharomyces cerevisiae* strain on survival of *Escherichia coli* O157:H7 in a dynamic gastrointestinal model. *Applied and Environmental Microbiology* 77: 1127-1131.

EC, European Comission, 2010. Functional Foods. *Studies and Reports*. http://ec.europa.eu/research/research-eu

FAO/WHO. 2001. Probiotics in food: Health and nutritional properties and guidelines for evaluation. *Report of joint FAO/WHO Expert Consultation on Evaluation of Health and Nutritional Properties of Probiotics in Food Including Powder Milk with Live Lactic Acid Bacteria.* Cordoba, Argentina, 1-4 October 2001.

Farnworth, E.R. 2005. Kefir – A complex probiotic. *Food Science and Technology Bulletin: Functional Foods* 2: 1-17.

Garrido-Fernández, A., M.J. Fernández-Díez and M.R. Adams. 1997. *Table Olives: Production and Processing.* Chapman & Hall. London, UK.

Galili, E., M. Weinstein-Evron and D. Zohary. 1989. Appearance of olives in submerged neolithic sites along the Carmel coast. *Journal of Israel Prehistoric Society* 22: 95-97. http://www.jstor.org/discover/10.2307/23373096?sid=21105962272443&uid=2&uid=4&uid=3738128

Ghanbari, R., F. Anwar, K.M. Alkhafry, A.H. Gilani and N. Saari. 2012. Valuable nutrients and functional bioactives in different parts of olive (*Olea europea* L.) – A Review. *International Journal of Molecular Science* 13: 3291-3340.

Gililland, J.R. and R.H. Vaughn. 1943. Characteristics of butyric acid bacteria from olives. *Journal of Bacteriology* 46: 315-322.

Grounta, A. and E.Z. Panagou. 2014. Mono and dual species biofilm formation between *Lactobacillus pentosus* and *Pichia membranifaciens* on the surface of black olives under different sterile brine conditions. *Annals of Microbiology* 64: 1757-1767.

Grounta, A., V. Iliopoulos, A.I. Doulgeraki, G-J.E. Nychas and E.Z. Panagou. 2014. Biofilm formation of *Lactobacillus pentosus* B281 and *Pichia membranifaciens* M3A during fermentation on naturally black conservolea olives. *In: Proceedings of the 24th International ICFMH Conference.* Nantes, France. Pp. 179.

Grounta, A., A.I. Doulgeraki and E.Z. Panagou. 2015. Quantification and characterization of microbial biofilm community attached on the surface of fermentation vessels used in green table olive processing. *International Journal of Food Microbiology* 203: 41-48.

Guerrieri, E., S. de Niederhäusern, P. Messi, C. Sabia, R. Iseppi, I. Anacarso and M. Bondi. 2009. Use of lactic acid bacteria (LAB) biofilms for the control of *Listeria monocytogenes* in a small-scale model. *Food Control* 20: 861-865.

Gupta, S. and N. Abu-Ghannam. 2012. Probiotic fermentation of plant based products: Possibilities and opportunities. *Critical Reviews in Food Science and Nutrition* 52: 183-199.

Hall-Stoodley, L., Costerton, J.W. and P. Stoodley. 2004. Bacterial biofilms: From the natural environment to infectious diseases. *Nature Reviews Microbiology* 2: 95-108.

Hasler, C.M. 2002. Functional foods: Benefits, concerns and challenges — A position paper from the American Council on Science and Health. *The Journal of Nutrition* 132: 3772-3781.

Heperkan, D. 2013. Microbiota of table olive fermentations and criteria of selection for their use as starters. *Frontiers in Microbiology* 4: Article 143.

Hernández, A., A. Martin, E. Aranda,F. Pérez-Nevado and M.G. Córdoba. 2007. Identification and characterization of yeast isolated from the elaboration of seasoned green table olives. *Food Microbiology* 24: 346-351.

Hunter, P. 2008. The mob response. *EMBO Reports* 9: 314-317.

Hurtado, A., C. Reguant, A. Bordons and N. Rozès. 2009. Influence of fruit ripeness and salt concentration on the microbial processing of Arbequina table olives. *Food Microbiology* 26: 827-833.

Hurtado, A., C. Reguant, A. Bordons and N. Rozès. 2010. Evaluation of single and combined inoculation of a *Lactobacillus pentosus* starter for processing *cv. Arbequina* natural green olives. *Food Microbiology* 27: 731-740.

Hurtado, A., C. Reguant, A. Bordons and N. Rozès. 2011. Expression of *Lactobacillus pentosus* B96 bacteriocin genes under saline stress. *Food Microbiology* 28: 1339-1344.

Hurtado, A., C. Reguant, A. Bordons and N. Rozès. 2012. Lactic acid bacteria from fermented table olives. *Food Microbiology* 31: 1-8.

IOC, International Olive Council, 2004. *Trade Standard Applying to Table Olives.* http://www.internationaloliveoil.org/estaticos/view/222-standards

IOC, International Olive Council, 2014. *World Table Olive Figures.* http://www.internationaloliveoil.org/estaticos/view/132-world-table-olive-figures

Jiménez-Díaz, R., R.M. Ríos-Sánchez, M. Desmazeaud, J.L. Ruiz-Barba and J.C. Piard. 1993. Plantaricins S and T, two new bacteriocins produced by *Lactobacillus plantarum* LPCO10 isolated from a green olive fermentation. *Applied and Environmental Microbiology* 59: 1416-1424.

Kailis, S.G. and D. Harris. 2007. *Producing Table Olives.* Landlinks Press. Australia.

Kaltsa, A., D. Papaliaga, E. Papaioannou and P. Kotzekidou. 2015. Characteristics of oleuropeinolytic strains of *Lactobacillus plantarum* group and influence on phenolic compounds in table olives elaborated under reduced salt conditions. *Food Microbiology* 48: 58-62.

Kaniewski, D., E. Van Campo, T. Boiy, J.F. Terral, B. Khadari and G. Besnard. 2012. Primary domestication and early uses of the emblematic olive tree: Palaeobotanical, historical and molecular evidence from the Middle East. *Biological Reviews* 87: 885-899.

Kawatomari, T. and R.H. Vaughn. 1956. Species of *Clostridium* associated with zapatera spoilage of olives. Journal of Food Science 21: 481-490.

Kountouri, A.M., A. Mylona, A.C. Kaliora and N.K. Andrikopoulos. 2007. Bioavailability of the phenolic compounds of the fruits (drupes) of *Olea europea* (olives): Impact on plasma antioxidant status in humans. *Phytomedicine* 14: 659-667.

Landete, J.M., J.A. Curiel, H. Rodríquez, B. De las Rivas and R. Muñoz. 2008. Study of the inhibitory activity of phenolic compounds found in olive products and their degradation by *Lactobacillus plantarum* strains. *Food Chemistry* 107: 320-326.

Lanza, B., M.G. Di Serio, E. Iannucci, F. Russi and P. Marfisi. 2010. Nutritional, textural and sensorial characterisation of Italian table olives (*Olea europea* L. cv. 'Intosso d'Abruzzo'). *International Journal of Food Science and Technology* 45: 67-74.

Lanza, B. 2013. Abnormal fermentations in table olive processing: Microbial origin and sensory evaluation. *Frontiers in Microbiology* 4: 1-7.

Lanza, B., M.G. Serio and E. Iannucci. 2013. Effects of maturation and processing technologies on nutritional and sensory qualities of Itrana table olives. *Grasas y Aceites* 64: 272-284.

Lavermicocca, P., F. Valerio, S.L. Lonigro, M. de Angelis, L. Morelli, M.L. Callegari, C.G. Rizzello, A. Visconti. 2005. Study of adhesion and survival of *Lactobacilli* and *Bifidobacteria* on table olives with the aim of formulating a new probiotic food. *Applied and Environmental Microbiology* 71: 4233-4240.

López-López, A., F. Rodríguez-Gómez, M. V. Ruíz-Méndez, A. Cortés-Delgado and A. Garrido-Fernández. 2009. Sterols, fatty alcohol and triterpenic alcohol changes during ripe olive processing. *Food Chemistry* 117: 127-134.

Marsilio, V., C. Campestre, B. Lanza and M. De Angelis. 2001a. Sugar and polyol compositions of some European olive fruit varieties (*Olea europea* L.) suitable for table olive purposes. *Food Chemistry* 72: 485-490.

Marsilio, V., C. Campestre, B. Lanza. 2001b. Phenolic compounds change during California-style ripe olive processing. *Food Chemistry* 74: 55-60.

Metchnikoff, É. 1907. *Essais optimistes. The Prolongation of Life: Optimistic Studies.* Translated and edited by P. Chalmers Mitchell. London: Heinemann, 1907.

Monds, R.D. and G.A. O'Toole. 2009. The developmental model of microbial biofilms: ten years of a paradigm up for review. *Trends in Microbiology* 17: 73-87.

Montaño, A., A.H. Sanchez and A. De Castro. 1993. Controlled fermentation of Spanish-type green olives. *Journal of Food Science* 58: 842-844.

Morikawa, M. 2006. Beneficial biofilm formation by industrial bacteria *Bacillus subtilis* and related species. Journal of Bioscience and Bioengineering 101: 1-8.

Nergiz, C. and Y. Engez. 2000. Compositional variation of olive fruit during ripening. *Food Chemistry* 69: 55-59.

Nortje, B. and R.H. Vaughn. 1953. The pectolytic activity of species of the genus *Bacillus*: Qualitative studies with *Bacillus subtilis* and *Bacillus pumilus* in relation to the softening of olives and pickles. *Journal of Food Science* 18: 57-59.

Nychas, G-J.E., E.Z. Panagou, M.L. Parker, K.W. Waldron and C.C. Tassou. 2002. Microbial colonization of naturally black olives during fermentation and associated biochemical activities in the cover brine. *Letters in Applied Microbiology* 34: 173-177.

Omar, S.H. 2010. Cardioprotective and neuroprotective roles of oleuropein in olive. *Saudi Pharmaceutical Journal* 18: 111-121.

Panagou, E.Z. 2006. Greek dry-salted olives: Monitoring the dry-salting process and subsequent physico-chemical and microbiological profile during storage under different packing conditions at 4 and 20°C. *LWT* 39: 322-329.

Panagou, E.Z. and C.C. Tassou. 2006. Changes in volatile compounds and related biochemical profile during controlled fermentation of cv. Conservolea green olives. *Food Microbiology* 23, 738-746.

Panagou, E.Z., U. Schillinger, C. M.A.P. Franz and G.J.E. Nychas. 2008. Microbiological and biochemical profile of cv. Conservolea naturally black olives during controlled fermentation with selected strains of lactic acid bacteria. *Food Microbiology* 25: 348-358.

Peres, C., L. Catulo, D. Brito and C. Pintado. 2008. *Lactobacillus pentosus* DSM 16366 starter added to brine as freeze-dried and as culture in the nutritive media for Spanish-style green olive production. *Grasas y Aceites* 59: 234-238.

Peres, C.M., C. Peres, A. Hernandez-Mendoza and F. Xavier Malcata. 2012. Review of fermented plant materials as carriers and sources of potentially probiotic lactic acid bacteria, with an emphasis on table olives. *Trends in Food Science and Technology* 26: 31-42.

Perricone, M., A. Bevilacqua, M.R. Corbo and M. Sinigaglia. 2010. Use of *Lactobacillus plantarum* and glucose to control the fermentation of 'Bella di Cerignola' table Olives, a traditional variety of Apulian region (Southern Italy). *Journal of Food Science* 75: M430-M436.

Pistarino, E., B. Aliakbarian, A.A. Casazza, M. Paini, M.E. Cosulich and P. Perego. 2013. Combined effect of starter culture and temperature on phenolic compounds during fermentation of Taggiasca black olives. *Food Chemistry* 138: 2043-2049.

Plastourgos, S. and R.H. Vaughn. 1957. Species of *Propionibacterium* associated with zapatera spoilage of olives. *Applied Microbiology* 5: 267-271.

Psani, M. and P. Kotzekidou. 2006. Technological characteristics of yeast strains and their potential as starter adjuncts in Greek-style black olive fermentation. *World Journal of Microbiology and Biotechnology* 22: 1329-1336

Ramírez, E., P. García-García, A. de Castro, C. Romero and M. Brenes. 2013. Debittering of black dry-salted olives. *European Journal of Lipid Science and Technology* 115: 1319-1324.

Randazzo, C.L., A. Todaro, A. Pinoa, I. Pitinoa, O. Corona, A. Mazzaglia and C. Caggia. 2014. *Giarraffa* and *Grossa di Spagna* naturally fermented table olives: Effect of starter and probiotic cultures on chemical, microbiological and sensory traits. *Food Research International* 62: 1154-1164.

Rickard, A.H., P. Gilbert, N.J. High, P.E. Kolenbrander and P.S. Handley. 2003. Bacterial coaggregation: An integral process in the development of multi-species biofilms. *Trends in Microbiology* 11: 94-100.

Riley, F.R. 2002. Olive oil production on Bronze Age Crete: Nutritional properties, processing methods and storage life of Minoan olive oil. *Oxford Journal of Archaeology* 21: 63-75.

Rodríguez-Gómez, F., J. Bautista-Gallego, F.N. Arroyo-López, V. Romero-Gil, R. Jiménez-Díaz, A. Garrido-Fernández and P. García-García. 2013. Table olive fermentation with multifunctional *Lactobacillus pentosus* strains. *Food Control* 34: 96-105.

Rodríguez-Gómez, F., V. Romero-Gil, J. Bautista-Gallego, P. García-García, A. Garrido-Fernández and F.N. Arroyo-López. 2014. Production of potential probiotic Spanish-style green table olives at pilot plant scale using multifunctional starters. *Food Microbiology* 44: 278-287.

Ross, R.P., S. Morgan and C. Hill. 2002. Preservation and fermentation: Past, present and future. *International Journal of Food Microbiology* 79: 3-16.

Ruiz-Barba, J.L., D.P. Cathcart, P.J. Warner and R. Jiménez-Díaz. 1994. Use of *Lactobacillus plantarum* LPCO10, a bacteriocin producer, as a starter culture in Spanish-style green olive fermentations. *Applied and Environmental Microbiology* 60: 2059-2064.

Ruiz-Barba, J.L., B. Caballero Guerrero, A. Maldonado-Barragán and R. Jiménez-Díaz. 2010. Co-culture with specific bacteria enhances survival of *Lactobacillus plantarum* NC8, an autoinducer-regulated bacteriocin producer, in olive fermentations. *Food Microbiology* 27: 413-417.

Sabatini, N., M.R. Mucciarella and V. Marsilio. 2008. Volatile compounds in uninoculated and inoculated table olives with *Lactobacillus plantarum* (*Olea europaea* L., cv. Moresca and Kalamata). *LWT — Food Science and Technology* 41: 2017-2022.

Sakouhi, F., S. Harrabi, C. Absalon, K. Sbei, S. Boukhchina and H. Kallel. 2008a. Tocopherol and fatty acids contents of some Tunisian table olives (*Olea europea* L.): Changes in their composition during ripening and processing. *Food Chemistry* 108: 833-839.

Salminen, S., A.C. Ouwehand and E. Isolauri. 1998. Clinical applications of probiotic bacteria. *International Dairy Journal* 8: 563-572.

Sánchez Gómez, A.H., P. García García and L. Rejano Navarro. 2006. Elaboration of table olives. *Grasas y Aceites* 57: 86-94.

Servili, M., L. Settanni, G. Veneziani, S. Esposto, O. Massitti, A. Taticchi, S. Urbani, G.F. Montedoro and A. Corsetti. 2006. The use of *Lactobacillus plantarum* 1MO to shorten the debittering process time of black table olives (Cv. Itrana and Leccino): A pilot-scale application. *Journal of Agricultural and Food Chemistry* 54: 3869-3875.

Silva, T., M. Reto, M. Sol, A. Peito, C.M. Peres, C. Peres and F. Xavier Malcata. 2011. Characterization of yeasts from Portuguese brined olives, with a focus on their potentially probiotic behavior. *LWT – Food Science and Technology* 44: 1349-1354.

Sisto, A. and P. Lavermicocca. 2012. Suitability of a probiotic *Lactobacillus paracasei* strain as a starter culture in olive fermentation and development of the innovative patented product 'probiotic table olives'. *Frontiers in Microbiology* 3: 1-5.

Spyropoulou, K.E., N.G. Chorianopoulos, P.N. Skandamis and G.-J.E. Nychas. 2001. Survival of *Escherichia coli* O157:H7 during the fermentation of Spanish-style green table olives (Conservolea variety) supplemented with different carbon sources. *International Journal of Food Microbiology* 66: 3-11.

Srey, S., I.K. Jahid and S.D. Ha. 2013. Biofilm formation in food industries: a food safety concern. *Food Control* 31: 572-585.

Sutherland, I.W. 2001. Biofilm exopolysaccharides: A strong and sticky framework. *Microbiology* 147: 3-9.

Tang, M.L.K., S.J. Lahtinen and R.J. Boyle. 2010. Probiotics and prebiotics: Clinical effects in allergic disease. *Current Opinion in Pediatrics* 22: 626-634.

Tassou, C.C. 1993. Microbiology of olives with emphasis on the antimicrobial activity of phenolic compounds. PhD Thesis, University of Bath, UK.

Tassou, C.C., E.Z. Panagou and K.Z. Katsaboxakis. 2002. Microbiological and physicochemical changes of naturally black olives fermented at different temperatures and NaCl levels in the brine. *Food Microbiology* 19: 605-615.

Vaughn, R.H., K.E. Stevenson, B.A. Davé and H.C. Park. 1972. Fermenting yeasts associated with softening and gas-pocket formation in olives. *Applied Microbiology* 23: 316-320.

Vitali, B., G. Minervini, C.G. Rizzello, E. Spisni, S. Maccaferri, P. Brigidi, M. Gobbetti and R. Di Cagno. 2012. Novel probiotic candidates for humans isolated from raw fruits and vegetables. *Food Microbiology* 31: 116-125.

Wodner, M., S. Lavee and E. Epstein. 1988. Identification and seasonal changes of glucose, fructose and mannitol in relation to oil accumulation during fruit development in *Olea europea* (L.). *Scientia Horticulturae* 36: 47-54.

Wollowski, I., G. Rechkemmer and B.L. Pool-Zobel. 2001. Protective role of probiotics and prebiotics in colon cancer. *The American Journal of Clinical Nutrition* 73: 451S-455S.

Zottola, E.A. and K.C. Sasahara. 1994. Microbial biofilms in the food processing industry – Should they be a concern? *International Journal of Food Microbiology* 23: 125-148.

9

Regional Fermented Vegetables and Fruits in Asia-Pacific

Manas Ranjan Swain[1]* and Marimuthu Ananadharaj[2]

1. Introduction

Fermentation is one of the oldest and cheapest methods to extend the shelf-life of perishable vegetables and fruits. The preparation methods of fermented foods have great heterogeneity as they are based on tradition, cultural preference and geographical area. Many methods for preparing traditional fermented foods are not known; however, it is believed that the methods are passed down to subsequent generations as a family tradition or secret (Swain et al. 2014). Our ancestors used fermented foods to survive during the winter and drought seasons. Drying, salting and fermentation processes were developed in order to preserve fruits and vegetables when they were available in plenty and also to impart desirable flavors, textures, increase nutritional value, reduce toxicity and decrease cooking time (Rolle and Satin 2002, Chapters 3 and 4 in this book). Further, this process is cheap as well as energy-efficient. Fermentation technologies play an important role in ensuring the food security of millions of people around the world, particularly of marginalized and vulnerable groups.

Fermentation is a slow decomposition process of organic substances brought about by various microorganisms or enzymes that convert carbohydrates into alcohols (i.e. ethanol) or organic acids (i.e. lactic acid, acetic acid, formic acid and propionic acid) (FAO 1998). Different types of fermentation methods are employed to preserve various food products, such as vegetables and fruits while maintaining and improving their safety, nutritional, sensory features and shelf-life. (Demir et al. 2006, Di Cagno et al. 2013). Carbohydrates in vegetables and fruits are converted to

[1] Department of Biotechnology, Indian Institute of Technology Madras, Chennai-600036, India
  E-mail: manas.swain@gmail.com
[2] Agricultural Biotechnology Research Center, Academia Sinica, Taipei, Taiwan
  E-mail: anandhraj49@gmail.com
* Corresponding author: manas.swain@gmail.com

lactic acid (LA) by a group of bacteria called Lactic Acid Bacteria (LAB), as already explained in the previous chapters and therefore the pH value of fermented products decreases (around 4.0). Lowering of pH value inhibits the growth of spoilage as well as pathogenic bacteria and fungi. These LAB also improve the human intestinal microbial balance and enhance one's health by inhibiting the growth of various intestinal pathogens, such as *Escherichia coli*, *Salmonella* and *Staphylococcus* (Ohmomo et al. 2000, Ross et al. 2002).

Fermented vegetables and fruits are found all over the globe, especially in the Asia-Pacific region, which provide a conducive atmosphere and social cultural practices for development of different types of fermented vegetable and fruits. In Asia-Pacific, the most commonly used vegetables and fruits for fermentation include mango, cabbage, radish, tomato, lemon, cauliflower, leafy mustard, carrot, beats, brinjal, chili, cucumber, ladies finger, sponge gourd, spinach, asparagus, and lettuce (Paramithiotis et al. 2010). Apart from these, in certain parts tender bamboo shoots are also preserved. A wide variety fermented vegetables and fruits are available in the Asia-Pacific region and though, their major ingredients are similar, the final products are quite unique and different from each other, specially due to specific culinary practices. In addition to this, fermented vegetables and fruits provide food security, health-promoting benefits particularly in underdeveloped countries and are considered a major source of income and employment (Montet et al. 2006). In this chapter, we discuss the fermented vegetables and fruits products of Asia-Pacific region which, divided into four major groups based on the geographical origin: (i) East Asia, (ii) South Asia, (iii) South East Asia and (iv) Oceania, while reviewing the research prospects with respect to human nutrition and health.

2. Fermented Fruits and Vegetables in South Asia

South Asia in the Asian continent, comprises the sub-Himalayan countries and, for some authorities, adjoining countries to the west and east of the Indian Plate. South Asia, which includes Bangladesh, Bhutan, India, Maldives, Nepal, Pakistan and Sri Lanka, is ethnically diverse with more than 2,000 ethnic groups constituting populations ranging from hundreds of millions to small tribal groups.

2.1. India and Nepal

In India, cultivation of fruits and vegetables is quite widespread because of the availability of vast expanses of cultivable lands. In each season, various crops are cultivated and preserved through ethnic methods, particularly in the Himalayan belt (Tamang et al. 2005).

2.1.1. Gundruk: Gundruk is an ethnic, non-salted, fermented and acidic vegetable product indigenous to the Himalayas, especially Sikkim.

2.1.1.1. Preparation method and consumption: Gundruk is prepared from the fresh leaves of local vegetables known as rayosag (*Brassica rapa* subsp. *campestris* var. cuneifolia), mustard leaves (*Brassica juncea* (L.) Czern), cauliflower leaves (*Brassica oleracea* L. var. *botrytis* L.) and cabbages (*Brassica* spp.) that are wilted for one to two days. They are then crushed mildly and pressed into a container or earthen pot, made airtight and kept in a warm place to ferment naturally for about 15–22 days. After the desirable fermentation, products are removed and sun dried for two to four days. Gundruk is consumed as pickle or soup (approximately 1.4 g/d) and has some similarities to other fermented acidic vegetable products such as kimchi of Korea, sauerkraut of Germany and sunki of Japan (Yan et al. 2008).

2.1.1.2. Microbiota analysis: Gundruk fermentation is carried out by diverse groups of LAB strains, predominantly by *Lactobacillus fermentum*, *Lb. plantarum*, *Lb. casei*, *Lb. casei* subsp. *pseudoplantarum and Pediococcus pentosaceus*. In some cases, *Staphylococcus aureus* and enterobacteria were isolated from some market samples of gundruk (Tamang et al. 2005).

2.1.2. Sinki: Sinki, an indigenous fermented radish taproot food, is traditionally prepared by pit fermentation, which is a unique type of bio-preservation of foods through LA fermentation in the Sikkim Himalayas.

2.1.2.1. Preparation method and consumption: Sinki is produced by the pit fermentation method; wherein a pit is dug with an area of 2–3 ft diameter in a dry place. The pit is cleaned, plastered with mud and warmed by burning. Then the ashes are removed and the pit lined with bamboo sheaths and paddy straw. Radish taproots are wilted for two to three days, crushed, dipped in lukewarm water, squeezed and pressed tightly into the pit, which is covered with dry leaves over which heavy planks or stones are placed. The top of the pit is plastered with mud and left to ferment for 22–30 days. After fermentation, fresh sinki is removed, cut into small pieces, sun-dried for two to three days, and stored at room temperature for future consumption (Tamang and Sarkar 1993). Sinki is used as a base for preparing soup. Sinki pickle is prepared by soaking it in water, squeezing it dry and mixing with salt, mustard oil and chilies.

2.1.2.2. Microbiota analysis: Fermentation of sinki is carried out by various LAB that include *Lb. plantarum*, *Lb. brevis*, *Lb. casei* and *Leuconostoc fallax* (Tamang et al. 2005).

2.1.3. Khalpi: Khalpi is fermented cucumber (*Cucumis sativus* L.), commonly prepared in the months of September/October. It is the only fermented cucumber product in the entire Himalayan region.

2.1.3.1. Preparation method and consumption: For the production of khalpi, ripe cucumbers are collected and cut into small pieces for sun drying for two days. They are then put into a bamboo vessel which is made airtight by covering it with dried leaves. Fermentation is carried out at room temperature for three to five days. After five days of fermentation, khalpi becomes sour in taste and is consumed as a pickle by adding mustard oil, salt and chili powder along with steamed rice (Tamang 2009, Tamang et al. 2005).

2.1.3.2. Microbiota analysis: Predominant microflora of khalpi include *Ln. fallax*, *Lb. plantarum* and *Lb. brevis* (Tamang et al. 2005). The microbial population of raw cucumber is very low but increases after 36 hours.

2.1.4. Inziangsang: In the northeast India, people of Nagaland and Manipur prepare inziangsang, a traditional fermented leafy vegetable product prepared from mustard leaves. Most of the properties of inziangsang are very similar to gundruk.

2.1.4.1. Preparation method and consumption: Mustard leaves, locally called hangam (*Brassica juncea* L. Czern) are collected, crushed and soaked in warm water. Excess water in the leaves is removed by squeezing and pressing into the container to provide anaerobic conditions for fermentation. The container is kept at ambient temperature (20–30°C) and allowed to ferment for seven to 10 days. Similar to gundruk, fresh inziangsang is sundried for four to five days and stored in a container for several years at room temperature. The Nagals consume inziangsang as a soup with steamed rice. In resident meals, the fermented extract of *ziang dui* is used as a condiment (Tamang 2009).

2.1.4.2. Microbiota analysis: *Lb. plantarum*, *Lb. brevis*, and *Pd. acidilactici* were isolated from inziangsang (Tamang et al. 2005).

2.1.5. Soidon: Soidon is a most popular fermented food product in Manipur and is prepared from the tips of mature bamboo shoots.

2.1.5.1. Preparation method and consumption: Soidon is prepared from tips or apical meristems of mature bamboo shoots (*Bambusa tulda*, *Dendrocalamus giganteus* and *Melocana bambusoides*). The outer case and lower parts are removed and submerged in water in an earthen pot. Sojim (sour liquid), which is previously prepared, is added as starter in 1:1 dilution. The preparation is covered and allowed to ferment at room temperature for three to seven days. To enhance the flavor of soidon, leaves of *Garcinia pedunculata* Roxb. (family *Guttiferae*), locally called heibungin, are added to the fermenting vessel. Fermented soidon were removed from the pot and stored in an air tight container for longer storage period. Soidon is consumed as a curry or pickle along with steamed rice (Tamang 2009, Swain et al. 2014).

2.1.5.2. Microbiota analysis: *Lb. brevis*, *Ln. fallax* and *Lactococcus lactis* are reported to contribute to the fermentation of soidon (Tamang et al. 2008).

2.1.6. Goyang: Goyang is a popular, ethnic fermented vegetable food product of Sikkim and Nepal prepared by Sherpa women in the Himalayas (Tamang 2010).

2.1.6.1. *Preparation method and consumption*: Goyang is prepared from the leaves of *magane-saag* (*Cardamine macrophylla* Willd.), belonging to the family *Brassicaceae*, which is abundant in eastern Himalayas. The leaves are collected and cut into small pieces, excess water removed by squeezing, and then tightly packed into a bamboo basket, which is lined with layers of fig plant. The basket is covered with fig leaves and allowed to ferment at room temperature (15–25°C) for 25–30 days. The shelf-life of goyang is two to three months, which is extended up to several months by sundrying. Goyang is consumed as a soup prepared by boiling it along with yak or beef meat and noodles (thukpa). Thupka is a common food among the Sherpas (Tamang 2009, Swain et al. 2014).

2.1.6.2. *Microbiota analysis*: Goyang fermentation is carried out by *Lb. plantarum*, *Lb. brevis*, *Lb. lactis*, *Ec. faecium*, *Pd. pentosaceus* and yeasts like *Candida* spp. (Tamang and Tamang 2007).

2.2. Fermented Fruits and Vegetables in Southeast Asia

Southeast Asia is a sub-region of Asia consisting of countries that are geographically south of China, east of India, west of New Guinea and north of Australia. The culture of Southeast Asia is very diverse: on mainland, the culture is a mix of Indian (Burma, Cambodia, Laos, Malaysia and Thailand) and Chinese (Vietnam and Singapore), while in Indonesia, the Philippines, Malaysia and Singapore, the culture is a mix of Chinese, Indian, Western and the indigenous Austronesian culture. Also Brunei shows a strong influence of Arabia; Vietnam and Singapore show more of Chinese influence.

2.2.1. Thailand. Traditional fermented products play an essential role in the Thai diet and are consumed either as main dishes or as condiments. A variety of fermented foods have gained popularity not only because of their own unique characteristics of sensorial quality, but also because of their nutritional value. Some of these foods have become more economically important as they are meant for both domestic consumption and export. Today, the interest in fermented items continues to contribute to the food industry by inspiring new products and lowering costs. As a result, an improvement is seen in the microbial processes on which food producers have long relied. In Thailand, there are over 60 traditional fermented products that are consumed all over the country (Phithakpol et al. 1995). Thai fermented products can be categorized according to their source, main processing techniques, ingredients added and the microorganisms associated with the food. Although some products are produced by similar processing, some ingredients can be varied, leading to differences in appearance, flavor, taste, as well as sensorial properties. In addition, several products are very similar to those found in other Asian countries.

2.2.1.1. Pak-Gard-Dong: Pak-gard-dong is a fermented mustard leaf (*Brassica juncea* L.) product made in Thailand.

2.2.1.1.1. Preparation method and consumption. To prepare pak-gard-dong, mustard leaves are washed and wilted in the sun. Salt is added to the leaves and packed into containers for 12 hours. Water is drained and sugar solutions are added at 3 per cent concentration. These preparations are then allowed to ferment for three to five days at room temperature (FAO 1998).

2.2.1.1.2. Microbiota analysis: Microorganisms associated with the fermentation of Pak-gard-dong include *Lb. brevis, Pd. cerevisiae* and *Lb. plantarum* (Boon-Long, 1992).

2.2.1.2. Pak-Sian-Dong: Pak-sian-dong is an indigenous fermented leafy vegetable product of Thailand (Dhavises 1972).

2.2.1.2.1. Preparation method and consumption: Pak-sian-dong is prepared from the leaves of pak-sian (*Gynandropis pentaphylla* DC) which are washed with fresh vegetables and spread out on a cloth under sunlight to dehydrate until the sample is distinctly flaccid. After drying, water, salt and sugar are added and filled in a closed container; this mixture is kept for two to three days for fermentation.

2.2.1.2.2. Microbiota analysis: *Ln. mesenteroides, Lb. fermentum, Lb. buchneri, Lb. plantarum, Lb. brevis* and *Pd. pentosaceus* are present in pak-sian-dong (Dhavises 1972).

2.2.1.3. Naw-mai-dong: Naw-mai-dong is an ethnic fermented bamboo shoot product of Thailand (Dhavises 1972).

2.2.1.3.1. Preparation method and consumption. For the preparation of naw-mai-dong, young bamboo shoots are picked and trimmed to remove the woody and defective portions before washing with water. The shoots are sliced into small pieces and mixed with 2 per cent salt. They are boiled in water and the bitter liquor discarded. This preparation is then packed into earthen jars, shielded with a plastic sheet, weighted on top and allowed to ferment for three to four weeks at room temperature.

2.2.1.3.2. Microbiota analysis: *Ln. mesenteroides, Lb. fermentum, Lb. buchneri, Lb. brevis,* and *Pd. pentosaceus* were isolated from naw-mai-dong (Dhavises 1972, Phithakpol et al. 1995).

2.2.3. Vietnam: An important part of daily nourishment in Vietnam comprises of fermented vegetables. Bacteria, especially LAB, play a central role in the production of many fermented vegetables. In Vietnam, in general, 'dua muoi' (fermented mustard or beet) and 'ca muoi' (fermented eggplant) are prepared in most areas throughout the year. In contrast, onion and cabbage fermentation is done only during the winter period.

2.2.3.1. *Dua muoi*: Dua muoi is a popular traditional fermented mustard or beet, which is prepared throughout the year in Vietnam. There are two types of dua muoi: 'Dua cai be' (fermented mustard) and 'Dua cu cai' (fermented beet).

2.2.3.1.1. Preparation method and consumption: Both types of dua muoi are prepared by using fresh mustard or beetroots mixed with onions, sugar and salt. The roots of fresh mustard plants are removed and the greens (for fermented beet both the green and tubers are used) are washed and dried in the sun for four hours. Brine is prepared from salt and sugar, the concentrations of which varies significantly between households. Then, they are mixed with onion, brined and kept for fermentation up to three days at 25–35°C (Nguyen et al. 2013).

2.2.3.1.2. Microbiota analysis: Predominant LAB associated with 'dua muoi' are *Lb. fermentum, Lb. pentosus, Lb. plantarum, Lb. paracasei, Lb. pantheris, Pd. acidilactici* and *Pd. pentosaceus* (Nguyen et al. 2013).

2.2.3.2. *Ca muoi*: Ca muoi is a delicious Vietnamese fermented food prepared from eggplant (*Solanum melongena*).

2.2.3.2.1. Preparation method and consumption. Fresh eggplants (white and small eggplants) are removed from the knob and are placed in a basket to dry in the sun for three to four hours. Then, the dried eggplants are mixed with garlic, chili and young galangals in a brine of salt; and occasionally sugar is added. Usually the fermentation is carried out for two to three days at 25–35°C (Nguyen et al. 2013).

2.2.3.2.2. Microbiota analysis: *Lb. fermentum, Lb. pentosus, Lb. plantarum* are reported to constitute the dominating lactic acid microbiota with *Pd. pentosaceus* and *Lb. brevis* forming a secondary one (Nguyen et al. 2013).

2.2.4. Indonesia

2.2.4.1. *Sayur asin*: Sayur asin is known as traditional fermented mustard (*Brassica juncea* var. rugosa) produced and consumed in many areas of Indonesia. Sayur asin is similar to hum choy produced in China and other Southeast Asian countries.

2.2.4.1.1. Preparation method and consumption: Mustard cabbage leaves (*Brassica juncea* var. rugosa) are wilted, rubbed or squeezed with 2.5–5 per cent salt. Mustard cabbage is very low in fermentable carbohydrates, so liquid from boiled rice is added to ensure that sufficient acid is produced during fermentation, which is carried out for one week at room temperature (Puspito and Fleet 1985).

2.2.4.1.2. Microbiota analysis: Fermentation is carried out by a sequential growth of *Ln. mesenteroides, Lb. confusus, Lc. curvatus, Pd. pentosaceus* and *Lb. plantarum*. Starch-degrading species of *Bacillus, Staphylococcus* and *Corynebacterium* exhibit limited growth during the first day of fermentation. The yeasts, *Candida sake* and *Candida guilliermondii*, contribute to fermentation (Puspito and Fleet 1985).

2.2.4.2. Growol: Growol is an Indonesian traditional food prepared from cassava (*Manihot esculenta* Crantz) by submerged lactic fermentation. Growol is a dough-like material with a sticky texture, a sour taste and a strong and pungent flavor. It becomes white in color after steaming and is primarily consumed as a substitute for rice.

2.2.4.2.1. Preparation method and consumption: The processing of cassava into growol is similar to that of West Nigerian fu-fu (Westby and Twiddy 1992). For the preparation of growol, one to two years old sweet cassava tubers are peeled and steeped in water in a clay pot and allowed to ferment naturally for four to six days at 30°C. The water is changed every day to avoid getting a pungent flavor. The fermented roots are sieved through a locally fabricated sieve to remove the inner cortical cellulosic materials. The starch is then separated and washed four to five times and drained. The cassava pulp is steamed for one hour.

2.2.4.2.2. Microbiota analysis: The main microorganisms present during fermentation are LAB and yeasts. The bacteria include *Streptococcus, Lactobacillus, Bacillus, Enterobacter, Acinetobacter, Moraxella* and *Micrococcus* species. (Endang et al. 1996).

2.2.5. Malaysia: Malaysia's cuisine reflects the multi-ethnic makeup of its population with several cultures from the surrounding regions having influenced the cuisine. Much of the influence comes from the Chinese, Indian, Thai, Javanese and Sumatran cultures largely due to the country being part of the ancient spice route. Similarly the fermented products normally adopted by Malaysia have a close similarity with surrounding countries and cultures.

2.2.5.1. Tempoyak: Tempoyak is a traditional Malaysian fermented condiment made from the pulp of durian fruit (*Durio zibethinus*).

2.2.5.1.1. Preparation method and consumption. Fermentation of durian into tempoyak is carried out in the traditional way to induce spontaneous fermentation. The durian fruit pulp is placed in jars and mixed with salt. Durian seeds are removed and mixed with small amounts of salt and allowed to ferment at ambient temperature in a tightly-closed container for four to seven days. The acidity of tempoyak is reported at approximately 2.8–3.6 per cent. The salt concentration significantly affect the sourness, sweetness, colour and aroma of tempoyak, with the most preferred tempoyak being that with 2 per cent salt (Leisner et al. 2001).

2.2.5.1.2. Microbiota analysis: The sour taste of tempoyak is brought about by the acid produced by LAB during fermentation. *Lb. brevis, Lb. mali, Lb. fermentum, Lb. durianis, Ln. mesenteroides* are reported to dominate this fermentation (Leisner et al. 2001).

2.2.5.2. *Jeruk*: Jeruk is a traditional homemade fermented pickle native to many places in Malaysia. It is prepared from common fruits and vegetables which are available in plenty during various seasons (Merican 1996). Most commonly used vegetables are gherkin and cucumber, ginger, onion, chili, bamboo shoot, mustard leaves, etc. Jeruk is a favourite and economic food for most rural people having a low income.

2.2.5.2.1. Microbiota analysis: Various species of *Leuconostoc, Lactobacillus, Pediococcus,* and *Enterococcus* are predominantly present in jeruk (Merican 1996).

2.2.6. Philippines: The Philippine archipelago has diverse ecosystems, organisms, peoples and cultures. Traditional or ethnic fermented foods are part of Philippino culture and intrinsic to their lifestyle. There are three major islands groups in Philippines — Luzon, Visayas and Mindanaow with each group having its own traditional way of preparing fermented food (Banaay et al. 2013).

2.2.6.1. *Burong mustasa*: Burong mustasa is made from fermented mustard leaves (*Brassica juncea*) and consumed by the Luzon region of Philippines.

2.2.6.1.1. Preparation method and consumption: Burong Mustasa is prepared from fresh mustard leaves, which are collected, washed with water and cut into small pieces (around 2″). Then rice broth (prepared by boiling one part rice with five parts of water) is added to the leaves along with 4.5 per cent of NaCl. The jar is covered and allowed to ferment for three to four days or until the desired acidity is reached (Sanchez 1999).

2.2.6.1.2. Microbiota analysis: Fermentation of burong mustasa is governed by *Ln. mesenteroides, Ec. faecalis* and *Lb. plantarum* (Banaay et al. 2013).

2.2.6.2. *Burong manga*: Burong manga is a traditional fermented green mango prepared and consumed by all the regions of Philippines.

2.2.6.2.1. Preparation method and consumption: Newly harvested green mangos are peeled, sliced to desired size and then packed into jars. The jars are filled with 3–4 per cent common salt (NaCl) solution. The mixture is allowed to ferment for three to five days at 30°C (Sanchez 1999).

2.2.6.2.2. Microbiota analysis: *Lb. brevis, Lb. plantarum Pd. cerevisiae* and *Ln. mesenteroides* were isolated from fermented burong manga (Banaay et al. 2013).

2.2.6.3. Atchara: Atchara (also spelled achara or atsara), is a pickle made from grated and raw papaya which is popular in the Philippines.

2.2.6.3.1. Preparation method and consumption: Atchara is prepared by overnight fermentation of salted green papaya to obtain the desired firm and crispy texture. Apart from papaya, several other vegetables, such as green pepper, carrot, ginger and garlic are sliced and mixed with it. The fermented pickle is packed in glass jars containing four parts vinegar and one part sugar. The mixture is allowed to stand for two days before consumption. This dish is habitually served as a side dish with fried or grilled foods, such as pork barbecue. The microbiological aspects of atchara are still unknown.

2.3. Fermented Fruits and Vegetables in East Asia

East Asia is the eastern sub-region of the Asian continent, which includes China, Japan, Korea, Mongolia and the Republic of China (Taiwan). Eastern Asia is one of the world's most populated areas, having diverse of cultural groups.

2.3.1. China and Taiwan: China is well known for its tradition and culture. Traditional or indigenous fermented foods and beverages are popular and integral part of the Chinese regime (Aidoo et al. 2006). The fermentation process was used to prepare various fermented foods before the science of microbiology came to be known. Fermented foods are mentioned in Chinese history and were used during festivals, celebrations, feasting and ritual activities. During the Shang dynasty (about 1600–1046 B.C.), three types of fermented beverages were made — *chang* (an herbal wine), *li* (probably a sweet, low-alcoholic rice or millet beverage), and *jiu* (a fully fermented and filtered rice or millet beverage or 'wine', with an alcoholic content of probably 10–15 per cent by weight). These fermented beverages were sacrificed to the royal ancestors in bronze vessels. However, the microorganism associated with Chinese fermented foods were documented only in mid 19th century (Xiao 1996).

2.3.1.1. Pao cai and suan cai: Pao cai is a most popular ethnic lactic acid fermented vegetable in northern and western China. A similar form of pao cai, called suan cai, is prepared and eaten in northeast China.

2.3.1.1.1. Preparation method and consumption: Surplus vegetables such as cabbage, celery, cucumber and radish are stored and collected for use in off season. These vegetables are cleaned, drained, cut into small pieces, and seasonings such as sugar, salt, vinegar and a little alcohol are added. Then this preparation is packed into jars, filled with water and allowed to ferment for more than one week. Usually pao cai is served as a side dish as a pickle, which varies in terms of taste and method of preparation in different areas. Taiwanese pao cai, which is made with different kinds of vegetables, spices and other ingredients by anaerobic fermentation in a special container, has a crunchy texture and tangy taste (Yan et al. 2008, Feng et al. 2012).

2.3.1.1.2. Microbiota analysis: Pao cai fermentation is facilitated by several microorganisms existing in the raw materials. *Lb. pentosus, Lb. plantarum, Lb. brevis, Lb. lactis, Lb. fermentum, Lb. harbinensis, Ln. mesenteroides* were isolated from various types of pao cai and suan cai (Miyamoto et al. 2005, Yan et al. 2008, Feng et al. 2012).

2.3.1.2. Yan-taozih: Yan-taozih is a pickled peach preparation of China and Taiwan and easily available in night or traditional markets.

2.3.1.2.1. Preparation method: Fresh peaches (*Prunus persica*) are harvested, and cleaned and mixed with 5–10 per cent salt. This mixture is gently shaken until water exudes from the peaches. The peaches are then washed and mixed with 5–10 per cent sugar and 1–2 per cent pickled plums. To obtain the sweet taste, licorice powder is added. All ingredients are mixed well and allowed to ferment at low temperature (6–10°C) for one day (Chen et al. 2013b, Swain et al. 2014).

2.3.1.2.2. Microbiota analysis: *Ln. mesenteroides, Lb. lactis, Weissella cibaria, W. paramesenteroides, W. minor, Ec. faecalis* and *Lb. brevis* were isolated from yan-taozih (Chen et al. 2013b).

2.3.1.3. Jiang-gua: Jiang-gua is a common ethnic fermented cucumber served as a side dish or a seasoning in Taiwan. Jiang-gua is available in almost every market of Taiwan, both traditional markets and supermarkets.

2.3.1.3.1. Preparation method: Fresh cucumber (*Cucumis sativus* L.) is harvested and washed to remove dirt, sliced and mixed with salt. Then they are layered in a bucket, sealed with heavy stones on the cover and allowed to ferment for four to five hours. some manufacturers resort to a longer processing time. Finally, water from the bucket is drained off and the cucumbers are mixed with sugar and vinegar. In some cases, soy sauce is also added. Fermentation usually continues for at least one day at low temperature (6–10°C). Jiang-gua is consumed as a seasoning or side dish wiht pork, fish, chicken and various other foods (Chen et al. 2012).

2.3.1.3.2. Microbiota analysis: Fermentation of jiang-gua depends upon *W. cibaria, W. hellenica, Lb. plantarum, Ln. lactis* and *Ec. casseliflavus* (Chen et al. 2012).

***2.3.1.4. Yan-tsai-shin*:** Broccoli (*Brassica oleracea* L.) is a cabbage family plant, the flower of which is widely used while the stems are discarded. In Taiwan, the broccoli stems are used for preparation of yan-tsai-shin, a traditional fermented food.

2.3.1.4.1. Preparation method: Freshly harvested broccoli is washed, peeled, cut, mixed with salt and kept in a bucket for approximately six hours. Excess amount of water is removed and various ingredients such as sugar, soy sauce and sesame oil added. Some manufacturers add rice wine or sliced hot pepper to obtain a distinctive flavor. All these ingredients are mixed well and fermented at low temperature (6–10°C) for one day (Chen et al. 2013a).

2.3.1.4.2. Microbiota analysis: *W. paramesenteroides*, *W. cibaria*, *W. minor*, *Ln. mesenteroides*, *Lb. plantarum*, and *Ec. sulphurous* are the predominant microbiota in yan-tsai-shin (Chen et al. 2013a).

***2.3.1.5. Pobuzihi*:** Pobuzihi is a widely used traditional fermented food prepared with cummingcordia in Taiwan. Two different types of pobuzihi (caked or granular) are available and easily differentiated by their appearance.

2.3.1.5.1. Preparation method: The initial process of caked or granular pobuzihi is similar to other fermented vegetables. Cummingcordia (*Cordia dichotoma* Forst. f.) is harvested, washed, boiled for several minutes and mixed with salt. To prepare caked pobuzihi, the boiled cummingcordia are filled into containers while hot; after cooling, they are removed from containers and placed into urns containing salt solutions. The granular pobuzihi is prepared by mixing the boiled cummingcordia with soy sauce, occasionally sugar, salt, licorice or sake (rice wine). Longer fermentation time is required for granular pobuzihi. The fermentation process of granular pobuzihi is around a week, but some manufacturers extend the fermentation periods up to 15 days or longer at room temperature.

2.3.1.5.2. Microbiota analysis: Chen et al. (2010) isolated and characterized the novel *Lb. pobuzihii*. Moreover, *Lb. plantarum*, *W. cibaria*, *W. paramesenteroides*, *Pd. pentosaceus* were also isolated from fermented pobuzihi (Chen et al. 2013c).

***2.3.1.6. Yan-dong-gua*:** In Taiwan, yan-dong-gua is a commonly used traditional food prepared from fermented wax gourd (*Benincasa hispida* Thunb.) and consumed as a seasoning due to its pleasant flavor.

2.3.1.6.1. Preparation method: Fresh wax gourds are harvested, washed, skins and seeds removed, sliced into tiny pieces and dried in sunlight for one to two days to remove the water content. After drying, it is combined with salt, sugar and fermented soybeans at final concentration of about 7–10 per cent and layered in a bucket. Typically, a small amount of mijiu (Taiwanese rice wine) is added before the bucket is sealed. Fermentation is carried out for one month or even longer. Yan-dong-gua is usually used as a seasoning for fish, pork, meatballs and various other foods.

2.3.1.6.2. Microbiota analysis: The LAB responsible for yan-dong-gua fermentation were *W. cibaria* and *W. paramesenteroides* (Lan et al. 2009).

2.3.1.7. Yan-jiang: Yan-jiang is traditional fermented ginger (*Zingiber officinale* Roscoe) that is widely used in Taiwan.

2.3.1.7.1. Preparation Method: Yan-jiang is prepared by two different methods — with and without addition of plums. Ginger is washed and skins are removed. Then it is shredded and mixed with salt. This mixture is layered in a bucket for two to six hours. After the removal of water, ginger is mixed with sugar and pickled plums. Salt and sugar are added to the final concentration of approximately 30–60 g kg/L and allowed to ferment for three to five days at low temperature (6–10°C), but some manufacturers maintain a fermentation time of one week or even longer. Yan-jiang is used as a seasoning due to its pleasing flavor (Chang et al. 2011).

2.3.1.7.2. Microbiota analysis: Initial fermentation of Yan-jiang is carried out by *Lb. sakei* and *Lc. lactis* but at the final stages of fermentation these species are replaced by *W. cibaria* and *Lb. plantarum* (Chang et al. 2011).

2.3.1.8. Fu-tsai and Suan-tsai: Fu-tsai and suan-tsai are traditional fermented mustard food products, majorly consumed by Hakka tribes of Taiwan (Chao et al. 2009).

2.3.1.8.1. Preparation method and consumption: To prepare fu-tsai and suan-tsai, freshly harvested mustard is cleaned and wilted, before being placed in a bucket in layers by alternating with 4 per cent salt and closed airtight. The bucket is allowed to ferment naturally for seven days (to prepare fu-tsai) to two months (to make suan-tsai) at ambient temperature. After the fermentation is complete, the product is cleaned with water, sun-dried for one to two days and then fermented again for two more days. This step of sun-drying during daytime and fermenting at night is repeated two or three times. In case of fu-tsai, partly dried mustards is tightly packed and sealed into bottles, placed upside down for maturation (three months). Fu-tsai and suan-tsai are consumed as a soup, or fried with shredded meat, or stewed with meat (Chao et al. 2009).

2.3.1.8.2. Microbiota analysis: Several LAB strains are isolated from fu-tsai and suan-tsai that include *Pd. pentosaceus*, *Tetragenococcus halophilus*, *Lb. farciminis*, *Ln. mesenteroides*, *Ln. pseudomesenteroides*, *W. cibaria* and *W. paramesenteroides* (Chao et al. 2009, Chen et al. 2006).

2.3.2. Japan: Fermented vegetables were introduced in Japan during the sixth century by Buddhist priests who began propagating Buddhism, according to which eating of meat and fish was prohibited. The Japanese adhere strictly to their Buddhist vegetarian diets (Fukushima 1985). In the next few paragraphs, some of the popular fermented vegetables of Japan are discussed.

2.3.2.1. Nozawana-zuke: Nozawana-zuke is an ethnic fermented low-salt pickle prepared by using field mustard, locally called Nozawana (*Brassica campestris* var. rapa), a leafy turnip plant, very popular among the Japanese.

2.3.2.1.1. Preparation method and consumption: Nozawana-zuke is prepared by lactic acid fermentation of nozawana along with inorganic salts and red pepper powder containing spicy components. The fermentation is carried out by various plant-based natural LAB, including *Lactobacillus* and *Leuconostoc*. These bacteria are responsible for the sensory properties of nozawana-zuke and inhibit the growth of unwanted bacteria by producing organic acids (Kawahara and Otani 2006).

2.3.2.1.2. Microbiota analysis: The microorganisms that are found include: *Lb. curvatus, Lb. brevis, Lb. delbrueckii, Lb. fermentum, Lb. plantarum* and *Ln. mesenteroides* (Kawahara and Otani 2006).

2.3.2.2. Sunki: Sunki is a traditional non-salted fermented vegetable prepared from the leaves of otaki-turnip in Kiso district of Japan.

2.3.2.2.1. Preparation method and consumption: Otaki turnip is boiled in water and mixed with wild small apples, locally called zumi. Then small amount of dried sunki preserved from the previous year is added and allowed to ferment for one to two months. Typically, sunki is manufactured under low temperatures (in winter season) and eaten with rice or in miso soup (Endo et al. 2008).

2.3.2.2.2. Microbiota analysis: The microorganisms involved include *Lb. plantarum, Lb. brevis, Bacillus coagulans* and *Pd. pentosaceus* (Endo et al. 2008).

2.3.3. Korea: Koreans are known for their special skill in preparing fermented food since more than 1500 years (Han et al. 1998). These foods, except for alcoholic beverages, are divided into three major categories, such as kimchi, soy-based products and fish- and shellfish-based products. Kimchi is a most abundantly used product that is prepared with various vegetables. Usually, these fermented foods are prepared by almost all households once a year and used for several years. Though different fermented foods are consumed in Korea, kimchi is the favourite and most studied one.

2.3.3.1. Kimchi: Kimchi is a well-known traditional fermented product of Korea, generally made from Chinese cabbage (beachu), radish, green onion, red pepper powder, garlic, ginger and fermented seafood (jeotgal), which is traditionally made at home and served as a side dish. Kimchi is a common term for more than 100 vegetable-based products fermented by LA fermentation (Lee et al. 2005).

2.3.3.1.1. Preparation method and consumption: Raw vegetables like cabbage or radish are pre-boiled then salted, blended with various ingredients like red pepper, garlic, green onion, ginger, etc. and other minor ingredients (seasonings, salted sea-food, fruits, vegetables, cereals, fish and meat, etc.). These mixtures are allowed to ferment at low temperature (2–5°C). To obtain authentic taste of kimchi, temperature control is an important factor; fermentation at 15°C may require one week, while at 25°C, three days are enough. However, a low temperature is always preferred in order to avoid over-ripening or strong acid production. Kimchi is characterised by its sour, sweet and carbonated taste and differs in flavor from sauerkraut and pickles that are also very popular fermented vegetables (Hong et al. 1999).

2.3.3.1.2. Microbiota analysis: The lactic acid bacteria responsible for kimchi fermentation include *Ln. mesenteroides, Ln. citreum, Ln. gasicomitatum, Ln. pseudomesenteroides, Lb. brevis, Lb. curvatus, Lb. plantarum, Lb. sakei, Lb. lactis, Pd. pentosaceus, W. confusa* and *W. koreensis* (Hong et al. 1999, Cho et al. 2006).

2.4. Fermented Fruits and Vegetables in Western Asia

Western Asia is the western sub-region of Asia and is a newer term for the area that encompasses Middle East and the Near East. The countries include Bahrain, Qatar, Oman, Saudi Arabia, Turkey, Syria, Iran, Iraq, etc.

2.4.1. Turkey

2.4.1.1. Shalgam juice: Shalgam, a famous traditional lactic acid fermented beverage, is a red colored, cloudy and sour soft drink, mainly consumed by people in some provinces of southern Turkey.

2.4.1.1.1. Preparation method and consumption: Shalgam juice (carrot juice) is prepared from a mixture of turnips (*Brassica rapa* L.), black carrot (*Daucus carota* ssp. sativus var. atrorubens Alef.), bulgur (broken wheat) flour, rock-salt and water by lactic acid fermentation (Tanguler and Erten 2012). Shalgam juices are prepared by two methods for commercial production — the traditional and the direct method. The traditional method has two stages of fermentation that include sour-dough fermentation (first fermentation) and carrot fermentation (second fermentation). The direct method includes only the second fermentation (Erten et al. 2008). Shalgam is widely consumed with food as a refreshing drink and also as a perfect accompaniment to the Turkish meatball called 'kebap'.

2.4.1.1.2. Microbiota analysis: Shalgam juice fermentation is mainly carried out by several LAB of the genera *Lactobacillus, Leuconostoc* and *Pediococcus*.

The predominant LAB species include *Lb. plantarum, Lb. brevis, Lb. paracasei, Lb. buchneri* and *Pd. pentosaceus* (Erten et al. 2008, Tanguler and Erten 2012).

2.4.1.2. *Tursu*: Tursu is one of the oldest fermented products consumed in Turkey. Tursu can be made from a wide variety of different vegetables and fruits. The word 'tursu' originates from the Persian word 'torsh' which means 'sour'.

2.4.1.2.1. Preparation method and consumption: Cucumber, cabbage, green tomato and green pepper are the most popular vegetables used to prepare tursu. In addition, melon (kelek), carrots, green beans, red beets and eggplants are used, depending on consumer preference. Garlic, parsley, fresh mint leaves, ginger, fresh dill leaves and bay leaves are also used as flavoring agents in the production of tursu. To prepare tursu, all vegetables and/or fruits together with the flavoring agents are filled in glass or plastic containers. Brine (a solution of 10–15 per cent NaCl in water) containing grape vinegar is added and the pickles are left to ferment at about 20°C for four weeks (Kabak and Dobson 2011).

2.4.1.2.2. Microbiota analysis: The LAB responsible for fermentation include *Lb. plantarum, Ln. mesenteroides, Lb. brevis, Pd. pentosaceus* and *Ec. faecalis* (Kabak and Dobson 2011).

2.4.1.3. *Hardaliye*: Hardaliye is an indigenous beverage in the Marmara region of Turkey where it has been produced for centuries mainly from red grape juice with the addition of crushed black mustard seeds.

2.4.1.3.1. Preparation method and consumption: To prepare hardaliye by the traditional method, the grapes are cleaned and pressed in wooden barrels; sometimes plastic barrels also used. These juices are mixed with crushed raw mustard seeds (0.2 per cent) and benzoic acid (0.1 per cent), and allowed to ferment for five to 10 days at room temperature. After fermentation, the hardaliye is stored in appropriate containers at 4°C. The pH of the finished product is around 3.21–3.97 (Kabak and Dobson 2011).

2.4.1.3.2. Microbiota analysis: The LAB counts of hardaliye range from 1×10^2 to 4×10^4 CFU/ml, with *Lb. paracasei* subsp. *paracasei, Lb. casei* subsp. *pseudoplantarum, Lb. pontis, Lb. brevis, Lb. acetotolerans, Lb. sanfranciscensis,* and *Lb. vaccinostercus* having been isolated from hardaliye samples (Arici and Coskun 2001).

2.5. Fermented Fruits and Vegetables from Oceania

Oceania includes all region cantered on the islands of the tropical Pacific Ocean. Oceania range from its three sub-regions of Melanesia, Micronesia, and Polynesia to, more broadly, the entire insular region between Asia and the Americas, including Australasia and the Malay Archipelago.

Table 1: Several Important Traditional Fermented Fruits and Vegetable Food Products, Commonly Used in Asian Subcontinent

Regions	Fermented Food Product	Fruit and Vegetables	Other Ingredients	Microorganisms	References
Fermented vegetables and fruits in South Asia					
India	Inziangsang	Mustard leaf	No	*Lb. plantarum, Lb. brevis, Pd. acidilactici*	Tamang (2009)
India	Soidon	Bamboo shoot	Water	*Lb. brevis, Lc. lactis, Ln. fallax*	Tamang et al. (2008)
India, Nepal and Bhutan	Sinki	Radish	No	*Lb. plantarum, Lb. brevis, Lb. fermentum, Lb. casei, Ln. fallax, Pd. pentosaceus*	Tamang et al. (2005)
Nepal	Khalpi	Cucumber	No	*Lb. plantarum, Lb. brevis, Ln. fallax, Pd. pentosaceus*	Tamang et al. (2005), Dahal et al. (2005)
Nepal, India	Gundruk	Cabbage, radish, mustard, cauliflower	No	*Pediococcus* spp. *and Lactobacillus* spp.	Tamang (2009), Dahal et al. 2005
Fermented Vegetables and Fruits in Southeast Asia					
Indonesia	Sayur Asin	Mustard, cabbage	Salt, liquid from boiled rice	*Ln. mesenteroides, Lb. confusus, Lb. plantarum, Lc. curvatus, Pd. pentosaceus*	Puspito and Fleet (1985)
Malaysia	Tempoyak	Durin (durio zibethinus)	Salt	*Lb. brevis, Ln. mesenteroides, Lb. mali, Lb. fermentum, Lb. durianis*	Leisner et al. (2001)
Philippines	Burong Mustasa	Mustard leaf	Rock salt	*Lb. plantarum, Ln. mesenteroides, Ec. faecalis*	Banaay et al. (2013)
Thailand	Dak-Gua-Dong	Mustard leaf	Salt	*Lb. plantarum*	Tanasupawat and Komagata (1995)
Thailand	Pak-Gard-Dong	Mustard leaf	Salt and sugar solution	*Lb. brevis, Pd. cerevisiae, Lb. plantarum*	Boon-Lung (1986, 1992)
Thailand	Pak-Sian-Dong	Leaves of pak-sian (Gynadropsis pentaphylla)	Brine	*Lb. brevis, Pd. cerevisiae, Lb. plantarum, Lb. buchneri, Lb. fermentum, Ln. mesenteroides*	Dhavises (1972), Tanasupawat and Komagata (1995)

Contd...

Table 1: (Contd.)

Regions	Fermented Food Product	Fruit and Vegetables	Other Ingredients	Microorganisms	References
Vietnam	Ca Muoi	Eggplant		*Lb. fermentum, Lb. pentosus, Lb. brevis, Lb. plantarum, Pd. pentosaceus*	Nguyen et al. (2013)
Vietnam	Dha Muoi	Cabbage, various vegetables		*Ln. mesenteroides, Lb. plantarum*	Steinkraus (1997)
Vietnam	Dua Muoi	Mustard or beet	Onion, sugar and salt	*Lb. fermentum, Lb. pentosus, Lb. plantarum, Lb. paracasei, Lb. pantheris, Pd. acidilactici, Pd. pentosaceus*	Nguyen et al. (2013)
Fermented Vegetables and Fruits in East Asia					
China	Pao Cai	Cabbage, celery, cucumber and radish	Ginger, salt, sugar, hot red pepper	*Lb. pentosus, Lb. plantarum, Ln. mesenteroides, Lb. brevis, Lb. lactis, Lb. fermentum, Lb. harbinensis*	Miyamoto et al. (2005), Yan et al. (2008), Feng et al. (2012)
China and Taiwan	Yan-Taozih	Peaches	Salt, sugar and pickled plums	*Ln. mesenteroides, W. cibaria, Lb. lactis subsp. lactis, W. paramesenteroides, Ec. faecalis, W. minor, Lb. brevis*	Chen et al. (2013b)
Japan	Nozawana-Zuke	Turnip		*Lb. curvatus, Lb. brevis, Lb. delbrueckii, Lb. fermentum, Lb. plantarum, Ln. mesenteroides*	Kawahara and Otani (2006)
Japan	Sunki	Leaves of otaki-turnip	Wild apple	*Lb. plantarum, Lb. brevis, Pd. pentosaceus, B. coagulans*	Battcock and Azam-Ali (2001), Endo et al. (2008)
Korea	Kimchi	Cabbage, radish, various vegetables	Garlic, red pepper, green onion, ginger and salt	*Ln. mesenteroides, Ln. citreum, Ln. gasicomitatum, Ln. pseudomesenteroides, Lb. curvatus, Lb. lactis, Lb. brevis, Lb. plantarum, Lb. sakei, W. confusa, W. koreensis, Pd. pentosaceus*	Hong et al. (1999), Lee et al. (2005), Cho et al. (2006)
Taiwan	Jiang-Gua	Cucumber	Salt	*W. cibaria, W. hellenica, Lb. plantarum, Ln. lactis, Ec. casseliflavus*	Chen et al. (2012)

Contd...

Table 1: (Contd.)

Regions	Fermented Food Product	Fruit and Vegetables	Other Ingredients	Microorganisms	References
Taiwan	Pobuzihi	Cummingcordia	Salt	*Lb. pobuzihii, Lb. plantarum, W. cibaria, W. paramesenteroides, Pd. pentosaceus*	Chen et al. (2010, 2013c)
Taiwan	Suan-Tsai	Chinese cabbage, cabbage, mustard leaves	Salt	*Pd. pentosaceus, Tc. halophilus*	Chen et al. (2006)
Taiwan	Yan-Dong-Gua	Wax gourd	Salt, sugar and fermented soybeans	*W. cibaria, W. paramesenteroides*	Lan et al. (2009)
Taiwan	Yan-Jiang	Ginger	Plums, salt	*Lb. sakei, Lc. lactis subsp. lactis, W. cibaria, Lb. plantarum*	Chang et al. (2011)
Taiwan	Yan-Tsai-Shin	Broccoli	Sugar, soy sauce and sesame oil	*W. paramesenteroides, W. cibaria, W. minor, Ln. mesenteroides, Lb. plantarum, Ec. sulfurous*	Chen et al. (2013a)
Fermented Vegetables and Fruits in West Asia					
Turkey	Shalgam Juice	Turnips, black carrot	Broken wheat, rock-salt, water	*Lb. plantarum, Lb. brevis, Lb. paracasei, Lb. buchneri, Pd. pentosaceus*	Erten et al. (2008), Tanguler and Erten (2012)
Turkey	Tursu	Cucumbers, cabbages, green tomatoes, carrots, green beans.	Garlic, parsley, fresh mint leaves, ginger	*Lb. plantarum, Lb. brevis, Ln. mesenteroides, Pd. pentosaceus, Ec. faecalis*	Kabak and Dobson (2011)
Turkey	Hardaliye	Grape	Mustard leaves, benzoic acid	*Lb. paracasei subsp. paracasei, Lb. casei subsp. pseudoplantarum, Lb. pontis, Lb. brevis, Lb. acetotolerans, Lb. sanfranciscensis, Lb. vaccinostercus*	Arici and Coskun (2001)

Lb.: Lactobacillus; Ln.: Leuconostoc; Lc.: Lactococcus; Pd.: Pediococcus; W.: Weissella; Ec.: Enterococcus; B.: Bacillus; Tc.: Tetragenococcus

2.5.1. Poi: Poi is a Polynesian staple food made from the underground plant stem or corm of the taro plant (known in Hawaiian as *kalo*). It is a traditional part of native Hawaiian cuisine (Matthews, 2000), produced by mashing the cooked corm (baked or steamed) until it turns into a highly viscous fluid. Water is added during mashing and again just before eating to achieve the desired consistency, which can range from liquid to dough-like (poi can be known as 'one-finger', 'two-finger' or 'three-finger' poi, alluding to how many fingers are required to scoop it up in order to eat it, which depends on the consistency). It can be eaten immediately, fresh and sweet, or left a bit longer to ferment. Although LAB are isolated from fermented poi, species level identification reports are not available in literature.

3. Conclusion

In the Asia-Pacific region, diverse groups of fermented foods have been used by diverse groups of ethnic communities. Understanding the population of microorganisms like bacteria, yeast and moulds in fermented fruits and vegetables is crucial to control the fermentation process. In Asia-Pacific, fermented foods help people survive in drastic climatic conditions and also provide several health benefits to the consumers. Apart from this, fermented food facilitates the preservation of easily perishable foods, removal of anti-nutritional compounds, bioavailability of nutrients and minerals, production of antioxidants and enhancement of immunity. However, consumption of most of the fermented foods is declining due to a modern lifestyle and availability of fast foods. This may lead to disappearance of information on the preparation and usage of ethnic fermented foods. This chapter tries to provide important facts about the preparation method and microbial population of several important fermented fruits and vegetables in the Asia-Pacific region.

Acknowledgement

The authors are thankful to Dr Ramesh C. Ray, Principal Scientist (Microbiology), ICAR- Central Tuber Crops Research Institute (Regional Centre), Bhubaneswar 751019, India for helpful suggestions and critically going through the manuscript.

References

Aidoo, K.E., R.M.J. Nout and P.K. Sarkar. 2006. Occurrence and function of yeasts in Asian indigenous fermented foods. *FEMS Yeast Research* 6: 30-39.

Arici, M. and F. Coskun. 2001. Hardaliye: Fermented grape juice as a traditional Turkish beverage. *Food Microbiology* 18: 417-421.

Banaay, C.G.B., M.P. Balolong and F.B. Elegado. 2013. Lactic acid bacteria in Philippine traditional fermented foods. *Lactic Acid Bacteria – R&D for Food, Health and Livestock Purposes*. Pp. 571- 588.

Battcock, M. and S. Azam Ali. 2001. *Fermented Fruits and Vegetables: A Global Perspective. In*: F. A. S. Bull (Ed.), vol. 134, Rome. Pp. 96

Boon-Long, N. 1986. Traditional technologies of Thailand: Traditional fermented food products, in *Traditional Foods: Some Products and Technologies. In*: India: Central Food Technological Institute.

Chang, C.H., Y.S. Chen and F. Yanagida. 2011. Isolation and characterization of lactic acid bacteria from yan-jiang (fermented ginger), a traditional fermented food in Taiwan. *Journal of the Science and Food Agriculture* 91(10): 1746-1750.

Chao, S.H., R.J. Wu, K. Watanabe and Y.C. Tsai. 2009. Diversity of lactic acid bacteria in suantsai and futsai, traditional fermented mustard products of Taiwan. *International Journal of Food Microbiology* 135: 203-10.

Chen, Y.S., F. Yanagida and J.S. Hsu. 2006. Isolation and characterization of lactic acid bacteria from suan-tsai (fermented mustard), a traditional fermented food in Taiwan. *Journal of Applied Microbiology* 101: 125-130.

Chen, Y.S., M.S. Liou, S.H. Ji, C.R. Yu, S.F. Pan and F. Yanagida. 2013a. Isolation and characterization of lactic acid bacteria from yan-tsai-shin (fermented broccoli stems), a traditional fermented food in Taiwan. *Journal of Applied Microbiology* 115(1): 125-132.

Chen, Y.S., M. Miyashita, K.I. Suzuki, H. Sato, J.S. Hsu and F. Yanagida. 2010. *Lactobacillus pobuzihii* sp. nov., isolated from pobuzihi (fermented cummingcordia). *International Journal of Systematic and Evolutionary Microbiology* 60(8): 1914-1917.

Chen, Y.S., H.C. Wu, H.Y. Lo, W.C. Lin, W.H. Hsu, C.W. Lin, P.Y. Lin and F. Yanagida. 2012. Isolation and characterization of lactic acid bacteria from jiang-gua (fermented cucumber), a traditional fermented food in Taiwan. *Journal of the Scientific and Food Agriculture* 92(10): 2069-2075.

Chen, Y.S., H.C. Wu, S.F. Pan, B.G. Lin, Y.H. Lin, W.C. Tung, Y.L. Li, C.M. Chiang and F. Yanagida. 2013b. Isolation and characterization of lactic acid bacteria from yan-taozih (pickled peaches) in Taiwan. *Annals of Microbiology* 63(2): 607-614.

Chen, Y.S., H.C. Wu, C.M. Wang, C.C. Lin, Y.T. Chen, Y.J. Jhong and F. Yanagida. 2013c. Isolation and characterization of lactic acid bacteria from pobuzihi (fermented cummingcordia), a traditional fermented food in Taiwan. *Folia Microbiologica* 58(2): 103-109.

Cho, J., D. Lee, C. Yang, J. Jeon, J. Kim and H. Han. 2006. Microbial population dynamics of kimchi, a fermented cabbage product. *FEMS Microbiology Letter* 257(2): 262-267.

Dahal, N.R., T.B. Karki, B. Swamylingappa, Q. Li and G. Gu. 2005. Traditional foods and beverages of Nepal — A review. *Food Reviews International* 21(1): 1-25.

Demir, N., K.S. Bahçeci and J. Acar. 2006. The effects of different initial *Lactobacillus plantarum* concentrations on some properties of fermented carrot juice. *Journal of Food Processing and Preservation* 30: 352-363.

Dhavises, G. 1972. Microbial studies during the pickling of the shoot of bamboo, *Bambusa arundinacea* Willd and of pak sian, Gynandropis pentaphylla DC. [MS thesis]. Bangkok: Kasetsart Univ.

Di Cagno, R., R., Coda, M. De Angelis and M. Gobbetti. 2013. Exploitation of vegetables and fruits through lactic acid fermentation. *Food Microbiology* 33: 1-10.

Endang, S.R., T.F. Djafar and D. Wibowo. 1996. Lactic acid bacteria from indigenous fermented foods and their anti-microbial activity. *Indonesian Food and Nutrition Progress* 3(2): 26-34.

Endo, A., H. Mizuno and S. Okada. 2008. Monitoring the bacterial community during fermentation of sunki, an unsalted, fermented vegetable traditional to the Kiso area of Japan. *Letters in Applied Microbiology* 47: 221-226.

Erten, H., H. Tanguler and A. Canbaş. 2008. A traditional Turkish lactic acid fermented beverage: Shalgam (Salgam). *Food Reviews International* 24(3): 352-359.

FAO. 1998. Fermented fruits and vegetables — A global perspective. *In*: (vol. 134). Rome: *FAO Agricultural Services Bulletin*.

Feng, M., X. Chen, C. Li, R. Nurgul and M. Dong. 2012. Isolation and identification of an exopolysaccharide-producing lactic acid bacterium strain from chinese paocai and biosorption of Pb (II) by its Exo-polysaccharide. *Journal of Food Sciences* 77(6): T111-T117.

Fukushima, D. 1985. Fermented vegetable protein and related food of Japan and China. *Food Reviews International* 1(1)0: 149-209.

Han, B.J., B.R. Han and H.S. Whang. 1998. Kanjang and doenjang, in one hundred Korean foods that Koreans must know, *Hyun Am Sa*, Seoul. pp. 500-507.

Hong, S.I., Y.J. Kimand and Y.R. Pyun. 1999. Acid tolerance of *Lactobacillus plantarum* from kimchi. *LWT - Food Science and Technology* 32(3): 142-148.

Kabak, B., and A.D.W. Dobson. 2011. An introduction to the traditional fermented foods and beverages of Turkey. *Critical Reviews in Food Science and Nutrition* 51(3): 248-60.

Kawahara, T. and H. Otani. 2006. Stimulatory effect of lactic acid bacteria from commercially available nozawana-zuke pickle on cytokine expression by mouse spleen cells. *Bioscience, Biotechnology and Biochemistry* 70(2): 411-417.

Lan, W.T., Y.S. Chen and F. Yanagida. 2009. Isolation and characterization of lactic acid bacteria from Yan-dong-gua (fermented wax gourd), a traditional fermented food in Taiwan. *Journal of Biosciences and Bioengineering* 108(6): 484-487.

Lee, J.S., G.Y. Heo, J.W. Lee, Y.J. Oh, J.A. Park, Y.H. Park, Y.R. Pyun and J.S. Ahn. 2005. Analysis of kimchi microflora using denaturing gradient gel electrophoresis. *International Journal of Food Microbiology* 102(2): 143-150.

Leisner, J.J., M. Vancanneyt, G. Rusul, B. Pot, K. Lefevre, A. Fresi and L.K Tee. 2001. Identification of lactic acid bacteria constituting the predominating microflora in an acid-fermented condiment (tempoyak) popular in Malaysia. *International Journal of Food Microbiology* 63(1-2): 149-157.

Matthews, P.J. 2000. An introduction to the history of taro as a food. *In*: *Potential of Root Crops for Food and Industrial Resources* (Eds.) M. Nakatani and K. Komaki, Twelfth Symposium of International Society of Tropical Root Crops (ISTRC), 10-16 Sept., 2000, Tsukuba, Japan. Pp. 484-511.

Merican, Z. 1996. Malayasian pickles. In. *Hand book of Indigenous Fermented Food* 2nd edn. Ed. K.H. Stein krous pp. 138-148. NewYork: Marcel Dekker. Inc.

Miyamoto, M., Y. Seto, D.H. Hao, T. Teshima, Y.B. Sun, T. Kabuki, L.B. Yao and H. Nakajima. 2005. *Lactobacillus harbinensis* sp. nov., consisted of strains isolated from traditional fermented vegetables 'suan cai' in harbin, northeastern China and *Lactobacillus perolens* DSM 12745. *Systematic Applied Microbiology* 28(8): 688-94.

Montet, D., G. Loiseau and N. Zakhia-Rozis. 2006. Microbial technology of fermented vegetables. *In*: R. C. Ray & O.P. Ward (Eds.), *Microbial Biotechnology in Horticulture*, vol. 1. pp. 309-343. Enfield, New Hampshire: Science Publishers Inc.

Nguyen, D.T.L., K. Van Hoorde, M. Cnockaert, E. De Brandt, M. Aerts, L. Binh Thanh and P. Vandamme. 2013. A description of the lactic acid bacteria microbiota associated with the production of traditional fermented vegetables in Vietnam. *International Journal of Food Microbiology* 163(1): 19-27.

Ohmomo, S., S. Murata, N. Katayama, S. Nitisinprasart, M. Kobayashi, T. Nakajima, M. Yajima and K. Nakanishi. 2000. Purification and some characteristics of enterocin ON-157, a bacteriocin produced by *Enterococcus faecium* NIAI 157. *Journal of Applied Microbiology* 88: 81-89.

Paramithiotis, S., O.L. Hondrodimou, and E.H. Drosinos. 2010. Development of the microbial community during spontaneous cauliflower fermentation. *Food Research International* 43(4): 1098-1103.

Phithakpol, B., W. Varanyanond, S. Reungmaneepaitoon and H. Wood. 1995. *The Traditional Fermented Foods of Thailand*. Kuala Lumpur: ASEAN Food Handling Bureau.

Puspito, H., and G. Fleet 1985. Microbiology of sayur Asian fermentation. *Applied Microbiology and Biotechnology* 22(6): 442-445.

Rolle, R. and M. Satin. 2002. Basic requirements for the transfer of fermentation technologies to developing countries. *International Journal of Food Microbiology* 75(3): 181-187.

Ross, R.P., S. Morgan and C. Hill. 2002. Preservation and fermentation: Past, present and future. *International Journal of Food Microbiology* 79 (1-2): 3-16.

Sanchez, C.P. 1999. Microorganisms and technology fermented foods of Philippine. *Japanese Journal of Lactic Acid Bacteria* 10(1): 19-28.

Steinkraus, K.H. 1997. Classification of fermented foods: Worldwide review of household fermentation techniques. *Food Control* 8 (5-6): 311-317.

Swain, M.R., M. Anandharaj, R.C. Ray, R. Parveen Rani. 2014. Fermented fruits and vegetables of Asia: a potential source of probiotics. *Biotechnology Research International*. 250424. doi: 10.1155/2014/250424.

Tamang B. and J.P. Tamang. 2007. Role of lactic acid bacteria and their functional properties in goyang, a fermented leafy vegetable product of the Sherpas. *Journal of Hill Research* 20(20): 53-61.

Tamang, B. and J.P. Tamang. 2010. *In situ* fermentation dynamics during production of gundruk and khalpi, ethnic fermented vegetables products of the Himalayas. *Industrial Journal of Microbiology* 50(1): 93-98.

Tamang, B., J.P. Tamang, U. Schilinger, C.M.A.P. Franz, M. Gores and W.H. Holzapfel. 2008. Phenotypic and genotypic identification of lactic acid bacteria isolated from ethnic fermented tender bamboo shoots of northeast India. *International Journal of Food Microbiology* 121: 35-40.

Tamang, J.P. 2009. *Himalayan Fermented Foods: Microbiology, Nutrition and Ethnic Values*. New Delhi: CRC Press.

Tamang, J.P. 2010. *Himalayan Fermented Foods: Microbiology, Nutrition, and Ethnic Values*. New York: CRC Press/Taylor & Francis Group.

Tamang, J.P. and P.K. Sarkar. 1993. Sinki: A traditional lactic acid fermented radish taproot product. *The Journal of General and Applied Microbiology* 39: 395-408.

Tamang, J.P., B. Tamang, U. Schillinger, C.M.A.P. Franz, M. Gores and W.H. Holzapfel. 2005. Identification of predominant lactic acid bacteria isolated from traditional fermented vegetable products of the Eastern Himalayas. *International Journal of Food Microbiology* 105(3): 347-356.

Tanasupawat, S. and K. Komagata. 1995. Lactic acid bacteria in fermented foods in Thailand. *World Journal of Microbiology and Biotechnology* 11(3): 253-256.

Tanguler, H. and H. Erten. 2012. Occurrence and growth of lactic acid bacteria species during the fermentation of shalgam (salgam), a traditional Turkish fermented beverage. *LWT-Food Science and Technology* 46(1): 36-41.

Xiao, Y. 1996. *Chinese Food Culture and Microorganism in Zhong Hua Shi Yuan*, Li, S. China Social Sciences Press, Beijing, China, Pp. 66-75.

Yan, P.M., W.T. Xue, S.S. Tan, H. Zhang and X.H. Chang. 2008. Effect of inoculating lactic acid bacteria starter cultures on the nitrite concentration of fermenting Chinese paocai. *Food Control* 19(1): 50-55.

Westby, A. and D.R. Twiddy. 1992. Characterization of the gari and fu-fu preparation procedure in Nigeria. *World Journal of Microbiology and Biotechnology* 8: 175-182.

10

Regional Fermented Vegetables and Fruits in Europe

Hüseyin Erten[1,*], Bilal Agirman[1], C. Pelin Boyaci Gunduz[1], and Akram Ben Ghorbal[1]

1. Introduction

Fermentation is one of the oldest manufacturing and preservation methods since ancient times (Hutkins 2006). Fermented foods are produced in different parts of the world, both at artisanal and industrial level. Vegetables and fruits are easily perishable, therefore difficult to store (Prajapati and Nair 2003). Fermentation plays an important role in the production of safe and stable foods with a longer shelf-life than its perishable raw material (Settanni and Moschetti 2014). Besides preserving, fermentation shows a strong potentiality to enhance the nutritional value, functionality and sensory properties of the food product (Hutkins 2006).

Lactic acid fermentation is a common method of preserving fresh vegetables and fruits. The micrsoorganisms mainly responsible for the lactic acid fermentation process are the lactic acid bacteria. They metabolize the sugars and produce lactic acid and other end-products. Two fermentative pathways metabolize the hexoses. In the homofermentative pathway, 90 per cent of the sugar is fermented exclusively to lactic acid by some *Lactobacillus*, *Pediococcus*, *Lactococcus* and *Streptococcus*. In heterofermentative pathway, 50 per cent of sugar is converted to lactic acid and the remaining sugar is fermented to CO_2 (carbon dioxide), ethanol and acetic acid by *Leuconostocs*, *Oenococcus*, *Weisella* and some *Lactobacillus* (Hutkins 2006).

Some lactic acid fermented foods, for example, pickles, table olives and sauerkraut are important industrial products produced mostly on a large scale. However, some fermented foods are specific to certain geographic regions. They are minor products in global terms, but are produced

[1] Cukurova University, Faculty of Agriculture, Department of Food Engineering, 01330, Balcali, Adana, Turkey.
* Corresponding author: E-mail: herten@cu.edu.tr

traditionally at home and also at small-scale level in industry. Pickled vegetables and fruits are used as appetizers in every meal. Lactic acid fermented beverages obtained from vegetables and fruits can be consumed as a refreshing drink. In this chapter, some of the regional lactic acid fermented vegetables and fruits in Europe are reviewed.

2. Almagro Eggplant Pickles

Almagro eggplant is a pickled product manufactured under the supervision of the 'Berenjena de Almagro' Protected Geographical Indication (PGI) since 1994. It is exclusive to Spain and more specifically to the town of Almagro and surrounding areas in the province of Ciudad Real. It is made from the fruits of a native variety of eggplant, *Solanum melongena* L. var. *esculetum depressum* or Almagro, without using starter cultures (Sánchez et al. 2000).

The Almagro eggplant pickle is highly seasoned and commonly used as an appetizer. It is considered by the Berenjena de Almagro PGI Regulatory Board as a product with economical importance in the region of Spain due to its high output, reaching 2,000 tons in 2009 (Seseña and Palop 2012).

This eggplant landrace has a small plant and fruit size with multiple inflorescences which often produce several fruits. The Almagro eggplant crop is not repeated in the same land for 10 or 15 years by farmers due to an ample availability of land and also due to isolation of Almagro eggplant fields from other eggplant or vegetable crop fields, leading to a very low incidence of pests and diseases. The fruits are harvested aged only two weeks after pollination and used as raw material in the pickling process (Prohens et al. 2009).

The chemical study of Almagro eggplants from differents cultivars shows that the total available carbohydrate fraction is in the average of 3.53 g/100 g and fiber in the average of 1.33 g/100 g (San José et al. 2013). The soluble sugars in the fruit reach the average of 1.45 g/100 g where the major compounds of this fraction are glucose and fructose, representing an average of 47.6 and 38.6 per cent, respectively. Moreover, fresh fruit materials have a high content of proteins with an average of 0.57 g/100 g, ascorbic acid with an average of 4.09 mg/100 g, as well as a high levels of total phenolic compounds with an average of 59.9 mg/100 g.

The technical aspects of the manufacturing process of Almagro eggplant pickle were first described by Morales Mocino (1985). Years later, regulations established by Berenjena de Almagro Regulating Board were made to describe aspects related to the characteristics of raw material and finished product as well as the production process necessary for any product to be admitted under the PGI. They summarized the pickling process in four steps (Seseña and Palop 2012). The first step is sorting of the fruits by removing the damaged ones and then grading them according to their size. Moreover, a peeling operation is involved to remove the stems and bracts. The second step is blanching of fruits in boiling water with sufficient time to destroy spoilage bacteria and preserve the texture of fruits from softening. In the third step, fermentation of fruits is performed in suitable containers, including brine at room temperature by their indigenous microflora. This step could take from 4 to 20 days and the brine is composed mostly of water, salt, vinegar and some seasonings like cumin and garlic. In the last step, the packaging of the fermented fruits is achieved under hygienic conditions in glass and metallic containers. Later, the packaged products are pasteurized at 65°C.

For preserving the fruit quality during fermentation and also for shortening of the fermentation period, sodium chloride (NaCl) is added to the brine. Seseña and Palop (2007) stated that the optimum salt level in the brine for indigenous fermentation of Almagro eggplant is 4 per cent. Moreover, this level should not exceed 6 per cent in fermentation when the starter culture is used (Ballesteros et al. 1999).

As an alternative to the Almagro fresh fruits, frozen fruits are used in the pickling process to overcome seasonal limitation of the Almagro eggplant fruits in the market. After thawing and blanching, the frozen materials undergo fermentation followed by packaging. The fermentation process can be extended by a few days in comparison to that of fresh materials since the freezing process of fruits affects their microbiota. (Seseña et al. 2002).

The concentration of reducing sugars (glucose and fructose) of fermented Almagro eggplants is related to the size of fruits and particularly the blanching step during processing. Thus the reducing sugars of big-size fruits did not get exhausted even after 19 days of fermentation where its level attains about 2 g/kg is opposite to the case of smaller-size fruits. Significant differences are observed in the lactic acid content, which is a major product of lactic acid fermentation. It is greater in the larger fruits, reaching 8 g/kg of fresh fruit at the end of the fermentation. The pH of the brine of big-size fruits is also higher during fermentation. At the end of the process, it decreases from 4.75 to 4.4 (Seseña and Palop 2012).

The industrial production of Almagro eggplant pickles is based on spontaneous fermentation achieved by the indigenous microbial flora of the fruit. This flora is dominated by *Lactobacillus* species, mainly *Lb. plantarum*, *Lb. fermentum* and *Lb. cellobiosus* and to a less extent by *Lb. pentosus* and *Lb. brevis* (Sánchez et al. 2000, Seseña et al. 2004). The noticeable events of this fermentation include domination of obligate heterofermentative bacteria *Lb. fermentum* in the initial stage and *Lb. plantarum* in the final stage with continuous increase in its population during the fermentation period (Sánchez et al. 2000, 2003). Moreover, Seseña and Palop (2007) showed a significant variation in composition of the lactic acid microbiota of Almagro eggplant fermentation in relation to the manufacturing environment and NaCl concentration. They identified the participation of *Lactococcus lactis, Leuconostoc mesenteroides* subsp. *mesenteroides, Lb. paracasei* subsp. *paracasei* and *Lb. acidophilus* in Almagro eggplant spontaneous fermentation.

In spite of conducting spontaneous fermentation in large-scale production of Almagro eggplant pickles, the process has numerous drawbacks which limits process yields and affects the final product quality. In fact, the microbial load of fresh fruits and the ambient temperature at harvest time, particularly at the end of the season, are factors exerting a major influence on the optimization of fermentation. As an alternative to uncontrolled fermentation and also to enable the use of brines with low salt concentrations without affecting the quality of the end-products, addition of starter is suggested in the pickling medium. A mixture of *Lactobacillus* species, composed mainly of *Lb. plantarum*, was used as a starter culture to produce Almagro eggplant pickles of good quality but with weak bitterness in a shorter fermentation period when compared to spontaneous fermentation (Ballesteros et al. 1999).

3. Brovada

Brovada, which is unique to the northeast region of Italy and known as Italian pickled turnips, is a conventional Italian product made by natural lactic acid fermentation of turnip (*Brassica rapa* L.). It is generally consumed after cooking but can be eaten raw (Maifreni et al. 2004).

The fermentable soluble sugars found in turnip are sucrose (0.21–3.42 per cent), glucose (0.20–1.41 per cent) and fructose (0.14–1.10 per cent) (Rodríguez-Sevilla et al. 1999, Erten and Tanguler 2012).

For the production of brovada, turnips are washed, cleaned and placed into barrels covered with a layer of grape skins. Prior to covering the barrels, water and sometimes salt are added. Fermentation is spontaneously carried out with the activity of lactic acid bacteria and yeasts to a less extent, and these come from turnips, grape skins and fermentation equipments (Maifreni et al. 1999). Fermentation of brovada usually continues for 30–40 days at 12–15°C before the turnips are taken from the barrels. They are peeled, sliced and wrapped in plastic bags to make them ready for the market.

The dominant lactic acid bacteria belong to the genera of *Lactobacillus* and *Pediococcus*. Moreover, the yeast *Candida* spp. also plays a role in the fermentation. However, dominant flora can change depending on the raw material, fermentation temperature and harvest conditions, because of which sensory characteristics of the product can show variations. At the beginning of fermentation, the heterofermantative bacterium *Lb. hilgardii* and the homofermantative bacterium *Pediococcus parvulus* are the main isolates of the fermentation media (Maifreni 1999, 2004). In addition to these bacteria, other lactic acid bacteria, such as *Lb. plantarum, Lb. coryniformis, Lb. maltaromicus, Lb. viridescens* and yeasts like *Hansenula* spp., *Candida* spp.

and *Issatchenkia* spp. were isolated in brovada fermentation (Vaughn 1985, Maifreni et al. 2004).

Lactic acid is the main end-product in brovada fermentation. Sugars are mainly fermented to lactic acid by the metabolism of lactic acid bacteria during fermentation. After four weeks of fermentation, the pH and concentration level of lactic acid are detected as 3.7 and 5 g/100 g, respectively. The flavor compounds determined during brovada fermentation are isoamyl-, propenyl- and 2-phenylethylesters and some other volatile compounds, especially isothiocyanates, which are responsible for the characteristic flavor and aroma of most *Brassica* vegetables (Maifreni et al. 2004).

4. Capers and Caper Berries

Caper is the common species of the genus *Capparis*, family Capparidaceae. It is known by various names in different countries and some of them are Alcaparro in Spain, Kapari in Turkey, Cappero in Italy, Kápparis in Greece, Kabbar in Arab countries and Câpres in France (Tlili et al. 2011b). Caper bush is a perennial shrub plant which is cultivated for its flower buds (capers) and fruits (caper berries) (Pulido et al. 2005, Tlili et al. 2011b, Pulido et al. 2012). The plant has a natural distribution in the coastal regions of the entire Mediterranean Sea basin, right from the Atlantic coast of the Canary Islands and Morocco to the Black Sea to the Crimea and Armenia and eastward to the Caspian Sea and into Iran (Romeo et al. 2007). *Capparis* L. has its maximum diversity in the Mediterranean region with species *Capparis spinosa, C. ovata, C. sicula, C decidua, C. masaikai, C. orientalis* and *C. zoharyi* (Inocencio et al. 2006, Tlili et al. 2011b). *C. spinosa* is the predominant and most important commercial species in the Mediterranean basin (Tlili et al. 2011b). The main producers of the plant are Mediterranean countries, especially Italy, Turkey, Morocco and Spain. The average annual production is estimated to be around 10,000 tons (Pulido et al. 2012).

Capers and caper berries are rich in unsaturated fatty acids and 58–63.5 per cent of total fatty acids found in flower buds and 73 per cent of total fatty acids in fruits are oleic, linoleic and linolenic acids (Vega and Ramos 1987, Rodrigo et al. 1992, Ozcan 1999b). Both capers and caper berries contain vitamins and minerals (Sozzi and Vicente 2006) and are good natural antioxidants with high levels of phenolic compounds. Total flavonoid content, as corresponding aglycones (quercetin and kaempferol), of capers are variable (1.82 and 7.85 mg/g) and the mean concentration is 3.86 mg/g fresh weight. The content of free aglycones, free quercetin and free kaempferol, are in the range of 0.03–1.45 mg/g and 0.1–2.8 mg/g, respectively (Inocencio et al. 2000). It was stated that the highest dietary sources of quercetin are capers (Andres-Lacueva et al. 2010). Tlili et al. (2011a) stated that the phenolic contents of commercial capers collected from different countries ranged from 1151.6 to 2243.96 mg/100 g while rutin contents were in the range of 150.62–732.61 mg/100 g. In the caper, β-carotene and lutein are also present at an important level in addition to the α-tocopherol (Tlili et al. 2010). In the study of Matthäus and Ozcan (2002), glucosinolates were investigated in the young shoots and buds of *C. spinosa* and *C. ovata* species. Glucocapperin was found to be the main glucosinolate with a proportion of 90 per cent of the total glucosinolates. Ozcan and Akgul (1998) referred to the reducing sugar levels of caper buds harvested in June. The results were different, depending on the diameter of the buds and ranging from 3.84 to 4.69 per cent for *C. spinosa* species and ranging from 4.16 to 4.97 per cent for the *C. ovata* species.

Fermented capers and caper berries are greatly appreciated for their unique organoleptic properties. They are traditionally produced at homes, besides being produced industrially on a small scale. Capers are fermented in brine and caper berries are pickled or fermented in brine or in water (Pulido et al. 2012). For the production of fermented caper berries, the harvested fruits in June and July are immersed in tap water and fermented for approximately four to seven days. Fermentation vessels are left out on sunny terraces where the fermentation takes place at a 20–45°C temperature range with a temperature rise and drop during the day and at night (Luna and Peres 1985). After this, fermented capers are placed in brine containing 10 per cent of NaCl and then distributed for consumption (Palomino et al. 2015).

Brine is also used for fermentation instead of plain water since immersion in tap water produces a strong fermentation and leads to color change (from green to yellow) and loss of texture due to flesh breakdown and gas accumulation, which negatively affect the value of the product (Sánchez et al. 1992, Pulido et al. 2012). Ozcan (1999a) suggests that the most suitable brine for lactic acid bacteria activity of caper berries fermentation is in the range of 5 to 10 per cent of NaCl. The best length of fermentation with respect to product color, flavor, acidity, pH and lactic acid bacteria activity in brine is 20 to 25 days. If the fermented product is given to the market directly, fermentations can be performed at low salt content. However, to keep the fermented product in good quality after fermentation in bulk for a long time, it is necessary to store the fermented caper berries in fresh brine containing up to 15 per cent of NaCl at equilibrium. But the desalting process should be done before putting out in the market since they cannot be consumed immediately due to the high salt content. In the study of Ozcan (1999a), pH of fermented caper berries of two species ranged from 4.73 to 4.90, depending on the salt concentration of brine (5 or 10 per cent) and the days of fermentation (20 or 25 days).

In the study of Ozcan and Akgul (1999a), different sized buds of *C. spinosa* and *C. ovata* species were pickled by putting into different salt concentrations of brine at 5, 10, 15 and 20 per cent, for caper fermentation. High salt concentrations inhibit the spoilage microorganisms, but have negative effects on the growth of lactic acid bacteria. It was reported that the most suitable salt concentrations are 5 per cent and partly 10 per cent for activity of lactic acid bacteria. In brines at 5 per cent NaCl, the fermentation must be continuously controlled due to the possibility of unwanted microorganisms growing, such as coliform bacteria, yeasts and moulds. It was stated that small buds of *C. ovata* for pickling have advantages of color and flavor. Pickling time of 40–50 days was suitable for both species regarding flavor, odor, brine acidity, pH and lactic acid bacteria activity (Ozcan and Akgul 1999a). The pH of fermented capers of both species ranged from 4.10 to 4.53, depending on the salt concentration of brine (5 or 10 per cent) and days of fermentation (40 or 50 days). The pH increased with increasing salt concentration (Ozcan and Akgul 1999a). A higher salt concentrations is applied for brining capers. It was reported that 20 per cent of NaCl can be used both as brine or dry salt for the fermentation of capers (Aktan et al. 1998). To decrease the cost of transportation and obtain a more intensive flavor, dry salt treatment is used in Italy. In Spain, large diameter capers are drained and mixed with dry salt in a maximum of 20 per cent (wt/wt) (Sozzi and Vicente 2006). Some researchers used pasteurization (80°C, 15 min.) of the final product to prevent the growth of certain spoilage-causing microorganisms (Alvarruiz et al. 1990, Ozcan and Akgul 1999b). However, pasteurization can cause loss of characteristic flavor and taste of capers and caper berries. In some cases, some bacteriostatics, like sorbic and benzoic acids and their salts, can be used as preservatives during final packaging but the product cannot be claimed to be naturally preserved. In order to overcome the effects of high salt concentration, at the beginning of fermentation, capers could be covered with diluted acetic acid or malt vinegar (4.3 to 5.9 per cent of acetic acid) (Sozzi and Vicente 2006).

Fermentation of capers and caper berries is a traditional process relying on spontaneous growth of naturally-occurring lactic acid bacteria. The predominant lactic acid bacterium of this fermentation is *Lb. plantarum* which is detected during the fermentation period of caper. The other species identified in the fermentation but to a lesser extent, include *Lb. paraplantarum*, *Lb. pentosus*, *Lb. brevis*, *Lb. fermentum*, *Pd. pentosaceus*, *Pd. acidilactici* and *Enterococcus faecium* (Pulido et al. 2005). It was reported that *Lb. fermentum* is detected towards the end of the fermentation process (Pulido et al. 2007).

There is always a risk of stuck fermentation and spoilage as a result of spontaneous fermentation and the starter cultures can be an issue to overcome these problems by controlling the fermentation process. Palomino et al. (2015) isolated *Lb. plantarum* Lb9 from spontaneous fermentation of caper berries and stated that this strain can be a potential candidate as starter culture for fermentation. Using *Lb. plantarum* Lb9 as starter leads to acceleration of lactic acid fermentation and reduces the fermentation risks in addition to standardization of the end-product quality. Consequently, homogeneous products with desirable properties and sensorial quality similar to the spontaneous fermented products can be produced.

5. Cauliflower

Cauliflower (*Brassica oleracea* var. *botrytis* L.) belongs to the family of Brassicaceae and it is one of the vegetables classified as subvarieties of the species *Brassica oleracea* (Beecher 1994, Köksal and Gülcin 2007, Stojceska et al. 2008). *Brassica oleracea* is an extremely diverse species which contains several important vegetable crops such as broccoli, Brussels sprouts, cabbage, kohlrabi, collards and kale other than cauliflower (Beecher 1994, Lo-Scalzo et al. 2008).

Cauliflower has been known for 2000 years in the Mediterranean and Asian regions. It is a cool-season vegetable adapted to humid climate. It is available throughout the year and is particularly abundant in spring and fall. Fermentation of cauliflower is not as popular as cucumber, sauerkraut and kimchi. However, cauliflower fermentation has a broad range of utilisation in some countries according to tradition and availability of the raw material (Paramithiotis et al. 2010, Sanz-Puig et al. 2015). Fermentation of cauliflower is popular in Spain, Italy, France and Greece (FAOSTAT 2007).

Raw cauliflower contains a high content of moisture and also some carbohydrates, fat, protein and dietary fiber. It also includes minerals, mainly potassium, sodium, calcium and iron and vitamins such as folic acid, ascorbic acid, vitamin A and niacin (Bhattacharjee and Singhal 2011). It is rich in antioxidant phytochemicals such as flavonols and hydroxycinnamic acids like other cruciferous vegetables. Cauliflower also has glucosinolates and caretonoids to a limited extent (Hertog et al. 1995, Heinonen and Meyer 2002, Podsedek 2007, Lo-Scalzo et al. 2008, Sikora et al. 2008, Volden et al. 2009). Some of the important phytochemicals found in cauliflower are; glucoraphanin, sulphoraphane, glucobrassicin, synapic acid, lutein, kaempferol, 3-methyl-thio-propyl isothiocyanate and 4-methyl-thio-butyl isothiocyanate (Picchi et al. 2012). Total phenol content of cauliflower is about 270 mg of gallic acid per 100 g of dried edible portion (Wu et al. 2004). The amount of ascorbic acid determined in different cauliflower genotypes is between 259.6–638.8 mg/100 g (Lo-Scalzo et al. 2007, Picchi et al. 2012).

To produce cauliflower pickle, whole cauliflower is chopped into small pieces and transferred to fermentation vessels filled with wheat grains. Then it is filled with brine of 8 per cent NaCl. Olive oil is added to the surface of the brine solution and left to ferment at 20°C for 20-30 days. Due to a fairly restricted market, cauliflower pickle is produced in household or by small entrepreneurs (Volden et al. 2009, Paramithiotis et al. 2010, 2012). The pH value of cauliflower pickle and brine is about 3.8–3.9 at the end of the spontaneous cauliflower fermentation. Total titratable acidity as lactic acid in cauliflower pickle and brine is in the percentage of 0.60 and 0.65 at the end of the fermentation, respectively (Paramithiotis et al. 2012).

Cauliflower fermentation is usually carried out spontaneously by lactic acid bacteria. The dominant bacteria in the beginning are mainly heterofermantative *Ln. mesenteroides*, followed by homofermantative strains like *Lb. plantarum*. In addition to the two bacteria, *Lb. pentosus, Lb. paraplantarum, Lb. hilgardii, Lb. spicheri* and *Lb. fabifermentans* participate in cauliflower fermentation (Anonymous 2005, Klaenhammer et al. 2005, Paramithiotis et al. 2010). Paramithiotis et al. (2010) stated that strains belonging to *Ec. faecium*-group and *Ec. faecalis*-group were also isolated in the early stages of fermentation.

6. Lactic Acid Fermented Vegetable Juices

Vegetables are important compounds of high nutritional value. By fermentation and processing of vegetables, vegetable juices can be preserved using lactic acid bacteria. Juices of vegetables, such as carrot, turnip, tomato, onion, sweet potato, cabbage, beet, pumpkin, cabbage and celery are produced (Aukrust et al. 1994, Montaňo et al. 1997, Karovicova et al. 1999, Karovicova and Kohajdova 2002, Karovicova et al. 2002a, 2002b, Rakin et al. 2003, 2004a, 2004b, 2005, 2007, Baráth et al. 2004, Klewicka et al. 2004, Yoon et al. 2004, 2005, 2006, Bergqvist et al. 2005, Kohajdová et al. 2006, Demir et al. 2006, Nazzaro et al. 2008, Buruleanu et al. 2009, Koh et al. 2010).

Lactic acid fermented vegetable juices are produced through two different methods. According to the first method, fermentation is carried out. Then the juice of the fermented product is obtained by pressing. Another method is firstly to process the vegetable to obtain the juice. Then it is pasteurized and fermented (Swain et al. 2014). Fermentation can be conducted spontaneously or by addition of the starter culture under controlled fermentation. Using starter culture is important, especially for fermentation of vegetable juices. The majority of lactic acid fermented vegetable juices are manufactured according to the 'lactoferment process' and lactic acid fermented juices are referred as 'lacto-juice' (Swain et al. 2014).

6.1. Beetroot Juice

Beetroot, *Beta vulgaris* L., is also known as red beet, table beet, garden beet or blood turnip. It is a well-known vegetable not only in Europe, but also all over the world. It is especially used as pickles, salad or juice. Red beet includes water-soluble nitrogenous substances, particularly betalain which is considered as a radical scavenger with antioxidant, anti-inflammatory and anti-carcinogenic activities (Kapadia and Rao 2013, Lee et al. 2014, Vanajakshi et al. 2015). The red color of beetroot is derived from betacyanin which belongs to the Betalain family. The Betalain family is divided in two groups according to the color: the betacyanins (red/purple) and betaxanthins (yellow) where betanin (red) and vulgaxanthine I (yellow) are the main Betalain components in beetroot, respectively (Sekiguchi et al. 2013). In red beet, the average content of Betalains is estimated as 1000 mg/100 g of total solids or 120 mg/100 g of fresh weight (Marmion 1991). The concentration of betanin, chemically betanidin 5-O-β-glucoside, which acts as an antioxidant in beetroot, varies from 300–600 mg/kg, whereas isobetanin, betanidin and betaxanthins are found in lower concentrations (Kanner et al. 1996, 2001, Shahidi et al. 2011). The other important compounds present in redbeet are sugars, vitamins, particularly folic acid, B_1, B_2, B_6 and ascorbic acid, minerals such as calcium, iron, potassium and magnesium and also carotenoids, saponins, polyphenols and flavanoids (Wang and Goldman 1997, Atamanova et al. 2005, Váli et al. 2007, Dias et al. 2009, Sárdi et al. 2009, Kazimierczak et al. 2014, Singh and Hathan 2014, Wruss et al. 2015). Red beet contains about 10 g/100 g carbohydrates which are mainly fermentable sugars (Titchenal and Dobbs 2004). The average concentration of fermentable sugars in different red beet varieties is 77.5 g/l. The concentrations of sucrose, glucose and fructose are 73.5, 2.62 and 1.51 g/l, respectively (Wruss et al. 2015).

The root of the plant is taken in its fresh form, pickled, cooked or used as a salad ingredient after it is sliced or diced, besides consumption as juice. Beetroot juice is produced as a probiotic beverage by small entrepreneurs and is referred as beet 'Kvass' in Russia, besides Europe (Anonymous 2012). Beetroot is also used as a natural colorant in the food, cosmetic and drug industries due to the Betalains (Klewicka and Czyżowska 2011, Zalán et al. 2012, Kapadia and Rao 2013).

Beetroot pickle is produced after cleaning the roots and then slicing and placing them in a container of brine. Beetroot or its juice can be fermented spontaneously or by addition of starter culture. Lactic acid bacteria are naturally found in beetroot. However, their number is usually very low during fermentation which could influence negatively the achievement of the process, thus affecting the quality of the end product. Therefore, for controlled fermentation, strains from the *Lactobacillus* genus are the most suitable starter cultures for lactic acid fermentation of beetroot (Zalán et al. 2012). It was reported that the cultures of *Lb. plantarum* and *Lb. acidophilus* have good biochemical activity on fermentation of the mixture of beetroot juice, brewer's yeast autolysate and carrot juice. However, the first strains show a better growth and lactic acid production than the second one. Enrichment of vegetable juices with brewer's yeast autolysate before lactic acid fermentation affects the number of lactic acid bacteria during fermentation (Rakin et al. 2004a, 2004b, 2007). Effect of starter culture on Betalain pigments is important since they are the most important bioactive component of red beet (Rakin et al. 2004a). Fermentation has an effect on the content and stability of Betalain and its derivative pigments, depending on the pH, fermentation conditions and the organism used (Manchali et al. 2013).

Four species of lactic acid bacteria (*Lb. acidophilus*, *Lb. casei*, *Lb. delbrueckii* and *Lb. plantarum*) were used as starters and all of them utilized the beet juice for cell growth and lactic acid production. However, *Lb.*

acidophilus and *Lb. plantarum* produced a greater amount of lactic acid than other cultures and both strains reduced the pH of beet juice from an initial value of 6.3 to lower than 4.5 after 48 hours of fermentation (Yoon et al. 2005). In the study of Klewicka and Czyżowska (2011), beet juice was fermented with probiotic *Lb. brevis* 0944 and *Lb. paracasei* 0920 bacteria. The number of cells for these starters exceeded the minimum number, 6-7 log cfu/ml of juice (Ishibashi and Shimamura 1993, Reid et al. 2001),and requested for probiotic juice label, by reaching 6.80 log cfu/ml of fermented beet juice after 180 days of storage (Klewicka and Czyżowska 2011).

6.2. Tomato Juice

Another commonly consumed juice in Europe is the juice of tomato, a plant that grow easily in the Mediterranean climate (Willcox et al. 2003, Ramandeep and Savage 2005).

Taste of tomato is mostly related to the concentration of sugars, acids and their ratio (Causse et al. 2010, Siddiqui et al. 2015). Glucose and fructose are the dominant sugars in tomato fruits and generally found at equal quantities (Beckles 2012). Because of the high activity of invertase, sucrose is not detectable or is present in low amounts in the cultivated tomato fruit (Beauvoit et al. 2014). Citric acid is the main organic acid of tomato followed by malic and oxalic acids (Fernandez-Ruiz et al. 2004, Siddiqui et al. 2015). The most characteristic antioxidant component of tomato is lycopene which is responsible for the red color of the mature fruit (Siddiqui et al. 2015). In addition to lycopene, another carotenoid present in tomato fruit is β-carotene (Cortes-Olmos et al. 2014). The mean glucose, fructose, citric acid, lycopene, β-carotene and ascorbic acid compositions of fresh tomato varieties are as follows — 8.84 g/kg, 12.96 g/kg, 3.56 g/kg, 45.71 mg/kg, 8.93 mg/kg and 190.8 mg/kg, respectively (Figas et al. 2015). Other health-promoting related components of tomato are flavonoids, vitamin E, folic acid, potassium and provitamin A (Lavelli et al. 2000, Sahlin et al. 2004, Sánchez-Moreno et al. 2006, Jacob et al. 2008).

Tomato juice production in the world is performed via thermal treatment without fermentation (Bates et al. 2001, Sánchez-Moreno et al. 2006, García-Alanso et al. 2009). On the other hand, some studies show use of starter cultures in the production of fermented tomato juice, though there is no industrial production of fermented product (Kohajdová et al. 2006, Tsen et al. 2008, Di Cagno et al. 2008, Koh et al. 2010).

For the production of tomato juice, fresh tomatoes are washed and dried at room temperature. Vapour is applied for 3–4 min. to make peeling process easier. After that the tomato fruit is grounded and filtered by pulper to separate the fruit seed and peel from the juice. Then sugars (sucrose, glucose, fructooligosaccharide) can be added at a certain concentrations prior to fermentation. Tomato juice is homogenized and then heated at 70°C for 5 min. (Kohajdová et al. 2006, Tsen et al. 2008). After it is cooled to 25°C, selected probiotic strains are added at inoculum level of 7 to 9 log cfu/ml. Fermentation can be carried out anaerobically for 17–80 hours at 25 to 37°C. This probiotic tomato juice is aseptically filled to the glass or tetrapak package materials (Plaza et al. 2003, Tsen et al. 2008, Koh et al. 2010).

It was reported that some probiotic species of lactic acid bacteria, such as *Lb. plantarum*, *Lb. acidophilus*, *Lb. casei*, *Lb. delbrueckii* are used for the production of probiotic tomato juice (Yoon et al. 2004, Tsen et al. 2008). Yoon et al. (2004) determined that *Lb. acidophilus* LA39, *Lb. plantarum* C3, *Lb. casei* A4 and *Lb. delbrueckii* D7 survived in fermented tomato juice with high acidity and low pH. Results indicated that viable cell counts of the four lactic acid bacteria in fermented tomato juice ranged from 6 to 8 log cfu/ml after four weeks of cold storage at 4°C and this beverage could serve as a probiotic beverage for consumers.

6.3. Some Other Vegetable Juices

Studies on other lactic acid fermented vegetable juices, such as carrot juice were conducted by some researchers (Demir et al. 2006, Kun et al. 2008, Nazzaro et al. 2008). Carrot (carotene group) is a rich source of carotene, especially β-carotene. Contents of β-carotene in different carrot varieties are in the range of 1.16–6.43 mg/100 g (Yahia and Ornelas-Paz 2010). Carrot contains carbohydrates up to 10.1 g/100 g and sugars is the main carbohydrate in the range of 5.4–7.5 g/100 g of raw vegetable (Titchenal and Dobbs 2004).

In a study by Demir et al. (2006), lactic acid fermented carrot juice was produced using *Lb. plantarum* starter culture. Optimum fermentation time and inoculum level of starter were investigated. Fermentation time of 15–16 hours is recommended for a pH value under 4.5. Starter culture addition (5.47 log cfu/g) produced better results for a fermented carrot juice.

Probiotic carrot juice was produced using probiotic strains of *Lb. rhamnosus* and *Lb. bulgaricus* for fermentation with prebiotic components, inulin and fructooligosaccharides. The resulting functional beverage showed a good antioxidant activity in comparsion to unfermented carrot juice. The results showed that fermented carrot juice was suitable as the basis of a complex functional product with higher added value (Nazzaro et al. 2008). In a study by Kun et al. (2008), probiotic carrot juice was produced using *Bifidobacterium* strains (*Bf. lactis* Bb-12, *Bf. bifidum* B7.1 and B3.2). Carrot juice contained 4.3–4.6 per cent of total sugars with more than 2 per cent (w/v) of sucrose, 1 per cent (w/v) of glucose and 0.8 per cent (w/v) of fructose. During fermentation the total soluble sugars decreased from 4.36 per cent (w/v) to 3.8 per cent (w/v) in the case of *Bf. lactis* Bb-12 and *Bf. bifidum* B3.2 and to 3.88 per cent (w/v) in the case of *Bf. bifidum* B7.1. Results showed that all the three *Bifidobacterium* strains had biochemical activities in carrot juice without addition of any nutrients.

Cabbage juice, including various levels of onion juice were fermented using *Lb. plantarum* strain. Most of lactic acid was produced in cabbage juice after the addition of 0.1 per cent of onion juice, giving the most acceptable combination for consumers depending on the sensory characteristics (Kohajdová and Karovičová 2005).

Gardner et al. (2001) stated that a starter mixture of *Lb. plantarum* NK-312, *Pd. acidilactici* AFERM 772 and *Ln. mesenteroides* BLAC can be used for fermentation of cabbage, carrot and beet-based vegetable products. In their study, starter addition gave a fermentation pattern similar to a spontaneous fermentation, but with a lower ethanol production and a faster acidification rate.

Di Cagno et al. (2008) identified *Ln. mesenteroides, Lb. plantarum, W. soli/W. koreensis, Ec. faecalis, Pd. pentosaceus* and *Lb. fermentum* from raw carrots, French beans and marrows. For fermentation of carrots, French beans and marrows, *Lb. plantarum, Ln. mesenteroides* and *Pd. pentosaceus* were selected as the autochthonous mixed starter due to the growth and acidification rates in vegetable juice media. Fermentation results showed that fermented vegetables with autochthonous starter were preferable (Di Cagno et al. 2008).

In a study by Yoon et al. (2006), probiotic cabbage juice was produced using three probiotic lactic acid bacteria (*Lb. plantarum* C3, *Lb. casei* A4 and *Lb. delbrueckii* D7). During cold storage at 4°C, both *Lb. plantarum* and *Lb. delbrueckii* were capable of surviving in low pH and high acidic conditions in fermented cabbage juice. On the other hand, *Lb. casei* did not survive in low pH and high acidity in fermented cabbage juice and died after only two weeks of cold storage (Yoon et al. 2006).

Cabbage, tomato, pumpkin and courgette were tested to prepare vegetable juices using *Lb. plantarum*. The pH value of resulted juice was low (3.6–3.75) to provide a good preservation effect. Due to lactic acid production and adequate pH reduction, all the vegetables were suitable substrates for the production of lactic acid fermented juice, especially, juice from cabbage and courgette due to their sensory results (Kohajdová et al. 2006).

In a study on the production of probiotic fruit juice, *Lb. casei* and *Lb. delbrueckii* were used for fermentation of peach juice. Results showed that *Lb. delbrueckii* was an appropriate starter culture to produce a probiotic beverage (Pakbin et al. 2014).

In Europe, there are some commercial probiotic juices, which are produced by adding probiotic lactic acid bacteria directly to the product. They are generally cereal-based products but some include fruits and vegetables too. Some of them are Gefilus® with *Lb. rhamnosus* GG (Valio Ltd., Finland), Bioprofit® with *Lb.* rhamnosus GG and *Propionibacterium freudenreichii* ssp. *shermanii* JS (Valio Ltd., Finland), Biola® with *Lb. rhamnosus* GG (Tine BA, Norway), Friscus® with *Lb. plantarum* HEAL9 and *Lb. paracasei* 8700:2 (Skånemejerier, Sweden) and Rela® with *Lb. reuteri* MM53 (Biogaia, Sweden) (Prado et al. 2008, Corbo et al. 2014).

7. Red Cabbage

Red cabbage (*Brassica oleracea* L. var. *capitata* f. *rubra*) is an indigenous vegetable grown in many regions, including the Mediterranean and southwest Europe. Red cabbage belongs to the family of Brassicaceae (FAOSTAT 2007, Arapitsas et al. 2008, Volden et al. 2008).

The main soluble fermentable sugars of fresh red cabbage are glucose (1.58 g/100 g), fructose (1.32 g/100 g) and followed by sucrose (0.76 g/100 g). Moreover, red cabbage is rich in phytochemicals, such as phenolics, vitamins, including vitamin A, ascorbic acid and vitamin K, glucosinolates and anthocyanin as well as minerals like potassium and manganese, which have a positive impact on human health. Also, red cabbage has high a amount of β-carotene, which has antioxidant benefits (Charron et al. 2007, Podsedek 2007, Volden et al. 2008, Wei et al. 2011).

Red cabbage is an attractive vegetable for consumers because of its taste and attractive red color with addition of nutritional value (Wiczkowski et al. 2015). Red cabbage has its own anthocyanin character and contains remarkable amounts of anthocyanin, ranging from 72 to 234 mg/100 g fresh weight with a mean level of 151 mg/100 g fresh weight (Tendaj et al. 2013). Red cabbage is rich in cyanidine with a content of around 72.9 mg/100 g fresh weight (Andres-Lacueva et al. 2010). Other major anthocyanins in red cabbage are pelargonidin glucoside, peonidin glucoside, cyanidin as aglycon, cyanidin-3-digucoside-5-glucoside and its acylated with sinapic acid forms (Arapitsas et al. 2008, Lin et al. 2008, Wiczkowski et al. 2013).

Red cabbage can be consumed fresh in salad or as sauerkraut. Sauerkraut is naturally lactic acid fermented cabbage. Even though sauerkraut traditionally is produced from white and green cabbage, it can also be made from red cabbage. The red cabbage has bluish-purple color at the beginning of the fermentation, but turns reddish-purple after processing due to acid formation (Ayotte 2013, Wiczkowski et al. 2013).

In the production of fermented red cabbage, first the outer leaves and core are removed, then red cabbage is divided into four segments and each segment is shredded by using the food processor or the food grinders. Then dry salt is sprinkled on the chopped cabbage at the percentage of 0.6–2.0, depending on the consumer demand. Then the mixture is placed in fermentation tanks and left overnight for complete removal of water till the cabbage is wholly soaked in the expelled water. The spontaneous fermentation is performed at 25–28°C for 15 days in fermentation vessels. After fermentation, the fermented red cabbage is stored at 5–6°C (Yoon et al. 2005, Ayotte 2013, Wiczkowski et al. 2015).

Red cabbage is a good source of dietary phenolic compounds among *Brassica oleracea* vegetables. The total phenolic content of red cabbage is sixfold higher than that of white cabbage (Podsedek et al. 2006). Phenolic acid contents are 0.03 and 0.05 mg/g dry weight for vanillic acid, 0.04 and 0.18 mg/g dry weight for chlorogenic acid, 0.04 and 0.18 mg/g dry weight for caffeic acid, 3.05 and 0.26 mg/g dry weight for p-coumaric acid, 2.15 and 1.37 mg/g dry weight for sinapic acid and 0.06 and 0.03 mg/g dry weight for rosmarinic acid in fresh and fermented red cabbage after 38 days of fermentation, respectively. Titratable acidity as lactic acid is at the percentage 1.5 and pH value ranges from 3.5 to 4 after 38 days of fermentation (Hunaefi et al. 2013).

Lactic acid bacteria that are involved in cabbage fermentation are *Lb. plantarum*, *Ln. mesenteroides*, *Lb. casei*, *Lc. lactis*, *Ec. faecalis* and *Lb. zeae* (Rodríguez et al. 2009, Buruleanu et al. 2013, Olszewska et al. 2015).

8. Sea Fennel

Sea fennel, *Crithmum maritimum* L., is a halophyte belonging to the Apiaceae family and known with different names as rock samphire, crest marine, marine fennel and sampier. It grows spontaneously on maritime rocks, rock crevices, rocky shores, piers, breakwaters and sandy beaches. It can be seen along the coast of Mediterranean and Black Sea and also along the Atlantic coast of Portugal, south and south-west England, Wales and southern Ireland (Renna and Gonnella 2012).

Sea fennel is a highly branched perennial herb growing up to 60 cm in height (Cornara et al. 2009). It has young green leaves and large yellow flowers. Young branches and fresh leaves of sea fennel are used in many European countries as pickle, condiment, seasoning and also as a salad ingredient because of its salty flavor and a pleasant taste (Ozcan 2000).

The plant contains high levels of ascorbic acid and has been traditionally eaten by sailors as a protection against scurvy (Cunsolo et al. 1993, Atia et al. 2011). Sea fennel is also rich in carbohydrates especially sucrose and glucose, phenolic compounds, minerals, organic acids and some bioactive substances; it is antioxidant and antibacterial (Ozcan 2000, Ruberto et al. 2000, Rossi et al. 2007, Meot-Duros et al. 2008, Meot-Duros and Magné 2009). The leaves contain high concentrations of fatty acids (Guil-Guerrero and Rodríguez-Garcia 1999).

To make production of pickles out of sea fennel, fresh leaves are washed and put into boiling water for a few minutes. Then the leaves are placed in a jar containing brines with water, vinegar, garlic, olive oil and salt. Fermentation is conducted at room temperature for one to three weeks. In the study by Ozcan (2000), two sea-fennel samples were fermented in two different brines. The first brine included 8 per cent of salt and the second contained 8 per cent salt, one per cent sugar and one per cent yogurt as starter. On the 25th day of fermentation, the pH fell to almost 5 and 3.7 in the first and second brines, respectively. It was reported that activity of lactic acid bacteria was significantly high in the second brine, containing yogurt, when compared to the brine without it. Acidity as lactic acid was almost 0.3 per cent after fermentation in the second brine. Moreover, it was determined that the best fermentation time was 15 to 25 days based on odor, flavor, pH and lactic acid bacteria activity in the end-product (Ozcan 2000).

9. Shalgam

Shalgam (şalgam) is a traditional lactic acid fermented Turkish beverage produced by lactic acid fermentation of black carrot (*Daucus carota* L.), sourdough, salt, bulgur flour, turnip (*Brassica rapa* L.) and sufficient quantity of drinking water. Shalgam has red color and is a cloudy and sour beverage. Shalgam is commonly produced in the Cukurova region of Turkey. Though traditionally made at home it is also manufactured commercially on small and large scales (Erten et al. 2014, Tanguler et al. 2014, 2015, Erten and Tanguler 2016).

During the shalgam production, roots of black carrot are used as a basic raw material for the production of lactic acid, which is the dominant end-product in sugar fermentation and also to obtain the red color of shalgam beverage (Canbas and Deryaoglu 1993, Erten and Tanguler 2012). Total sugar percentage varies from 5.12 to 7.09 in black carrot. The major soluble fermentable sugars of black carrot grown in Turkey are sucrose (1.20–4.36 g/100 g), glucose (1.10–5.60 g/100 g) and fructose (1.00–4.36 g/100 g) (Kammerer et al. 2004, Erten et al. 2008, Tanguler 2010).

Commercially there are two main processing methods for production of shalgam — the traditional and the direct method. The traditional method includes two stages; the first (dough) fermentation and the second (carrot or main) fermentation, while in the direct method, the first fermentation is not carried out (Erten and Tanguler 2012, 2016).

In the traditional method, the purpose of first fermentation (sourdough fermentation) is to provide enrichment of lactic acid bacteria and yeasts. The dough is prepared from a mixture of bulgur flour (at the percentage of 3), rock salt (at the percentage of 0.2), sourdough (at the percentage of 0.2) and adequate water. Then it is kneaded and left to ferment for three to five days at room temperature. After first fermentation, the dough is extracted with adequate water about three to five times. During the first fermentation, acidity rises significantly and pH drops as a result of the activities of lactic acid bacteria mainly and yeasts to a lesser extent. The extract obtained from the first fermentation makes a good start for the second (carrot, main) fermentation. On the other hand, some producers do not extract the fermented mixture of bulgur flour and sourdough with water. After performing the first fermentation, the dough is placed inside a fabric bag and hung in the fermentation tank to start the second fermentation. If extraction is done, the extracts are combined with sorted and chopped black carrots (at the percentage of 15–20), salt (at the percentage of 1.5), sliced turnip (at the percentage of 2) and adequate water in a tank to perform the second fermentation. At the end of fermentation, a red colored, sour beverage is obtained. Fermented juice can also be seasoned by adding chili powder. Generally fiberglass, plastic or stainless steel tanks are used for fermentation of shalgam (Erten et al. 2008, Erten and Tanguler 2012, Tanguler and Erten 2012a, b).

In the direct method, sourdough fermentation is not carried out. Black carrots are sorted out and cut into small pieces. The pieces of black carrots (at the percentage of 15–20), salt (at the percentage of 1–2), sliced turnips (at the percentage of 1–2), bakers' yeast (*Saccharomyces cerevisiae*) or sourdough (at the percentage of 0.2) and adequate water are transferred to a tank to facilitate lactic acid fermentation, which is called the main (carrot) fermentation (Erten and Tanguler 2012, 2016).

Generally, the total fermentation period varies between 13–25 days according to the temperature which too can vary between 10–35°C and other factors. The fermented product is removed from the vessel and marketed in sealed bottles or in plastic containers (Tanguler and Erten 2012a, Tanguler et al. 2014).

The microbiology of shalgam is complex and fermentation occurs naturally wherein lactic acid bacteria and yeasts to a lesser extent, are involved (Erginkaya and Hammes 1992, Arici 2004, Erten and Tanguler 2012, 2016). Lactic acid bacteria and yeasts come from the surface of raw materials, surface of vessels where it is produced and stored and also from the sourdough extract (Erten et al. 2008, Erten and Tanguler 2012). During the shalgam fermentation, the counts in total mesophilic aerobic bacteria, yeasts and non-*Saccharomyces* yeasts usually decline, while the dominant microflora, lactic acid bacteria, incease (Tanguler 2010, Agirman 2014).

Lb. plantarum is the dominant lactic acid bacteria species during fermentation in the traditional method. In addition, *Lb. paracasei* subsp. *paracasei* is dominant bacterium with a lesser growth rate when compared to *Lb. plantarum*. *Lb. fermentum*, *Lc. lactis*, *Lb. brevis*, *Lb. arabinosus*, *Ln. mesenteroides* subsp. *mesenteroides*, *Pd. pentosaceus* and *Lb. delbrueckii* subsp. *delbrueckii* are also involved in shalgam fermentations. The most promising bacteria for industrial usage as a starter culture are *Lb. plantarum*, *Lb. fermentum* and *Lb. paracasei* subsp. *paracasei* (Tanguler 2010, Tanguler and Erten 2011, 2013, Erten and Tanguler 2012, 2016).

The basic end-product of shalgam fermentation is lactic acid, which is responsible for the total acidity. The lactic acid helps to preserve the beverage and provides taste and aroma. Its concentration ranges from 5.18 to 8.26 g/l in shalgam (Canbas and Deryaoglu 1993, Tanguler 2010, Erten and Tanguler 2012, 2016). During fermentation, mainly lactic acid bacteria give shalgam its typical taste and flavor while producing lactic acid, ethanol and other organic compounds. Some volatile compounds have been identified in shalgam and these include carbonyl compounds, volatile acids, higher alcohols, esters, terpenols, norisoprenoids, lactones and volatile phenols, which contribute to the overall flavor of shalgam (Tanguler 2010, Erten and Tanguler 2012).

10. Cranberrybush

Viburnum opulus L., which is widely known as cranberrybush in the world, belongs to the plant family Adoxaceae (formerly Caprifoliaceae). It is called 'Gilaburu' in Turkey and is also known by different names, such as Guelder rose, Crampbark and European Cranberrybush (Kalyoncu et al. 2013). *Viburnum opulus* L. is native to eastern, northeastern, western and central part of Europe and Anatolia (Altan and Maskan 2004, Cesoniene et al. 2012, Yilmaztekin and Sislioglu 2015). The genus *Viburnum* L. comprises over 230 species and *Viburnum opulus* is primarily known through its various cultivars grown in north Asia, northwest Africa, North America and the central zone of western Russia (Velioglu et al. 2006). In Turkey, it is mostly grown around Kayseri city in central Anatolia (Cam et al. 2007, Dinc et al. 2012, Yilmaztekin and Sislioglu 2015). Cranberrybush grows fast to reach a height of 1.3–4 meters. It has showy creamy-white flowers in spring and red berries throughout the autumn. Some species have edible fruits which are red drupes and utilized in the form of dried fruits, pickle and jam (Sonmez et al. 2007, Ozrenk et al. 2011, Kalyoncu et al. 2013).

The fruit is a source of various biochemical components, which include phenolic compounds like flavanoids, including (+) catechin, (–) epicatechin, quercetin glycosides, anthocyanins and proanthocyanins and chlorogenic and 3-methyl-butanoic acids. The other compounds identified are ascorbic acid, some organic acids like malic acid and tartaric acid, minerals and carotenoids (Velioglu et al. 2006, Zayachkivska et al. 2006, Cam et al. 2007, Gavrilin et al. 2007, Cesoniene et al. 2008, Rop et al. 2010, Ozrenk et al. 2011). In the study of Ozrenk et al. (2011) tartaric and malic acid contents of cranberrybush fruits were determined at 1.24–1.41 and 1.21–1.37 g/kg, respectively. Catechin is the most abundant component among all the phenolic compounds and ranges from 284.96 to 352.04 mg/kg (Ozrenk et al. 2011). In the study by Akbulut et al. (2008), the energy level and the reducing sugar, protein, cellulose, oil, ash, acidity, ascorbic acid, total phenolics, total anthocyanin and soluble solid matter contents of cranberrybush fruits were found to be 256.56 kJ/g, 63.46 g/kg, 64.85 mg/kg, 180.71 g/kg, 6.70 g/kg, 12.83 g/kg, 17.92 g/kg, 595.24 mg/kg, 3253.87 mg/kg, 654.23 mg/kg and 104.31 g/kg, respectively. Glucose, fructose and sucrose contents of fruits were in the range of 2.34–2.42, 1.59–1.67 and 0.064–0.069 g/100 g, respectively (Ozrenk et al. 2011).

The juice obtained from the spontaneously fermented cranberrybush fruits is a traditional drink in central Anatolia region. After collection, the fruits, are washed and placed in plastic jars containing water and allowed to ferment in the dark at room temperature for about three to five months. The fruits ripen at the end of this process. After fermentation, the fruit juice is obtained by squashing the fruits and diluting it with water (1:4) before drinking. The juice is not acceptable organoleptically due to its astringent taste which leaves a tactile sensation (Velioglu et al. 2006) and consumed by adding some water and sugar, if desired. The long fermentation and additon of water or sugar reduce the astringency. It is known that the fermentation conditions, especially time and temperature, directly affect the taste of the juice. Therefore, there is a need to standardize the production process to enhance the taste and quality of the product (Soylak et al. 2002, Sonmez et al. 2007, Sagdic et al. 2014).

It was reported that the number of lactic acid bacteria vary and are quite high in the juice samples. Sagdic et al. (2014) identified 332 isolates belonging to *Lactobacillus* and *Leuconostoc* species in the fermented cranberrybush drink. The lactic acid bacteria counts in the fermented fruit juice are in the range of 3.92-8.30 log cfu/ml. The major lactic acid bacteria are *Lb. plantarum*, *Lb. casei* and *Lb. brevis*; the other lactic acid bacteria strains isolated include *Ln. mesenteroides*, *Lb. hordei*, *Lb. paraplantarum*, *Lb. cornyformis*, *Lb. buchneri*, *Lb. parabuchneri*, *Lb. pantheris*, *Ln. pseudomesenteroides* and *Lb. harbinensis*.

The pH and total acidity of fermented juices range between 2.40–2.69 and 1.59–2.23 per cent as lactic acid, respectively (Yapar 2008). Reducing sugar content decreases in fermented samples from 60.39–77.42 to 13.52–55.14 g/kg. After fermentation, 58 volatile compounds including acids (10 compounds), alcohols (27 compounds), aldehydes (3 compounds), esters (3 compounds), ketones (12 compounds) and lactones (3 compounds) were determined in the fermented juice samples (Yilmaztekin and Sislioglu 2015).

11. Hardaliye

Hardaliye is a traditional grape-based non-alcoholic beverage. It is very popular in Thrace of Marmara region of Turkey (Altay et al. 2013). It is produced by fermentation of red grapes (*Vitis vinifera* L.) or grape juice which are rich in highly bioavailable antioxidants, including phenolic compounds, pigments and ascorbic acid (Arici and Coskun 2001). Phenolic compounds in grapes are found in high amounts which are in the range of 1.61–10.85 g/kg (Kelebek 2009). Phenolic acids, stilbenes, principally resveratrol and flavonoids, which include flavonols, tannins and anthocyanins are found in grapes and grape products (Rice-Evans et al. 1995, Downey et al. 2006, Yahia 2010, Amoutzopoulos et al. 2013). The prime sugars in grapes are glucose, fructose and to a lesser extent sucrose while in ripe grapes, the sugar content can vary from 18 to 26 g/100 g, depending on the cultivar and degree of ripeness. In unripe grapes, glucose is found in high concentrations but during ripening, its amount decreases and fructose concentration increases. In ripe grapes, the fructose and glucose concentrations are close to each other and the ratio of these two sugars is about 1 (Kelebek 2009). Organic acids are found around 0.5–2 per cent and tartaric acid is the main organic acid followed by malic acid in grapes (Coskun 2001).

In addition to grapes, which are the main ingredient of hardaliye, crushed mustard seeds and sometimes sour cherry leaves are also used. Addition of black mustard seeds and sometimes preservatives leads to decreased alcohol formation by yeasts. Mustard seeds and cherry leaves (if added) give flavor and taste. Black mustard (*Brassica nigra* L. *Koch*) seeds contain 35–40 per cent oil, 20 per cent protein and thioglucosides, for example sinigrin, and the enzyme, myrosinase. Isothiocyanates, also called mustard oils, are the products of hydrolysis of thioglucosides in the presence of water and enzyme. In this case, the sinigrin is broken down to allyl isothiocyanate which is responsible for the characteristic pungency of hardaliye (Cleenewerck and Martin 2006). Moreover, due to its inhibitory effect on yeasts, allyl isothiocyanate decreases the alcohol production during hardaliye processing (Flamini and Vedova 2007). Tsao et al. (2002) reported that the allyl isothiocyanate contents of various grounded mustard seeds were in the range of 0.7–3.9 g/kg and that use of black mustard seeds is better for the production of hardaliye as they impart both flavor and taste of hardaliye, and have a good inhibitory effect on microorganisms by preventing alcohol production in the final product when compared to white mustard seeds, which have low levels of volatile oils (Coskun and Arici 2011).

There is not a standard production procedure for hardaliye. In general, grapes are washed and pressed. Crushed raw mustard seeds and usually benzoic acid, as a preservative, are added in percentages of 0.2 and 0.1, respectively. Pressed grapes and crushed mustard seeds are placed in wooden or plastic barrels in alternate layers, one by one. Then the barrels are closed and the grapes fermented for five to 10 days at room temperature. During that time, the lowermost layer is taken from the bottom of the tank and put on top once a day. After fermentation, it is filtered and kept in cold storage for three months (Coskun and Arici 2011).

Lactic acid bacteria are the dominant microorganisms responsible for fermentation of hardaliye. The lactic acid bacteria counts were reported in the range of 2–4.60 log cfu/ml. The major lactic acid bacteria isolated from spontaneously fermented hardaliye are *Lb. paracasei* subsp. *paracasei* and *Lb. casei* subsp. *pseudoplantarum*. Other lactic acid bacteria, such as *Lb. pontis*, *Lb. brevis*, *Lb. acetotolerans*, *Lb. sanfranciscensis* (formerly *Lb. sanfrancisco*) and *Lb. vaccinostercus* also contribute to the fermentation process (Arici and Coskun 2001). It was reported that *Lb. plantarum* addition as starter culture to pasteurized grapes for the production of hardaliye gives different organoleptic results when compared to traditional production (Coskun et al. 2012).

The pH of hardaliye is in the range of 3.21-3.97 (Arici and Coskun 2001). Amoutzopoulos et al. (2013) stated that hardaliye is relatively rich in phenolic compounds. The mean level of total phenolic compounds and total anthocyanins was found to be 2128 mg/l as gallic acid equivalent and 41.7 mg/l as cyanidin-3-glucoside equivalent, respectively. Total phenolic acids were determined at 397 mg/l. Quercetin, gallic acid and *trans*-resveratrol were found at important levels in hardaliye samples—65.5 mg/l, 47 mg/l and 2.72 mg/l, respectively.

12. Others

Besides the common fermented vegetables, other vegetables and fruits are fermented in Europe and these include radish, celery, garlic, broccoli, mustard, green tomatoes, pepper, artichoke, jerusalem artichoke, okra, green beans, green peas, asparagus, baby corn, mushroom, kohlrabi, leek, onion, swedes, apple, plum and mixed vegetables at homes traditionally (Di Cagno et al. 2013, Buckenhueskes 2015).

References

Agirman, B. 2014. Using different chloride salts to reduce the amount of sodium chloride in the production of shalgam juice. M.S. Thesis, Cukurova University, Adana, Turkey (*in Turkish*).

Akbulut, M., S. Causir, T. Marakoglu and H. Coklar. 2008. Chemical and technological properties of European cranberrybush. *Asian Journal of Chemistry* 20: 1875-1885.

Aktan, N., U. Yücel and H. Kalkan. 1998. *Pickle Technology*. Ege University Press, İzmir, Turkey (*in Turkish*).

Altan, A. and M. Maskan. 2004. Studies on the production of ready fruit juice powders from gilaburu (*Viburnum Opulus* L.). *Traditional Symposium on Foods*, 23-24 September, Van, 18-23. Pp. (*in Turkish*).

Altay, F., F. Karbancıoğlu-Güler, C. Daşkaya-Dikmen and D. Heperkan. 2013. A review on traditional Turkish fermented non-alcoholic beverages: Microbiota, fermentation process and quality characteristics. *International Journal of Food Microbiology* 167: 44-56.

Alvarruiz, A., M. Rodrigo, J. Miquel, V. Giner, A. Feria and R. Vila. 19 90. Influence of brining and packing conditions on product quality of capers. *Journal of Food Science* 55: 196-198.

Amoutzopoulos, B., G.B. Löker, G. Samur, S.A. Çevikkalp, M. Yaman, T. Köse and E. Pelvan. 2013. Effects of a traditional fermented grape-based drink 'hardaliye' on antioxidant status of healthy adults: A randomized controlled clinical trial. *Journal of the Science of Food and Agriculture* 93: 3604-3610.

Andres-Lacueva, C., A. Medina-Remon, R. Llorach, M. Urpi-Sarda, N. Khan, G. Chiva-Blanch, R. Zamora-Ros, M. Rotches-Ribalta and R. M. Lamuela Raventos. 2010. Phenolic compounds: Chemistry and occurrence in fruits and vegetables. Pp. 53-88. *In*: L.A. de la Rosa, E. Alvarez-Parrilla, G.A. Gonzalez-Aguilar (Eds.). *Fruit and Vegatable Phytochemicals Chemistry, Nutritional Value, and Stability*. Blackwell Publishing, Singapore.

Anonymous. 2005. Health Protection Agency Advisory Committee on the microbiological safety of food information paper. *Microbiological Status of Ready-to-eat Fruit and Vegetables*. ACM/745. London, UK: Food Standarts Agency from http://food.gov.uk/multimedia/pdfs/acm745amended.pdf. Accessed 14.06.2011.

Anonymous. 2012. Beet Kvass: *What is It and Why am I Drinking It?* http://livingmaxwell.com/beet-kvass. Accessed 16.06.2015.

Arapitsas, P., P.C.R. Sjöberg and C. Turner. 2008. Characterisation of anthocyanins in red cabbage using high resolution liquid chromotography coupled with photodiode array detection and electrospray ionization-linear ion trap mass spectrometry. *Food Chemistry* 109: 219-226.

Arici, M. and F. Coskun. 2001. Hardaliye: Fermented grape juice as a traditional Turkish beverage. *Food Microbiology* 18: 417-421.

Arici, M. 2004. Microbiological and chemical properties of a drink called shalgam. *Ernahrungs-Umschau* 51(1):10-11.

Atamanova, A., T.A. Brezhneva, A.I. Slivkin, V.A. Nikolaevskii, V.F. Selemenev and N.V. Mironenko. 2005. Isolation of saponins from table beetroot and primary evaluation of their pharmacological activity. *Pharmaceutical Chemistry Journal* 39: 650-652.

Atia, A., Z. Barhoumi, R. Mokded, C. Abdelly and A. Smaoui. 2011. Environmental eco-physiology and economical potential of the halophyte *Crithmum maritimum* L. (Apiaceae). *Journal of Medicinal Plants Research* 5(16): 3564-3571.

Aukrust, T.W., H. Blom, F. Sandtorv and E. Slinde. 1994. Interaction between starter culture and raw material in lactic acid fermentation of sliced carrot. *Lebensmittel-Wissenschaft und-Technologie* 27: 337-341.

Ayotte, E. 2013. Sauerkraut. http://www.uaf.edu/files/ces/publications- db/catalog/hec/FNH-00170.pdf Accessed 22.06.2015.

Ballesteros, C., M.L. Palop and I. Sanchez. 1999. Influence of sodium chloride concentration on the controlled lactic acid fermentation of 'Almagro' eggplants. *International Journal of Food Microbiology* 53: 13-20.

Baráth, A., A. Halász, E. Neméth and Z. Zalán. 2004. Selection of LAB strains for fermented red beet juice production. *European Food Research and Technology* 218: 184-187.

Bates, R.P., J.R. Morris and P.G. Crandall. 2001. Principles and practises of small and medium scale fruit juice processing. *FAO Agricultural Services Bulletin*, Rome, Italy.

Beauovit, B.P., S. Colombie, A. Monier, M.H. Andrieu, B. Blais, C. Benard, C. Cheniclet, M. Dieuaide-Noubhani, C. Nazaret, J.P. Mazat and Y. Gibon. 2014. Model-assisted analysis of sugar metabolism throughout tomato fruit development reveals enzyme and carrier properties in relation to vacuole expansion. *The Plant Cell* 26: 3222-3223.

Beckles, D.M. 2012. Factors affecting the post-harvest soluble solids and sugar content of tomato (*Solanum lycopersicum* L.) fruit. *Post-harvest Biology and Technology* 63: 129-140.

Beecher, C.W.W. 1994. Cancer preventive properties of varieties of *Brassica oleracea*: A review. *The American Journal of Clinical Nutrition* 59: 1166-1170.

Bergqvist, S.W., A.S. Sandberg, N.G. Carlsson and T. Andlid. 2005. Improved iron solubility in carrot juice fermented by homo- and hetero-fermentative lactic acid bacteria. *Food Microbiology* 22: 53-61.

Bhattacharjee, P. and R.S. Singhal. 2011. Asparagus, broccoli and cauliflower: Production, quality, and processing. Pp. 507-524. *In:* N.K. Sinha, Y.H. Hui, E.O. Evranuz, M. Siddiq, J. Ahmed (Eds.). *Handbook of Vegetables and Vegetable Processing*. Blackwell Publishing, Singapore.

Buckenhueskes, H.J. 2015. Quality improvement and fermentation control in vegetables. Pp. 515-539. *In:* W. Holzapfel (Ed.). *Advances in Fermented Foods and Beverages Improving Quality, Technologies and Health Benefits*. Elsevier, Cambridge.

Buruleanu, L., I. Manea, M.G. Bratu, D. Avram and C.L. Nicolescu. 2009. Effects of prebiotics on the quality of lactic acid fermented vegetable juices. *Ovidius University Annals of Chemistry* 20: 102-107.

Buruleanu, L.V., M.G. Bratu, I. Manea, D. Avram and C.L. Nicolescu. 2013. Fermentation of vegetable juices by *Lactobacillus acidophilus* LA-5. Pp. 173-194. *In:* M. Kongo (Eds.). *Lactic Acid Bacteria – R&D for Food, Health and Livestock Purposes*. InTech, Rijeka.

Cam, M., Y. Hisil and A. Kuscu. 2007. Organic acid, phenolic content, and antioxidant capacity of fruit flesh and seed of *Viburnum opulus*. *Chemistry of Natural Compounds* 43(4): 460-461.

Canbas, A. and A. Deryaoglu. 1993. A research on the processing techniques and characteristics of shalgam beverage. *Doğa-Turkish Journal of Agricultural and Forestry* 17: 119-129 (*in Turkish*).

Causse, M., C. Friguet, C. Coiret, M. Lepicier, B. Navez, M. Lee, N. Holthuysen, F. Sinesio, E. Moneta and S. Grandillo. 2010. Consumer preferences for fresh tomato at the European scale: A common segmentation on taste and firmness. *Journal of Food Science* 75: 531-541.

Cesoniene, L., R. Daubaras and P. Viskelis. 2008. Evaluation of productivity and biochemical components in fruit of different *Viburnum* accessions. *Biologija* 54(2): 93-96.

Cesoniene, L., R. Daubaras, P. Viskelis and A. Sarkinas. 2012. Determination of the total phenolic and anthocyanin contents and anti-microbial activity of *Viburnum opulus* fruit juice. *Plant Foods for Human Nutrition* 67: 256-61.

Charron, C.S., B.A. Clevidence, S.J. Britz and J.A. Novotny. 2007. Effect of dose size on bioavailability of acylated and nonacylated anthocyanins from red cabbage (*Brassica oleracea* L. var. *capitata*). *Journal of Agricultural and Food Chemistry* 55(13): 5354-5362.

Cleenewerck, M.B. and P. Martin. 2006. *Food.* Pp. 285-304. *In*: A. L. Chew and H.I. Maibach (Eds.). *Irritant Dermatitis.* Springer, Germany.

Corbo, M.R., A. Bevilacqua, L. Petruzzi, F.P. Casanova and M. Sinigaglia. 2014. Functional beverages: The emerging side of functional foods commercial trends, research and health implications. *Comprehensive Reviews in Food Science and Food Safety* 13(6): 1192-1206.

Cornara, L., C. D'Arrigo, F. Pioli, B. Borghesi, C. Bottino, E. Patrone and M.G. Mariotti. 2009. Micromorphological investigation on the leaves of the rock samphire (*Crithmum maritimum* L.): Occurrence of hesperidin and diosmin crystals. *Plant Biosystems* 143: 283-292.

Cortes-Olmos, J., M. Leiva-Brondo, J. Rosello, M.D. Raigon and J. Cebolla-Cornejo. 2014. The role of traditional varieties of tomato as sources of functional compounds. *Journal of the Science of Food and Agriculture* 94: 2888-2904.

Coskun, F. 2001. A study on the production technology of Hardaliye. PhD Thesis, Trakya University, Tekirdag, Turkey (*in Turkish*).

Coskun, F. and M. Arici. 2011. Effect of the use of different mustard seeds and grape varieties on some properties of Hardaliye. *Academic Food Journal (Akademik Gıda)* 9(3): 6-11 (*in Turkish*).

Coskun, F., M. Arici, G. Çelikyurt and M. Gülcü. 2012. Changes occuring at the end of storage in some properties of hardaliye produced by using different methods. *Journal of Tekirdag Agricultural Faculty* 9(3): 62-67 (*in Turkish*).

Cunsolo, F., G. Ruberto, V. Amico and M. Piattelli. 1993. Bioactive metabolites from Sicilian marine fennel *Crithmum maritimum. Journal of Natural Products* 56(9): 1598-1600.

Demir, N., K.S. Bahceci and J. Acar. 2006. The effects of differential initial *Lactobacillus plantarum* concentrations on some properties of fermented carrot juice. *Journal of Food Processing and Preservation* 30(3): 352-363.

Di Cagno, R., R.F. Surico, S. Siragusa, M. De Angelis, A. Paradiso, F. Minervini, L. De Gara and M. Gobbetti. 2008. Selection and use of autochthonous mixed starter for lactic acid fermentation of carrots, French beans or marrows. *International Journal of Food Microbiology* 127: 220-228.

Di Cagno, R., R. Coda, M. De Angelis and M. Gobbetti. 2013. Exploitation of vegetables and fruits through lactic acid fermentation. *Food Microbiology* 33: 1-10.

Dias, M.G., M.F.G.F.C. Camoes and L. Oliveira. 2009. Carotenoids in traditional Portuguese fruits and vegetables. *Food Chemistry* 113: 808-815.

Dinc, M., D. Aslan, N.C. Içyer and M. Çam. 2012. Microencapsulation of gilaburu juice. *Electronic Journal of Food Technologies* 7(2): 1-11.

Downey, M.O, N.K. Dokoozlian and M.P. Krstic. 2006. Cultural practice and environmental impacts on the flavonoid composition of grapes and wine: A review of recent research. *American Journal of Enology and Viticulture* 57: 257-268.

Erginkaya, Z. and W.P. Hammes. 1992. A study on the development of microorganisms during the shalgam fermentation and identification of isolated lactic acid bacteria. *Food (Gıda)* 17(5): 311-314 (*in Turkish*).

Erten, H., H. Tanguler and A. Canbaş. 2008. A traditional Turkish lactic acid fermented beverage: Shalgam (Salgam). *Food Reviews International* 24: 352-359.

Erten, H. and H. Tanguler. 2012. Şalgam (Shalgam). Pp. 657-664. *In*: Y.H. Hui, E.O. Evranuz, F.N. Arroyo-López, L. Fan, Å.S. Hansen, M.E. Jaramillo-Flores, M. Rakin, R.F. Schwan and W.Zhou (Eds.). *Handbook of Plant-Based Fermented Food and Beverage Technology.* CRS Press, Boca Raton.

Erten, H., B. Agirman, C.P. Boyaci-Gündüz, E. Carsanba, S. Sert, S. Bircan and H. Tanguler. 2014. Importance of yeast and lactic acid bacteria in food processing. Pp. 351-378. *In*: A.

Malik, Z. Erginkaya, S. Ahmad and H. Erten (Eds.). *Food Processing: Strategies for Quality Assesment*. Springer, Newyork.

Erten, H. and H. Tanguler. 2016. Shalgam (şalgam): A traditional Turkish lactic acid fermented beverage obtained mainly from black carrot. *In*: Y.H. Hui, E.O. Evranuz, H. Erten, G. Bingöl, M.E.J. Flores (Eds.). *Handbook of Vegetable Preservation and Processing*, second edition. CRC Press, Boca Raton (*in Press*).

FAOSTAT. 2007. http://top5ofanything.com/list/6be531bb/Cauliflower-Producing-Countries. Accessed 09.06.2015.

Fernandez-Ruiz, V., M.C. Sanchez-Mata, M. Camara, M.E. Torija, C. Chaya, L. Galiana-Balaguer, S. Rosello and F. Nuez. 2004. Internal quality characterization of fresh tomato fruits. *Hortscience* 39: 339-345.

Figas, M.R., J. Prohens, M.D. Raigon, P. Fernandez-de-Cordova, A. Fita and S. Soler. 2015. Characterization of a collection of local varieties of tomato (*Solanum lycopersicum* L.) using conventional descriptors and the high-throughput phenomics tool Tomato Analyzer. *Genetic Resources and Crop Evolution* 62: 189-204.

Flamini, R. and A.D. Vedova. 2007. Preservation of *Cabernet Sauvignon* grape must samples destined for chemical analysis: Addition of sodium azide, allyl isothiocyanate, octanoic acid and ethyl bromoacetate, and effect of pasteurization. *Journal of Food Processing and Preservation* 31:345-355.

García-Alanso, F.J., S. Bravo, J. Casas, D. Perez-Conesa, K. Jacob and M.J. Periago. 2009. Changes in antioxidant compounds during the shelf-life of commercial tomato juices in different packaging materials. *Journal of Agricultural and Food Chemistry* 57(15): 6815-6822.

Gardner, N.J., T. Savarda, P. Obermeierb, G. Caldwellb and C.P. Champagne. 2001. Selection and characterization of mixed starter cultures for lactic acid fermentation of carrot, cabbage, beet and onion vegetable mixtures. *International Journal of Food Microbiology* 64: 261-275.

Gavrilin, M.V., M. Markova, T. Likhota and A. Izmailova. 2007. Optimization of the procedure of vitamin determination in *Viburnum* oil. *Pharmaceutical Chemistry Journal* 41: 101-104.

Guil-Guerrero, J.L. and I. Rodríguez-Garcia. 1999. Lipid classes, fatty acids and carotenes of the leaves of six edible wild plants. *European Food Research and Technology* 209: 313-316.

Heinonen, I.M. and A.S. Meyer. 2002. Antioxidants in fruits, berries and vegetables. Pp. 23-51. *In*: W. Jongen (Ed.). *Fruit and Vegetable Processing: Improving Quality*. CRC Press, Boca Raton.

Hertog, M.G.L., D. Kromhout, C. Aravanis, H. Blackburn, R. Buzina, F. Fidanza, S. Giampaoli, A. Jansen, A. Menotti, S. Nedeljkovic, M. Pekkarinen, B.S. Simic, H. Toshima, E.J.M. Feskens, P.C.H. Hollman and M.B. Katan. 1995. Flavanoidintake and long-term risk of coronary-heart-disease and cancer in the 7 countries study. *Archives of Internal Medicine* 155: 381-386.

Hunaefi, D., D.N. Akumo and I. Smetanska. 2013. Effect of fermentation on antioxidant properties of red cabbages. *Food Biotechnology* 27: 66-85.

Hutkins, R.W. 2006. *Microbiology and Technology of Fermented Foods*. Blackwell Publishing IFT Press, Iowa.

Inocencio, C., D. Rivera, F. Alcaraz and F.A. Tomas-Barberan. 2000. Flavonoid content of commercial capers (*Capparis spinosa*, *C. sicula* and *C. orientalis*) produced in Mediterranean countries. *European Food Research and Technology* 212: 70-74.

Inocencio, C., D. Rivera, M.C. Obon, F. Alcaraz and J.A. Barrena. 2006. A systematic revision of Capparis section *Capparis* (Capparaceae). *Annals of the Missouri Botanical Garden* 93(1): 122-149.

Ishibashi, N. and S. Shimamura. 1993. Bifidobacteria: Research and development in Japon. *Food Technology* 47: 126-135.

Jacob, K., M.J. Periago, V. Böhm and G.R. Berruezo. 2008. Influence of lycopene and vitamin C from tomato juice on biomarkers of oxidative stress and inflammation. *British Journal of Nutrition* 99: 137-146.

Kalyoncu, I.H., N. Ersoy, A.Y. Elidemir and M.E. Karalı. 2013. Some physico-chemical characteristics and mineral contents of Gilaburu (*Viburnum opulus* L.) fruits in

Turkey. *International Journal of Biological, Food, Veterinary and Agricultural Engineering* 7(6): 169-171.

Kammerer, D., R. Carle and A. Schieber. 2004. Quantification of anthocyanins in black carrot extracts (*Daucus carota* ssp. *sativus* var. *atrorubens* Alef.) and evaluation of their color properties. *European Food Research and Technology* 219(5): 479-486.

Kanner, J., S. Harel and R. Granit. 1996. Pharmaceutical compositions containing antioxidants betalains and a method for their preparation. *Israel Patent* 119-872.

Kanner, J., S. Harel and R. Granit. 2001. Betalains, a new class of dietary cationized antioxidants. *Journal of Agricultural and Food Chemistry* 49: 5178-5185.

Kapadia, G.J. and G.S. Rao. 2013. Anticancer effects of red beet pigments. Pp. 125-154. *In*: B. Neelwarne (Ed.). *Red Beet Biotechnology Food and Pharmaceutical Applications.* Springer, New York.

Karovicova, J., M. Drdak, G. Greif and E. Hybenova. 1999. The choice of strains of *Lactobacillus* species for the lactic acid fermentation of vegetable juices. *European Food Research and Technology* 210: 53-56.

Karovicova, J. and Z. Kohajdova. 2002. The use of PCA, FA, CA for the evaluation of vegetable juices processed by lactic acid fermentation. *Czech Journal of Food Sciences* 20: 135-143.

Karovicova, J., Z. Kohajdova, M. Greifova, D. Lukacova and G. Grief. 2002a. Porovnanie fermentacii zeleninovych stiav. *Bulletin of Food Research* 41: 197-213.

Karovicova, J., Z. Kohajdova, M. Greifova, D. Lukacova and G. Grief. 2002b. Using of multivariate analysis for evaluation of lactic acid fermented cabbage juices. *Chemical Papers* 56: 267-274.

Kazimierczak, R., E. Hallmann, J. Lipowski, N. Drela, A. Kowalik, T. Pussa, D. Matt, A. Luik, D. Gozdowski and X.K.E. Rembia. 2014. Beetroot (*Beta vulgaris* L.) and naturally fermented beetroot juices from organic and conventional production: Metabolomics, antioxidant levels and anti-cancer activity. *Journal of the Science of Food and Agriculture* 94: 2618-2629.

Kelebek, H. 2009. Researches on the phenolic compounds profile of Okuzgozu, Bogazkere and Kalecik Karası cultivars grown in different regions and their wines. PhD Thesis, Cukurova University, Adana, Turkey (*in Turkish*).

Klaenhammer, T.R., R. Barrangou, B.L. Buck, M.A. Azcarate-Peril and E. Altermann. 2005. Genomic features of lactic acid bacteria effecting bioprocessing and health. *FEMS Microbiology Reviews* 29: 393-409.

Klewicka, E., I. Motyl and Z. Libudzisz. 2004. Fermentation of beet juice by bacteria of genus *Lactobacillus* spp. *European Food Research and Technology* 218: 178-183.

Klewicka, E. and A. Czyżowska. 2011. Biological stability of lactofermented beetroot juice during refrigerated storage. *Polish Journal of Food and Nutrition Sciences* 61(4): 251-256.

Koh, J.H., Y. Kim and J.H. Oh. 2010. Chemical characterization of tomato juice fermented with *Bifidobacteria. Journal of Food Science* 75: 428-432.

Kohajdová, Z. and J. Karovičová. 2005. Sensory and chemical evaluation of lactic acid–fermented cabbage-onion juices. *Chemical Papers* 59: 55-61.

Kohajdová, Z., J. Karovičová and M. Greifová. 2006. Lactic acid fermentation of some vegetable juices. *Journal of Food and Nutrition Research* 45(3): 115-119.

Köksal, E. and I. Gülcin. 2007. Antioxidant activity of cauliflower (*Brassica oleracea* L.). *Turkish Journal of Agriculture and Forestry* 32: 65-78.

Kun, S., J.M. Rezessy-Szabo, Q.D. Nguyen and A. Hoschke. 2008. Changes of microbial population and some components in carrot juice during fermentation with selected *Bifidobacterium* strains. *Process Biochemistry* 43: 816-821.

Lavelli, V., C. Peri and A. Rizzola. 2000. Antioxidant activity of tomato products as studied by model reactions using xanthine oxidase, myeloperoxidase and copper-induced lipid peroxidation. *Journal of Agricultural and Food Chemistry* 48: 1442-1448.

Lee, E.J., D. An, C.T.T. Nguyen, B.S. Patil, J. Kim and K.S. Yoo. 2014. Betalain and betaine composition of greenhouse- or field-produced beetroot (*Beta vulgaris* L.) and inhibition of HepG2 cell proliferation. *Journal of Agricultural and Food Chemistry* 62: 1324-1331.

Lin, J.Y., C.Y. Li and I.F. Hwang. 2008. Characterization of the pigment components in red cabbage (*Brassica oleracea* L. var.) juice and their anti-inflammatory effects on LPS-stimulated murine splenocytes. *Food Chemistry* 109: 771-781.

Lo-Scalzo, R., G. Bianchi, A. Genna and C. Summa. 2007. Antioxidant properties and lipidic profile as quality indexes of cauliflower (*Brassica oleracea* L. var. *botrytis*) in relation to harvest time. *Food Chemistry* 100: 1019-1025.

Lo-Scalzo, R., A. Genna, F. Branca, M. Chedin and C.H. Chassaigne. 2008. Anthocyanin composition of cauliflower (*Brassica rapa* L. var. *botrytis*) and cabbage (*Brassica oleracea* L. var. *capitata*) and its stability in relation to thermal treatments. *Food Chemistry* 107: 136-144.

Luna, F. and M. Perez. 1985. *La tapanera o alcaparra. Cultivo y aprovechamiento.* Ministerio de Agricultura, Pesca y Alimentacion, Madrid, Spain.

Maifreni, M., M. Marino, G. Rondinini. 1999. Microbiological aspects (processing and storage) of a fermented vegetable (brovada). *Mitteilungen aus Lebensmitteluntersuchung und Hygiene* 90: 211-221.

Maifreni, M., M. Marino and L. Conte. 2004. Lactic acid fermentation of *Brassica rapa*: Chemical and microbial evaluation of a typical Italian product (Brovada). *European Food Research and Technology* 218: 469-473.

Manchali, S., N.K.C. Murthy, S. Nagaraju and B. Neelwarne. 2013. Stability of betalain pigments of red beet. pp 55-74. *In*: B. Neelwarne (Ed.). *Red Beet Biotechnology Food and Pharmaceutical Applications.* Springer, New York.

Marmion, D.M. 1991. *Handbook of US Colorants: Foods, Drugs, Cosmetics and Medicinal Devices.* Wiley-Interscience, New York.

Matthäus, B. and M. Ozcan 2002. Glucosinolate composition of young shoots and flower buds of capers (*Capparis* species) growing wild in Turkey. *Journal of Agricultural and Food Chemistry* 50(25): 7323-7325.

Meot-Duros, L., F.G. Le and C. Magné. 2008. Radical scavenging, antioxidant and anti-microbial activities of halophytic species. *The Journal of Ethnopharmacology* 116: 258-262.

Meot-Duros, L. and C. Magné. 2009. Antioxidant activity and phenol content of *Crithmum maritimum* L. leaves. *Plant Physiology and Biochemistry* 47: 37-41.

Montaño, A., A.H. Sánchez, L. Rejano and A. De Castro. 1997. Processing and storage of lye-treated carrots fermented by a mixed starter culture. *International Journal of Food Microbiology* 35: 83-90.

Morales Mocino, A. 1985. *El cultivo y principios de comercializacion de la berenjena de Almagro.* Proyecto Fin de Carrera E.U.I.T.A. Ciudad Real.

Nazzaro, F., F. Fratianni, S. Alfonso and P. Orlando. 2008. Synbiotic potential of carrot juice supplemented with *Lactobacillus* spp. and inulin or fructooligosaccharides. *Journal of the Science of Food and Agriculture* 88: 2271-2276.

Olszewska, M.A., A.M. Kocot and L.L. Trokenheim. 2015. Physiological functions at single-cell level of *Lactobacillus* spp. isolated from traditionally fermented cabbage in response to different pH conditions. *Journal of Biotechnology* 200: 19-26.

Ozcan, M. and A. Akgul 1998. Influence of species, harvest date and size on composition of capers (*Capparis spp.*) flower buds. *Nahrung* 42(2). 102-105.

Ozcan, M. 1999a. Pickling and storage of caperberries (*Capparis* spp.). *Zeitschrift fuer Lebensmittel Untersuchung und Forschung* 208: 379-382.

Ozcan, M. 1999b. The physical and chemical properties and fatty acid compositions of raw and brined caperberries (*Capparis* spp.). *Turkish Journal of Agriculture and Forestry* 23: 771-776.

Ozcan, M. and A. Akgul. 1999a. Pickling process of capers (*Capparis* spp.) flower buds. *Grasas Aceites* 50(2): 94-99.

Ozcan, M. and A. Akgul. 1999b. Storage quality in different brines of pickled capers (*Capparis* spp.). *Grasas Aceites* 50: 269-274.

Ozcan, M. 2000. The use of yogurt as starter in rock samphire (*Crithmum maritimum* L.) fermentation. *European Food Research and Technology* 210: 424-426.

Ozrenk, K., M. Gündoğdu, N. Keskin and T. Kaya. 2011. Some physical and chemical characteristics of Gilaburu (*Viburnum opulus* L.) fruits in Erzincan region. *Iğdır University Journal of the Institute of Science and Technology* 1(4): 9-14.

Pakbin, B., S.H. Razavi, R. Mahmoudi and P. Gajarbeygi. 2014. Producing probiotic peach juice. *Biotechnology and Health Sciences* 1(3): e24683.

Palomino, J.M., J. Toledo del Árbo, N. Ben Omar, H. Abriouel, M. Martínez Canamero, A. Galvez and R.P. Pulido. 2015. Application of *Lactobacillus plantarum* Lb9 as starter culture in caper berry fermentation. *Food Science and Technology* 60: 788-794.

Paramithiotis, S., O.L. Hondrodimou and E.H. Drosinos. 2010. Development of the microbial community during spontaneous cauliflower fermentation. *Food Research International* 43: 1098-1103.

Paramithiotis, S., A.I. Doulgeraki, I. Tsilikidis, G.J.E. Nychas and E.H. Drosinos. 2012. Fate of *Listeria monocytogenes* and *Salmonella Typhimurium* during spontaneous cauliflower fermentation. *Food Control* 27(1): 178-183.

Picchi, V., C. Migliori, R. Lo Scalzo, G. Campanelli, V. Ferrari and L.F. Di Cesare. 2012. Phytochemical content in organic and conventionally grown Italian cauliflower. *Food Chemistry* 130: 501-509.

Plaza, L., M. Munoz, B. De Ancos and M.P. Cano. 2003. Effect of combined treatments of high pressure, citric acid and sodium chloride on quality parameters of tomato puree. *European Food Research and Technology* 216: 514-519.

Podsedek, A., D. Sosnowska, M. Redzynia and B. Anders. 2006. Antioxidant capacity and content of *Brassica oleracea* dietary antioxidants. *International Journal of Food Science and Technology* 41: 49-58.

Podsedek, A. 2007. Natural antioxidants and antioxidant capacity of *Brassica* vegetables: A review. *LWT- Food Science and Technology* 40: 1-11.

Prohens, J., J.E. Munoz-Falcon, A. Rodriguez-Burruezo, F. Ribas, A. Castro and F. Nuez. 2009. 'H15', an Almagro-type pickling eggplant with high yield and reduced prickliness. *Horticultural Science* 44: 2017-2019.

Prado, F.C., J.L. Parada, A. Pandey and C.R. Soccol. 2008. Trends in non-dairy probiotic beverages. *Food Research International* 41: 111-123.

Prajapati, J.B. and B.M. Nair. 2003. The history of fermented foods. Pp. 1-24. *In*: E.R. Farnworth (Ed.). *Handbook of Fermented Functional Foods*. CRC Press, Boca Raton.

Pulido, R.P., N. Ben Omar, H. Abriouel, R.L. López, M.M. Canamero and A. Gálvez. 2005. Microbiological study of lactic acid fermentation *of* caper berries by molecular and culture dependant methods. *Applied and Environmental Microbiology* 71: 7872-7879.

Pulido, R.P., N. Ben Omar, H. Abriouel, R.L. López, M.M. Canamero, J-P. Guyot and A. Gálvez. 2007. Characterization of *Lactobacilli* isolated from caperberry fermentations. *Journal of Applied Microbiology* 102: 583-590.

Pulido, R.P., N. Ben Omar, M.M. Cañamero, H. Abriouel and A. Gálvez. 2012. Fermentation of caper products. Pp. 201–208. *In*: Y.H. Hui, E.O. Evranuz, F.N. Arroyo-López, L. Fan, Å.S. Hansen, M.E. Jaramillo-Flores, M. Rakin, R.F. Schwan and W.Zhou (Eds.). *Handbook of Plant-based Fermented Food and Beverage Technology*, second edition. *CRC Press, Boca Raton, FL.*

Rakin, M., J. Baras, M. Vukašinović and D. Mijuca. 2003. Optimization of lactic-acid fermentation of beetroot juice and brewer's yeast autolysate. *Romanian Biotechnological Letters* 8(5-6):1421-1430.

Rakin, M., J. Baras, M. Vukasinovic and M. Maksimovic. 2004a. The examination of parameters for lactic acid fermentation and nutritive value of fermented juice of beetroot, carrot and brewer's yeast autolysate. *Journal of the Serbian Chemical Society* 69 (8-9): 625-634.

Rakin, M., J. Baras and M. Vukašinović. 2004b. The influence of brewer's yeast autolysate and lactic acid bacteria on the production of a functional food additive based on beetroot juice fermentation. *Food Technology and Biotechnology* 42(2): 109-113.

Rakin, M., J. Baras and M. Vukašinović. 2005. Lactic acid fermentation in vegetable juices supplemented with different content of brewer's yeast autolysate. *Acta Periodica Technologica* 36: 71-80.

Rakin, M., M. Vukašinović, S. Siler-Marinkovic and M. Maksimovic. 2007. Contribution of lactic acid fermentation to improved nutritive quality vegetable juices enriched with brewer's yeast autolysate. *Food Chemistry* 100: 599-602.

Ramandeep, K.T. and G.P. Savage. 2005. Antioxidant activity in different fractions of tomatoes. *Food Research International* 38: 487-494.

Reid, G., D. Beuerman, C. Heinemann and A.W. Bruce. 2001. Probiotic *Lactobacillus* dose required to restore and maintain a normal vaginal flora. *FEMS Immunology and Medical Microbiology* 32: 37-41.

Renna, M. and M. Gonnella. 2012. The use of the sea fennel as a new spice-colorant in culinary preparations. *International Journal of Gastronomy and Food Science* 1: 111-115.

Rice-Evans, C.A., N.J. Miller, P.G. Bolwell, R.M. Bramley and J.B. Pridham. 1995. The relative antioxidant activities of plant-derived polyphenolic flavonoids. *Free Radical Research* 22: 375-383.

Rodrigo, M., M.J. Lazaro, A. Alvarruiz and V. Giner. 1992. Composition of capers (*Capparis spinosa*): Influence of cultivar, size and harvest date. *Journal of Food Science* 57: 1152-1154.

Rodríguez, H., A.J. Curiel, J.M. Landete, B. de las Rivas, F.L. de Felipe, C. Gomez-Cordoves, J.M. Manchenoc and R. Munoz. 2009. Food phenolics and lactic acid bacteria. *International Journal of Food Microbiology* 132: 79-90.

Rodríguez-Sevilla, M.D., M.J. Villanueva-Suarez and A. Redondo-Cuenca. 1999. Effects of processing conditions on soluble sugar content of carrot, beetroop and turnip. *Food Chemistry* 66: 81-85.

Romeo, V., M. Ziino, D. Giuffrida, C. Condurso and A. Verzera. 2007. Flavour profile of capers (*Capparis spinosa* L.) from the Eolian Archipelago by HS-SPME/GC-MS. *Food Chemistry* 101: 1272-1278.

Rop, O., V. Reznicek, M. Valsikova, T. Jurikova, J. Mlcek and D. Kramarova. 2010. Antioxidant properties of European cranberrybush fruit (*Viburnum opulus* var. *edule*). *Molecules* 15: 4467-4477.

Rossi, P.G., L. Berti, J. Panighi, A. Luciani, J. Maury and A. Muselli. 2007. Anti-bacterial action of essential oils from Corsica. *Journal of Essential Oil Research* 19: 176-182.

Ruberto, G., M.T. Baratta, S.G. Deans and H.J.D. Dorman. 2000. Antioxidant and antimicrobial activity of *Foeniculum vulgare* and *Crithmum maritimum* essentials oils. *Planta Medica* 66: 687-693.

Sagdic, O., I. Ozturk, N. Yapar and H. Yetim. 2014. Diversity and probiotic potentials of lactic acid bacteria isolated from gilaburu, a traditional Turkish fermented European cranberrybush (*Viburnum opulus* L.) fruit drink. *Food Research International* 64: 537-545.

Sahlin, E., G.P. Savage and C.E. Lister. 2004. Investigation of the antioxidant properties of tomatoes after processing. *Journal of Food Composition and Analysis* 17: 635-647.

San José, R., M.C. Sánchez, M.M., Cámara and J. Prohens. 2013. Composition of eggplant cultivars of the Occidental type and implications for the improvement of nutritional and functional quality. *International Journal of Food Science and Technology* 48: 2490-2499.

Sánchez, A.H., A. De Castro and L. Rejano 1992. Controlled fermentation of caperberries. *Journal of Food Science* 57: 675-678.

Sánchez, I., L. Palop and C. Ballesteros. 2000. Biochemical characterization of lactic acid bacteria isolated from spontaneous fermentation of 'Almagro' eggplants. *International Journal of Food Microbiology* 59(1-2): 9-17.

Sánchez, I., L. Palop and C. Ballesteros. 2003. Identification of lactic acid bacteria from spontaneous fermentation of 'Almagro' eggplants by SDS-PAGE whole cell protein fingerprinting. *International Journal of Food Microbiology* 82(2): 181-189.

Sánchez-Moreno, C., L. Plaza, B. de Ancos and M.P. Cano. 2006. Nutrional characterisation of commercial traditional pasteurised tomato juice: Caretenoids, vitamin C and radical-scavenging capacity. *Food Chemistry* 98: 749-756.

Sanz-Puig, M., M.C. Pina-Perez, D. Rodrigo and A. Martinez-Lopez. 2015. Anti-microbial activity of cauliflower (*Brassica oleracea* L.var. *botrytis*) by-product against *Listeria monocytogenes*. *Food Control* 50: 435-440.

Sárdi, É., É. Stefanovits-Bányai, I. Kocsis, M. Takács-Hájos, H. Fébel and A. Blázovics. 2009. Effect of bioactive compounds of table beet cultivars on alimentary induced fatty livers of rats. *Acta Alimentaria* 38: 267-280.

Sekiguchi, H., Y. Ozeki and N. Sasaki. 2013. Biosynthesis and regulation of betalains in red beet. Pp. 45-54. *In*: B. Neelwarne (Ed.). *Red Beet Biotechnology Food and Pharmaceutical Applications*. Springer, New York.

Seseña, S., I. Sánchez, M.A. González Vinas and L. Palop. 2002. Effect of freezing on the spontaneous fermentation and sensory attributes of 'Almagro eggplants'. *International Journal of Food Microbiology* 77: 155-159.

Seseña, S., I. Sánchez- Hurtado and L. Palop. 2004. Genetic diversity (RAPD-PCR) of *Lactobacilli* isolated from 'Almagro' eggplant fermentations from two seasons. *FEMS Microbiology Letters* 238: 159-165.

Seseña, S. and M.L. Palop. 2007. An ecological study of lactic acid bacteria from Almagro eggplant fermentation brines. *Journal of Applied* Microbiology 103: 1553-1561.

Seseña, S. and M.L. Palop. 2012. Almagro eggplant: From homemade tradition to small-scale industry. Pp. 407-418. *In*: Y.H. Hui, E.O. Evranuz, F.N. Arroyo-López, L. Fan, Å.S. Hansen, M. E. Jaramillo-Flores, M. Rakin, R. F. Schwan and W.Zhou (Eds.). *Handbook of Plant-Based Fermented Food and Beverage Technolog,.* second edition. CRC Press, New York.

Settanni, L. and G. Moschetti. 2014. New trends in technology and identity of traditional dairy and fermented meat production processes: Preservation of typicality and hygiene. *Trends in Food Science and Technology* 37 (1): 51-58.

Shahidi, F., A. Chandrasekara and Y. Zhong. 2011. Bioactive phytochemicals in vegetables. Pp. 125 158. *In*: N.K. Sinha, Y.H. Hui, E.O. Evranuz, M. Siddiq, J. Ahmed (Eds.). *Handbook of Vegetables and Vegetable Processing*. Blackwell Publishing, Singapore.

Siddiqui, M.W., J.F. Ayala-Zavala and R.S. Dhua. 2015. Genotypic variation in tomatoes affecting processing and antioxidant properties. *Critical Reviews in Food Science and Nutrition* 55: 1819-1835.

Sikora, E., E. Cieslik, T. Leszczynska, A. Filipiak-Florkiewicz and P.M. Pisulewski. 2008. The antioxidant activity of selected cruciferous vegetables subjected to aquathermal processing. *Food Chemistry* 107: 55-59.

Singh, B. and B.S. Hathan. 2014. Chemical composition, functional properties and processing of Beetroot — A review. *IJSER Journal* 5(1): 679-684.

Sonmez, N., H.H.A. Alizadeh, R. Öztürk and A.İ. Acar. 2007. Some physical properties of gilaburu seed. *Agricultural Science Journal* (*Tarım Bilimleri Dergisi*) 13 (3): 308-311 (*in Turkish*).

Soylak, M., L. Elci, S. Saracoglu and U. Divrikli. 2002. Chemical analysis of fruit juice of European cranberry bush (*Viburnum opulus*) from Kayseri, Turkey. *Asian Journal of Chemistry* 14: 135-138.

Sozzi, G.O. and A.R. Vicente. 2006. Caper and caperberries. Pp. 230-256. *In*: K.V. Peter (Eds.). *Handbook of Herbs and Spices*, volume 3. Woodhead Publishing Limited and CRC Press, Cambridge.

Stojceska, V., P. Ainsworth, A. Plunkett, E. Ibanoğlu and S. Ibanoğlu. 2008. Cauliflower by-products as a new source of dietary fibre, antioxidants and proteins in cereal based ready-to-eat expanded snacks. *Journal of Food Engineering* 87: 554-563.

Swain, M.R., M. Anandharaj, R.C. Ray and R.P. Rani. 2014. Fermented fruits and vegetables of Asia: A Potential Source of Probiotics. *Biotechnology Research International* 250-424.

Tanguler, H. 2010. Identification of predominant lactic acid bacteria isolated from shalgam beverage and improvement of its production technique. PhD Thesis, Cukurova University, Adana, Turkey (*in Turkish*).

Tanguler, H. and H. Erten. 2011. The microbiological investigation of dough fermentation in the production of shalgam beverage. *Journal of Agricultural Faculty of Cukurova University* 26(2): 15-22 (*in Turkish*).

Tanguler, H. and H. Erten. 2012a. Occurrence and growth of lactic acid bacteria species during the fermentation of shalgam (şalgam), a traditional Turkish fermented beverage. *LWT-Food Science and Technology* 46: 36-41.

Tanguler, H. and H. Erten. 2012b. Chemical and microbiological characteristics of shalgam (şalgam); A traditional Turkish lactic acid fermented beverage. *Journal of Food Quality* 35: 298-306.

Tanguler, H. and H. Erten. 2013. Selection of potential autochtonous starter cultures through lactic acid bacteria isolated and identificated from shalgam: A traditional Turkish fermented beverage. *Turkish Journal of Agriculture and Forestry* 37: 212-220.

Tanguler, H., G. Günes and H. Erten. 2014. Influence of addition of different amounts of black carrot (*Daucus carota*) on shalgam quality. *Journal of Food Agriculture and Environment* 12(2): 60-65.

Tanguler, H., Saris Per. E.J. and H. Erten. 2015. Microbial, chemical and sensory properties of shalgams made using different production methods. *Journal of the Science of Food and Agriculture* 95: 1008-1015.

Tendaj, M., K. Sawicki and B. Mysiak. 2013. The content of some chemical compounds in red cabbage (*Brassica oleracea* l.var. *capitata* f. *rubra*) after harvest and long-term storage. Electronic Journal of Polish Agricultural Universities. 16(2) http://www.ejpau.media.pl/volume16/issue2/art-02.html Accessed 20.06.2015

Titchenal, C.A. and J. Dobbs. 2004. Nutritional value of vegetables. Pp. 39-53. *In*: Y. H. Hui, S. Chazala, D.M. Graham, K.D. Murrell and W.K. Nip (Eds.). *Handbook of Vegetable Preservation and Processing*. Marcel Dekker, New York.

Tlili N., A. Khaldi, S. Triki and S. Munne-Bosch. 2010. Phenolic compounds and vitamin antioxidants of caper (*Capparis spinosa*). *Plant Foods for Human Nutrition* 65: 260-265.

Tlili N, N. Nasri, A. Khaldi, S. Triki and S. Munne-Bosch. 2011a. Phenolic compounds, tocopherols, carotenoids and vitamin C of commercial caper. *Journal of Food Biochemistry* 35: 472-483.

Tlili, N., W. Elfalleh, E. Saadaoui, A. Khaldi, S. Triki and N. Nasri. 2011b. The caper (*Capparis* L.): Ethnopharmacology, phytochemical and pharmacological properties. *Fitoterapia*. 82: 93-101.

Tsao, R., Q. Yu, J. Potter and M. Chiba. 2002. Direct and simultaneous analysis of sinigrin and allyl isothiocyanate in mustard samples by high-performance liquid chromatography. *Journal of Agricultural and Food Chemistry* 50: 4749-4753.

Tsen, J.H., Y.P. Lin, H.Y. Huang and V.E. King. 2008. Studies on the fermentation of tomato juice by using K-Carrageenan immobilized *Lactobacillus acidophilus*. *Journal of Food Processing and Preservation* 32: 178-189.

Váli, L., E. Stefanovits-Banyai, K. Szentmihalyi, H. Febel, E. Sardi, A. Lugasi, I. Kocsis and A. Blazovics. 2007. Liver-protecting effects of table beet (*Beta vulgaris* var. *rubra*) during ischemia-reperfusion. *Nutrition* 23: 172-178.

Vanajakshi, V., S.V.N. Vijayendra, M.C. Varadaraj, G. Venkateswaran and R. Agrawal. 2015. Optimization of a probiotic beverage based on *Moringa* leaves and beetroot. *LWT-Food Science and Technology* 63(2): 1268-1273.

Vaughn, R. 1985. The microbiology of vegetable fermentations. Pp. 45-72. *In*: B.J.B. Wood (Ed.). *Microbiology of Fermented Foods*. Elsevier, London.

Vega, M.N. and C. Ramos. 1987 . Constituents of capers and changes during pickling. *Grasas Aceites*, 38: 173-5.

Velioglu, Y.S., L. Ekici and E.S. Poyrazoglu. 2006. Phenolic composition of European cranberrybush (*Viburnum opulus* L.) berries and astringency removal of its commercial juice. *International Journal of Food Science and Technology* 41: 1011-1015.

Volden, J., G.I.A. Borge, G.B. Bengtsson, M. Hansen, I.E. Thygesen and T. Wicklund. 2008. Effect of thermal treatment on glucosinolates and antioxidant-related parameters in red cabbage (*Brassica oleracea* L. spp. *capitata* f. *rubra*). *Food Chemistry* 109: 595-605.

Volden, J., G.I.A. Borge, M. Hansen, T. Wicklund and G.B. Bengtsson. 2009. Processing (blanching, boliling, steaming) effects on the content of glucosinolates and antioxidant-related parameters in cauliflower (*Brassica oleracea* L. spp. *botrytis*). *LWT-Food Science and Technology* 42: 63-73.

Wang, M. and I.L. Goldman. 1997. Accumulation and distribution of free folic acid content in red beet (*Beta vulgaris* L.). *Plant Foods for Human Nutrition* 50: 1-8.

Wei, J., H. Miao and Q. Wang. 2011. Effect of glucose on glucosinolates, antioxidants and metabolic enzymes in *Brassica* sprouts. *Scientia Horticulturae Amsterdam* 129: 535-540.

Wiczkowski, W., D. Szawara-Novak and J. Topolska. 2013. Red cabbage anthocyanins: Profile, isolation, identification and antioxidant activity. *Food Research International* 51: 303-309.

Wiczkowski, W., D. Szawara-Novak and J. Topolska. 2015. Changes in the content and composition of anthocyanins in red cabbage and its antioxidant capacity during fermentation, storage and stewing. *Food Chemistry* 167: 115-123.

Willcox, J.K., G.L. Catignani and S. Lazarus. 2003. Tomatoes and cardiovascular health. *Critical Reviews in Food Science and Nutrition* 43: 1-18.

Wruss, J., G. Waldenberger, S. Huemer, P. Uygun, P. Lanzerstorfer, U. Müller, O. Höglinger and J. Weghuber. 2015. Compositional characteristics of commercial beetroot products and beetroot juice prepared from seven beetroot varieties grown in Upper Austria. *Journal of Food Composition and Analysis* 42: 46-55.

Wu, X., G.R. Beecher, J.M. Holden, D.B. Haytowitz, S.E. Gebhardt and R.L. Prior. 2004. Lipophilic and hydrophilic antioxidant capacities of common foods in the United States. *Journal of Agricultural and Food Chemistry* 52(12): 4026-4037.

Yahia, E.M. 2010. The contribution of fruit and vegetable consumption to human health. Pp. 3-52. *In*: L.A. Rosa, E. Alvarez-Parrilla and G.A. Gonzalez-Aguilar (Eds.). *Fruit and Vegetable Phytochemicals Chemistry, Nutritional Value and Stability*. Blackwell Publishing, USA.

Yahia, E.M. and J.J. Ornelas-Paz. 2010. Chemistry, stability and biological actions of carotenoids. Pp. 177-222. *In*: L.A. Rosa, E. Alvarez-Parrilla and G.A. Gonzalez-Aguilar (Eds.). *Fruit and Vegetable Phytochemicals Chemistry, Nutritional Value, and Stability*. Blackwell Publishing, USA.

Yapar, N. 2008. The possibility of using lactic acid bacteria isolated from traditionally fermented European Cranberrybush (*Viburnum opulus* L.) juice in the manufacture of industrial fruit juice. M.S. Thesis, Erciyes University, Kayseri, Turkey (*in Turkish*).

Yilmaztekin, M. and K. Sislioglu. 2015. Changes in volatile compounds and some physicochemical properties of European cranberrybush (*Viburnum opulus* L.) during ripening through traditional fermentation. *Journal of Food Science* 80(4): 687-694.

Yoon, K.Y., E.E. Woodams and Y.D. Hang. 2004. Probiotication of tomato juice by lactic acid bacteria. *Journal of Microbiology* 42(4): 315-318.

Yoon, K.Y., E.E. Woodams and Y.D. Hang. 2005. Fermentation of beet juice by benefical lactic acid bacteria. *LWT-Food Science and Technology* 38: 73-75.

Yoon, K.Y., E.E. Woodams and Y.D. Hang. 2006. Production of probiotic cabbage juice by lactic acid bacteria. *Bioresource Technology* 97: 1427-1430.

Zalán, Z., A. Halász and Á. Baráth. 2012. Fermented red beet juice. Pp. 373-384. *In*: Y.H. Hui, E.O. Evranuz, F.N. Arroyo-López, L. Fan, Å.S. Hansen, M.E. Jaramillo-Flores, M. Rakin, R.F. Schwan and W. Zhou (Eds.). *Handbook of Plant-Based Fermented Food and Beverage Technology*, second edition. CRC Press, Boca Raton.

Zayachkivska, O.S., M.R. Gzhegotsky, O.I. Terletska, D.A. Lutsyk, A.M. Yaschenko and O.R. Dzhura. 2006. Influence of *Viburnum opulus* proanthocyanidins on stress-induced gastrointestinal mucosal damage. *Journal of Physiology and Pharmacology* 57: 155-167.

11

Regional Fermented Fruits and Vegetables in Africa

Folarin A. Oguntoyinbo[1,2]* and Charles M.A.P. Franz[2]

1. Introduction

While the human population in the world is increasing, there is a corresponding reduction in the agricultural land for farming. Additionally, global warming is having a deleterious impact on agricultural productivity with dire consequences on the food supply for both developed and developing countries (Rosenzweig and Parry 1994). Luckily for Africa, it is particularly rich in traditional fermented foods, particularly plant-based fermented foods, which are produced using minimal technology and inputs (Odunfa 1985). However, sub-Saharan Africa has the highest percentage of chronically malnourished people in the world (OECD-FAO 2011). The reasons for this are mainly agronomic constraints and lack of appropriate local processing techniques. Accordingly, a huge proportion of ca. 30 to 50 per cent of harvest is lost at the postharvest stage (Shiundu and Oniang'o 2007). The main causes for this are inadequate production conditions (Abukutsa-Onyango 2007, Diwani and Janssens 2001) as well as rapid product decay during transport, storage and marketing (Muchoki et al. 2007).

Future issues surrounding global food supply and security are posing a challenge if adequate attention is not placed on traditional food production, sustainable preservation and reduction of food waste in different continents. The endeavor to provide sufficient food on a world-wide scale will require a proper understanding of the lesser known crops, low-cost preservation techniques and new methodologies of production and processing that can guarantee adequate food availability (Huang et al. 2002). For example, Africa's large quantities of different agricultural

[1] Department of Microbiology, Faculty of Science, University of Lagos, Akoka, Lagos, Nigeria.
[2] Department of Microbiology and Biotechnology, Max Rubner-Institut, Hermann-Weigmann-Straße 1, 24103 Kiel, Germany.
* Corresponding author: E-mail: foguntoyinbo@unilag.edu.ng

harvests can meet the increasing needs of other parts of the world. Many African produce are drought-resistant and these tropical crops are grown in abundance. Therefore, effective post-harvest strategies based on sound scientific principles need to be developed for efficient crop utilization. The strategies must be applicable and adaptable to different situations in African countries where there are varying levels of infrastructure and technology. Africa also is a region where a very high variety of diverse food crops is grown. Vegetables and fruits are produced throughout the continent, but due to limited industrial-scale processing of agricultural products, large economic losses of up to 40 per cent are experienced, and as a consequent, poverty and hunger (Gustavsson et al. 2011).

Traditional methods of processing and value addition to vegetables and fruits have a long history throughout Africa (Steinkraus 1985). Odunfa (1985) identified food processing that involved fermentation as an important method to facilitate the availability of food and support food security throughout the continent. Available information shows that archeological and anthropological records, as well as the available documented history of African foods and diets show that food fermentation originated by trial and error, before it became an art. Subsequently techniques were transferred from one generation to the other (Odunfa 1985, Stanton 1985, Battcock and Azam-Ali 1998, Deshpande et al. 2000).

Climatic conditions characterized by high humidity and temperatures contribute to the diversity and quantities of agricultural produce in Africa. The need to adopt preservation and optimize traditional techniques for this has been emphasized to greatly influence the type of food that can be fermented, as well as the scale at which it should be done (Sanni 1993). Cereals and tubers, as well as legumes, fruits and vegetables are produced in large quantities in many parts of Africa and because of their perishable nature, they are subject to post-harvest processing. Post-harvest processing based on fermentation is used to produce and increase the shelf-life of a variety of foods at either household or small-scale, cottage-type businesses in Africa since decades (Odunfa 1985, Steinkraus 1995). The many advantages of fermenting agricultural produce have been recognized throughout the continent as an important strategy for improving palatability, detoxification, mineralization, as well as increasing shelf-life and digestibility. The significance of food fermentation as a sustainable post-harvest technology, especially in developing countries, has become well-recognized by FAO published global perspectives (Battcock and Azam-Ali 1998, Haard et al. 1999, Deshpande et al. 2000). Apart from contributing to the dietary intake of the people, it improves safety, quality, availability of foods and generates income for the food processors.

The aim of this chapter is to describe different lactic acid fruit and vegetable fermentation processes that are currently in use in Africa. In addition, the involvement of different lactic acid bacteria (LAB) populations associated with fermentation, and possibly other microorganisms associated with such fermentations will be discussed. The beneficial roles, which traditional fermented foods play in the diet and health of the consumer will also be addressed, and so will be the development of concepts that could facilitate process optimization, safety and quality, which may potentially lead to the development of functional foods.

Fruits and vegetables produced in different regions of Africa are classified as fruits, starchy vegetables, protein-oil seeds and leafy vegetables. Very high percentages of fruits and vegetables are consumed after harvest in Africa. In many countries, traditional processing of fruits and vegetables play an important role in the food supply, especially during off-seasons and harvest.

2. Role of Fruits and Vegetables in African Diet, Nutrition and Health

Plant products including fruits and vegetables, legumes, seeds, roots and tubers are an important source of fiber, carbohydrate, protein, as well as amino acids, fatty acids, minerals and vitamins. They also serve as important antioxidants (Willcox et al. 2003). A shift in the oxidative potential in the human body is due to the limitation of antioxidants, which lead to oxidative stress and cellular oxidative damage. Antioxidants from fruits and vegetables are essential for balancing of oxidative stress (Rautenbach et al. 2010) by way of supplying antioxidants, such as vitamin E and C, carotenoids, tocopherols and polyphenols, which are important to human health. Antioxidants play an important role in preventing development of chronic diseases such as cancer, cardiovascular disease (hypertension) and pathogenesis of immune deficiency virus (Willcox et al. 2004). Some fermented plant products possess higher vitamins than the unfermented foods; for instance, fermented vegetable proteins, such as *iru* or *dawadawa* contain higher riboflavin than unfermented seeds (Odunfa 1986). Methionine- and lysin- producing *lactobacilli* strains has been isolated from traditional fermented *ogi* (Odunfa et al. 2001). A novel *Lactobacillus rossiae* DSM15814[T] species was shown to possess a complete *de novo* biosynthetic pathway for synthesis of riboflavin, vitamin B$_{12}$ and other B vitamins (De Angelis et al. 2014) and an *in situ* study showed the relevance of such strains in cereal fermentation (Capozzi et al. 2012).

3. Food Fermentation as a Post-harvest Strategy for Food Security in Africa

In Africa, traditional processing of fruits and vegetables is an important component of agricultural post-harvest value addition and essential to the food security identified by WHO/FAO. This traditional processing contributes to human energy food requirements, protein intake, fatty acids and trace element such as vitamins, flavonoid, polyphenol, carotenoids and minerals such as iron, zinc, calcium, magnesium, etc. It has been reported that lactic fermentation used as a traditional food processing technique includes general methods, such as mechanical dehulling of seeds, peeling of tubers, grating, boiling, soaking and pressing, as well as the usual fermentation stage, where microbial biochemical changes based on LAB sugar metabolism enhance flavor and aroma besides product acidification. This has significantly impacted the availability of a variety of foods in Africa and contributed to food preservation with a concomitant reduction in agricultural waste. Traditional processes that involve fermentation of agricultural products are a common practice throughout Africa with a long history of household and small-scale, cottage-type production (Holzapfel 2002, Kimaryo et al. 2000). Many methods were developed to meet the need for food preservation and for adequate nutrition intake. Furthermore, fermentation processes result in acceptable organoleptic flavor and aromas or detoxification of product, all of which improve the raw material sensory characteristics or make them edible. The major types of fruits and vegetables fermentation processes identified in different regions of Africa are classified here on the basis of LAB either dominating or occurring in co-metabolism with other microbes as to impact biochemical transformation of different vegetable raw materials and fruits in Africa. These include (i) lactic acid fermented starchy vegetables (root crops and cereals) (ii) lactic-acid fermented leafy vegetables (iii) alkaline-fermented vegetable proteins containing lactic acid bacteria (iv) mixed lactic acid and alcohol fermented starchy-vegetable substrates and fruits (v) mixed lactic, acetic and alcohol-fermented fruits and vegetables. These are discussed with different examples in the following paragraphs.

4. Fermented Vegetables

4.1. Lactic Acid Fermented Starchy Vegetables (*Root Crops and Cereals*)

Lactic acid fermented starchy vegetables are consumed in many African countries as staple foods by over ca. 800 million people. Starchy cereals and tubers are processed and made into a dough or gruel or porridge. As many of these crops are drought resistant, their production needs limited inputs of irrigation and they are processed into a variety of foods by utilizing low-cost technologies. Some are consumed as the main daily meal, while others are served as a beverage, gruel or complementary food for infants during weaning. Some gruels are also consumed by adults for breakfast. Examples of these foods produced in different regions of Africa are shown in Tables 1 and 2. Their processing involves the initial preparation of substrates, i.e. tubers, such as cassava are peeled, washed and grated, whereas cereals such as maize, sorghum, millet and *tef* are pre-soaked before grating or occasionally mixed with malted grains in the case of *hussuwa*. Then follows lactic acid fermentation which relies on autochthonous LAB stemming

Table 1: Examples of Lactic Fermented Cereal Vegetable Starchy Foods in Africa

Fermented food product	Country	Vegetable substrate	Microorganisms	References
Ogi	Nigeria/ Benin	Maize, sorghum	*Lb. plantarum, Lb. fermentum, Pd. pentosaceus* and yeasts	Oguntoyinbo and Narbad (2012)
Kunu-zaki	Nigeria	Millet, sorghum	*Lb. plantarum, Lb. fermentum, Str. gallolyticus* subsp. *macedonicus* and *Pd. pentosaceus*	Oguntoyinbo and Narbad (2012)
Togwa	Tanzania	Millet, sorghum	*Lb. plantarum, Lb. fermentum, Issatchenkia orientalis, S. cerevisiae, C. pelliculosa* and *C. tropicalis*	Mugula et al. (2003a, b)
Agbelima	Ghana	Cassava	*Lb. plantarum, Lb. fermentum*	Amoa-Awua et al. (1996)
Kenkey	Ghana	Maize, sorghum	*Pd. pentosaceus, Lb. fermentum/ reuteri, Lb. brevis* and yeasts	Halm et al. (1993), Olsen et al. (1995)
Koko	Ghana	Maize, sorghum	*W. confusa* and *Lb. fermentum*, followed by *Lb. salivarius* and *Pediococcus* spp, yeasts	Lei and Jakobsen (2004)
Ben-saalga	Burkina faso	Millet, sorghum	*Lb. plantarum, Lb. fermentum, W. confusa*	Nout (2009), Songre-Ouattara et al. (2008)
Ingera	Ethiopia	*Tef*	*Lb. buchneri* and *Pd. pentosaceus*	(Nigatu 2000).
Kisra	Sudan	Sorghum	*Pd. pentosaceus, W. confusa, Lb. brevis, Lactobacillus* sp., *Erwinia ananas, Klebsiella pneumoniae, Enterobacter cloacae* and yeasts (*C. intermedia* and *Debaryomyces hansenii*), and molds (*Aspergillus* sp., *Penicillium* spp., *Fusarium* spp., and *Rhizopus* spp.).	Mohammed et al. (1991)
Hussuwa	Sudan	Sorghum	*Lb. plantarum, Lb. fermentum, Enterococcus* spp.	Yousif et al. (2010)
Uji	Kenya and Tanzania	Sorghum Sorghum bicolor, S. vulgare	*Lactobacillus* sp.	Mbugua (1985), Mbugua and Njenga (1992)
Obushera	Uganda	Millet, sorghum	*Lb. plantarum, Lb. fermentum, Str. gallolyticus, Str. infantarius* subsp. *infantarius*	Mukisa et al. (2012)
Ting (*motogo* and *bogobe*)	Botswana	Sorghum	*Lb reuteri, Lb. fermentum, Lb. harbinensis, Lb. plantarum, Lb. parabuchneri, Lb. casei* and *Lb. coryniformis*	Sekwati-Monang et al. (2012)
Degue	Burkina faso	Millet	*Lb. gasseri, Enterococcus* spp., *Escherichia coli, Lb. plantarum/ paraplantarum, Lb. delbrueckii, Bacillus* spp., *Lb. reuteri* and *Lb. casei*	Abriouel et al. (2006)

Contd...

Table 1: (Contd.)

Fermented food product	Country	Vegetable substrate	Microorganisms	References
Poto poto	Congo	Maize	*Lb. gasseri, Enterococcus* spp., *E. coli, Lb. plantarum/ paraplantarum, Lb. acidophilus, Lb.delbrueckii, Bacillus* spp., *Lb. reuteri* and *Lb. casei*	Abriouel et al. (2006)

Table 2: Lactic Acid Fermented African Vegetable Starchy Foods from Root Crops

Fermented food product	Country	Vegetal substrate	Microorganisms	References
Fufu or *foofoo*	Nigeria	Cassava	*Lactococcus lactis, Ln. mesenteroides, Lb. plantarum*	Oyewole (1997) Brauman et al. (1996)
Lafu	Nigeria	Cassava	*Lb. fermentum Lb. plantarum, W. confusa, S. cerevisiae, P. scutulata, K. marxianus, H. guilliermondii, P. rhodanensis, C. glabrata, P. kudriavzevii, C. tropicalis, T. asahii*	Padonou et al. (2009)
Gari	Nigeria	Cassava	*Lb. plantarum, Lb. fermentum, W. confusa, C. tropicalis*	Huch Nee Kostinek et al. (2008), Oguntoyinbo and Dodd (2010)
Kocho	Ethiopia	Enset	*Lb. plantarum, Lb. fermentum, Lb. rhamnosus, Lb. sakei, Lb. parabuchneri, Lb. gallinarum, Lb. casei, Weissella* spp.	Nigatu (2000)
Agbelima	Ghana	Cassava	*Lb. plantarum, Lb. fermentum*	Amoa-Awua et al. (1996)
Candi	Nigeria	Cassava	LAB, yeasts	Odunfa (1988)
Kpokpogari	Nigeria	Cassava	LAB, yeasts	Odunfa (1988)
Akyeke	Ghana		*Lb. plantarum, Lb. brevis, Ln. mesenteroides* subsp. *cremoris*	Obilie et al. (2003, 2004)
Bikedi	Congo		*Lactobacillus* spp. *Bacillus* spp.	Kobawila et al. (2005)
Elubo		Yam	*Lb. plantarum, Lb. brevis, Lb. delbrueckii* and *B. subtilis, Citrobacter freundii, P. burtonii* and *C. krusei*	Achi and Akubor (2000)
Iyan ewu		Yam	—	

from plant surfaces for fermentation. In some cases, two-stage fermentation processed used besides addition of other products such as spices, groundnut, ginger, pepper, sweetener and salts.

The representative flow diagrams showing the processing of starchy vegetables into dough during production of *fufu*, *gari*, *agalima* and *kocho* from root crops, as well as *ingera*, *kisara*, *hussuwa* and *ting* (*motogo* and *bogobe*) from cereals, are shown in Figure 1a, b. Also, gruel, porridge and

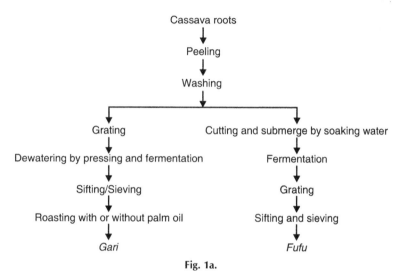

Fig. 1a.

cooked gel formation during processing of *ogi*, *kunu-zaki*, *togwa*, *koko* or *akasa* and *kenkey* are shown in Figures 2 and 3. The different traditional fermentations practised during production include solid state fermentation with consistent removal of water, e.g. cassava (*Manihot esculenta* Crantz) processing into *gari* in Nigeria, Ghana, Benin, or *chikawangue* in Congo, or *tef* fermentation for production of *ingera* in Ethiopia. Submerged fermentations, on the other hand, are performed by submerging substrates in water, as is the case for maize, sorghum and millet for production of *ogi*, *togwa*, *kunu-zaki*, and *hussawa* processing in Nigeria, Tanzania and Sudan, respectively, as well as for cassava fermentation for *fufu* and *lafun* in Nigeria. A further method is pit fermentation in processing of false banana (*Ensete ventricosum*) into *kocho* as is popular in Ethiopia. Pits are first dug with caution to prevent flooding and soil contamination of the fermenting mash by using layers of banana leaves. Pits vary in size and depth, but an average-size pit for family use was reported to be about 0.6 to 1.5 meters deep and 1.2 to 2 meters wide. The pit facilitates fermentation, storage and subsequent starter culture inoculation similar to back-slopping (Nigatu 2000). It was reported that all the above fermentations in different countries were mainly lactic fermentations, in which LAB are the major microbiota responsible for biochemical changes during fermentation and leading to a

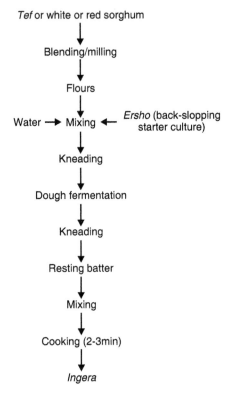

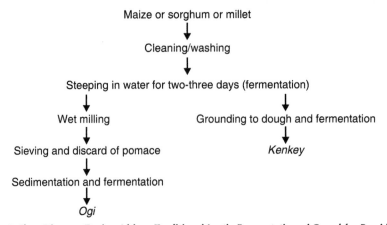

Fig. 1b: Flow Diagram showing Processing of Starchy Vegetables into Dough (a) Cassava Processing for *gari* and *fufu* Production (b) Processing of *tef* and Sorghum Processing for *ingera* Production

Fig. 2: Flow Diagram During African Traditional Lactic Fermentation of Cereal for Porridge Production

rapid acidification from ca. pH 6.5 to 3.5. The acidic pH contributes to the flavor characteristics and the preservation of the final product (Nout et al. 1989, Nout and Sarkar 1999). Many of these products show a long shelf-life, while maintaining their nutritional composition with improved protein digestibility (Axtell et al. 1981), thus supporting the dietary requirements of the people of Africa.

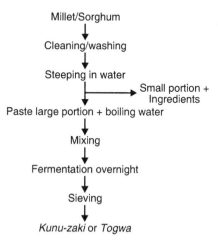

Fig. 3: Flow Diagram of African Traditional Lactic Fermentation of Cereal for Gruel Production

Studies on microbial dynamics during fermentation showed that traditional fermentation as practised in different countries usually involves the occurrence of successions of different LAB. Furthermore, recent microbial diversity studies using denaturing gradient gel electrophoresis (DGGE), clone libraries and high throughput-sequencing techniques showed a considerable agreement with culture-dependent methods and that different species of LAB were the dominant microorganisms during traditional fermentation (Humblot and Guyot 2009, Mukisa et al. 2012, Oguntoyinbo et al. 2011). Table 1 shows a list of bacteria identified from different fermented vegetable starchy foods in Africa. There is a general agreement that LAB including *Lb. plantarum*, *Lb. fermentum*, *Weissella* spp., *Leuconostoc mesenteroides*, *Enterococcus* spp. and *Pediococcus* spp. are the major bacteria associated with different products in Africa. Also, studies show that *Lb. plantarum*, *Lb. fermentum* and *Leuconostoc* species as predominant in most of the products and information is available on the succession of these bacteria, their complementary roles, as well as acidification during fermentation (Efiuvwevwere and Akona 1995, Nout 2009).

Different studies identified the important role played by *Leuconostoc* in initiating the fermentation of plant materials. However, amylase activity has also been identified as essential enzymatic activity during fermentation of starchy vegetables — a trait which occurs especially among some amylolytic *Lactobacillus* spp. (Calderon Santoyo et al. 2003, Humblot et al.

2014, Maktouf et al. 2013). Well-studied amylase-producing strains, such as *Lb. plantarum* A6 and *Lb. plantarum* ULAG11 were identified with properties that are essential for starting of the fermentation with a rapid drop in the pH, especially during the first 24 hours of lactic acid fermentation (Oguntoyinbo and Narbad 2012). The *in situ* expression of amylase genes during fermentation of starchy vegetable was recently reported (Oguntoyinbo and Narbad 2012, Humblot et al. 2014).

Also, the occurrence of heterofermentative LAB that produce CO_2 (carbon dioxide) in plant fermentations could be very important in the prevention of growth of yeast and oxidase-positive and Gram-negative bacteria such as *Aeromonas* and *Pseudomonas*. Co-inoculation of *Lb. plantarum* with complementary strains, such as *Lb. fermentum* and *Leuconostoc*, can be essential to a safe fermentation process development.

Microbial phytase plays an important role during vegetable starch fermentation in Africa. Recently, *Lb. buchneri* strain MF58 and *Pd. pentosaceus* strain MF35 were isolated from *tef-ingera* demonstrated phytate hydrolysis in a pilot fermentation study (Fischer et al. 2014). Emerging information shows that yeast, such as *Pichia kudriavzevii* TY13 and *Hanseniaspora guilliermondii* TY14, also showed strong phytate degradation in a model *togwa* production from maize (Hellstrom et al. 2012). This is relevant to the mineralization of fermented cereals particularly rich in phytate, but low in iron and zinc.

Natural fermentation practices, i.e. non-utilization of starter cultures for the production of these foods in Africa, are common. Possibly as a consequence of this, different potentially pathogenic bacteria were isolated from some of these fermented plant foods (Oguntoyinbo 2014, Schoustra et al. 2013). Toxinogenic bacilli strains, for example, were isolated from *ben-saalga*, *ogi* and *kunu-zaki* (Humblot et al. 2012, Oguntoyinbo and Narbad 2012), *Streptococcus galactolyticus* and *Str. infantarius* subsp. *infantarius* with undefined virulence and pathogenic traits were isolated from *kunu-zaki* and *obushera* (Mukisa et al. 2012, Oguntoyinbo and Narbad 2012). Also, species of enterococci with anti-microbial resistances or presence of virulence determinants were detected in *hussuwa* (Yousif et al. 2005, Yousif et al. 2010).

Investigations on the shelf-life of cassava flour also showed that the storage quality is subject to microbial contamination by *Aspergillus niger* and *Rhizopus* spp. (Obadina et al. 2007). Issues of mycotoxin contamination of traditional fermented foods and its challenges in Africa have been reported (Bankole et al. 2006, Bankole and Mabekoje 2004). The use of defined starter cultures capable of fast acid production and the possible industrialization of such traditional fermentations of starchy vegetables could therefore increase the safety and prevent adulteration by spoilage microorganisms.

In many parts of West Africa, yam tubers (*Dioscorea rotundata*) are produced and consumed after boiling directly, or are pounded with a pestle and mortar to produce a gel that is consumed with different soups and sauces. The remaining portions are fermented by some tribes for preservation and generation of the characteristic flavour that is very pleasing to taste. The fermented product is called *ewu iyan* by the Igbomina land-people in Nigeria or *iyan anan* by Ikere-Ekiti people in south-western Nigeria. Also, yams are directly processed into chips or cut into small sizes, blanched in water at 60°C for 30 min and fermented naturally for 24 hours. After this, they are sun-dried, pulverised and reconstituted in hot water to form a gelatinous product, known as *elubo. Elubo* is an important food product consumed by millions of people in Nigeria, especially amongst the Yoruba ethnic group. The flow diagram of *elubo* production is shown in Figure 4. Achi and Akubor (2000) reported an underlying lactic fermentation and identified *Lb. plantarum, Lb. brevis* and *Lb. delbrueckii* as dominating bacteria responsible for rapid acidification. Also, yeast species were identified to occur during fermentation and included yeasts such as *P. burtonii* and *Candida krusei*. It appears that different yeast strains belonging to *Saccharomyces cerevisiae, P. scutulata, Kluyveromyces marxianus, H. guilliermondii, P. rhodanensis* and *C. glabrata*, as well as *P. kudriavzevii, C. tropicalis* and *Trichosporon asahii* are commonly associated with lactic fermentation in Africa (Padonou et al. 2009).

Yam tuber
↓
Washing
↓
Peeling
↓
Cutting to uniform sizes
↓
Blanching for 10 min in boiling water
↓
Fermentation
↓
Drying
↓
Milling into (*elubo*) flour

Fig. 4: Flow Diagram of Lactic Acid Fermentation of Yam During *elubo* Production

4.2. *Lactic Fermented Leafy Vegetables*

The tropical climate and agricultural land in Africa supports the growth of different leafy vegetables. Some of these plants are traditional to Africa and only successfully grow in this continent; these are listed in Table 3. Leafy vegetables have a short shelf-life and are highly perishable, and different African leafy vegetables (ALVs) have been identified as being indigenous to different regions in Africa (Shiundu and Oniang'o 2007) (Table 3). Traditional processing of ALVs immediately after harvest includes

Table 3: Distribution of Some Regional and Common African Leafy Vegetables*

All over the sub-continent	West/East and Central Africa	West and Southern Africa	East/Central and Southern Africa
Abelmoschus esculentus (ladies' fingers)	*Basella alba* (vine spinach)	*Amaranthus caudatus* (Aluma)	*Solanum nigrum* (black nightshade)
Amaranthus cruentus (amaranth)	*Citrullus lunatus* (watermelons)	*Amaranthus hybridus* (amaranth)	*Bidens pilosa* (black-jack)
Corchorus olitorius (jute mallow)	*Colocasia esculenta* (cocoyam)	*Portulaca oleracea* (purslane)	*Cleome gynandra* (African cabbage)
Cucurbita maxima (pumpkins)	*Hibiscus sabdariffa* (zobo)		
Vigna unguiculata (cowpea)	*Ipomea batatas* (sweet potato)		
Solanum macrocarpon (African eggplant)	*Manihot esculenta* (cassava)		
	Solanum aethiopicum (Mock Tomato)		
	Solanum scarbrum (Garden Huckleberry)		
	Talinium triangulare (waterleaf)		
	Vernonia amygdalina (ewuro)		
	Moringa oleifera (moringa or drumstick tree)		
	Solanecio biafrae (Worowo)		

*Adapted from Smith and Eyzaguirre (2007)

washing, shredding and drying. Sun-drying and fermentation are the two most important processing techniques used for processing of ALVs (Ayua and Omware 2013). Some ALVs are also fermented after shredding; an example is the production of *kawal* in the Sudan, where the fresh leaves of the leguminous plant *Cassia obtusifolia* are fermented and consumed as meat or fish protein substitutes in soups and sauces (Suliman et al. 1987). The leaves are abundantly available and serve as a cheap source of proteins and amino acids, with a high composition of oxalate (Dirar et al. 1985). Production of *kawal* involves solid state fermentation of the leguminous leaves that involves bacterial species, such as *B. subtilis*, *Propionibacterium* and *Staphylococcus sciuri*, with participation of LAB, such as *Lb. plantarum* (Dirar et al. 1985).

In Congo, *ntoba mbadi* is a fermented leafy vegetable consumed as a condiment (Sanni and Oguntoyinbo 2014). Kobawila et al. (2005)

reported a flow diagram for the fermentation processing of *ntoba mbadi*. The processing involves sun-drying cassava leaves for two to three hours to wilt the leaves, which allow easier removal of stalks and petioles. The lamina are cut into fragments and then washed with water, packed and wrapped in papaya (*Carica papaya*) leaves and are then left to ferment for two to four days in a basket. The fermentation is a semi-solid process that involves microbial metabolic activities leading to the change to alkalinity with a steady increase to pH 8.5. Bacteria reported to be involved in this alkaline fermentation include the *Bacillus* spp. *B. macerans*, *B. subtilis* and *B. pumilus*. Other bacteria, such as *St. xylosus* and *Erwinia* spp., as well as LAB such as *Ec. faecium*, *Ec. hirae*, *Ec. casseliflavus*, *W. confusa*, *W. cibaria* and *Pediococcus* spp. have also been reported to occur (Sanni and Oguntoyinbo 2014, Ouoba et al. 2010). Apart from the effect of lactic acid preservative influence, reduction of cyanogenic acid in the leaves and mineralization are further beneficial changes brought about by the fermentation process (Ouoba et al. 2010, Sanni and Oguntoyinbo 2014). In Kenya, cowpea leaves (*Vigna unguiculata* syn. *Vigna sinensis*) are part of the diet of the people and a recent study showed that natural fermentation can improve the storage quality, retain beta-carotene by 91 per cent and ascorbic acid by 15 per cent, while a sensory evaluation showed consumer acceptance of the fermented cowpea (Muchoki et al. 2007).

In Kenya, African kale leaves are processed in a fermentation-like manner by soaking the vegetables in milk for a few days to remove the bitter taste. Unconfirmed information (Mathara, personal communication) reported LAB in the milk was responsible for the fermentation that led to the removal of bitterness. Recently, LAB fermentation of ALV in a 2.8–3.0 per cent NaCl solution was shown to impact the preservative influence on some ALVs, increase shelf-life due to acidity of the medium that prevented spoilage bacteria and limit yeast proliferation in the menstruum through production of CO_2. This process involved rapid metabolism of LAB within the fermenting menstruum with the dominance of LAB throughout the process being essential. Therefore, NaCl could facilitate osmotic pressure that releases metabolizable carbohydrate for enzymatic breakdown by LAB to support their rapid growth and produce lactic acid that reduces pH of the medium from 6.0 to 3.8 (Halász et al. 1999).

4.3. Alkaline Fermented Vegetable Proteins Involving Lactic Acid Bacteria in the Fermentation

A significant proportion of the protein intake in African countries is through vegetable-plant-protein sources, notably the proteinacous seeds (oilseeds), which are consumed in the form of fermented vegetable proteins (Odunfa 1988). The seeds bearing the cotyledon used in production of condiments are produced in large quantities in Africa, especially by members of the family *Malvaceae*, such as *Adansonia digitata*, *Parkia biglobosa, Prosopis africana, Hibiscus sabdariffa* and *Fabaceae,* leguminous bean-producing plants, e.g. cowpea (*Vigna unguiculata*) and soy beans (*Glycine max*) (Parkouda et al. 2009). Some of the African fermented vegetable-protein seeds and their corresponding condiments produced and consumed in different regions of Africa are shown in Table 4.

Table 4: African Fermented Vegetable Proteins with Reported Lactic Acid Bacteria

Fermented food product	Country	Vegetable substrate	Microorganisms	References
Iru or *Dawadawa*	Nigeria	*Pakia biglobosa*	*B. subtilis, B. amyloliquifaciens,* other LAB	Adewumi et al. (2012)
Okpehe	Nigeria	*Prosopis africana*	*B. subtilis, B. amyloliquefaciens, B. cereus,* and *B. licheniformis, Enterococcus* spp.	Oguntoyinbo et al. (2010)
Maari	Burkina Faso	*Adansonia digitata*	*B. subtilis, Ec. faecium, Ec. casseliflavus, Pd. acidilactici*	Sanni and Oguntoyinbo (2014), Parkouda et al. (2010)
Bikalga	Burkina faso	*Hibiscus sabdariffa*	*B. subtilis, B. licheniformis, B. cereus, B. pumilus, B. badius, W. confusa, W. cibaria, Lb. plantarum, Pd. pentosaceus, Ec. casseliflavus, Ec. faecium, Ec. faecalis, Ec. avium* and *Ec. hirae, Brevibacillus bortelensis, Br. sphaericus and Br. fusiformis.*	Ouoba et al. (2010, 2008)

The climatic conditions in Africa favor a wide diversity and distribution of *Malvaceae* across the continent. The seeds are not directly consumed because of their anti-nutritional compounds, such as proteinase inhibitors, amylase inhibitors, metal chelators, flatus factors, hemagglutinins, saponins, cyanogens, lathyrogens, tannins, allergens, acetylenic furan and isoflavonoid phytoalexins (Pariza 1996). *Parkia biglobosa* and soybean contain trypsin inhibitors, which reduce the digestibility of proteins (Collins and Saunders 1976) and molecular weight carbohydrate fractions, which are responsible for flatulence after ingestion (Fleming 1981). Also, soybean contains high level (120–150 g kg^{-1} dry wt) of α-galactosides of sucrose, causing gastrointestinal gas production in humans (Rackis et al. 1970). Kawamura (1954) observed that over 90 per cent of the sugars present in ripe soybeans comprise sucrose, raffinose and stachyose. Cottonseed also contains gossypol, an anti-nutritional factor, while mesquite seeds, *Prosopis Africana*, could cause fetal abortion in domestic animals. However, there is long history of consumption of these seeds in Africa (Odunfa 1985). Processing and fermentation therefore contributes significantly to the extensive hydrolysis of the seeds and concomitant detoxification. Different communities have developed strategies for processing of the seeds for food, especially through natural fermentation to produce foods which are rich in vegetable proteins and are used as seasoning agents, or as meat or fish substitutes (Odunfa 1985, Steinkraus 1996).

Traditional processing of seeds includes wet dehulling, boiling and fermentation. There are similar fermented vegetable proteins with different names in Africa; even the processing techniques often follow a similar methodology. The common examples of fermented vegetable proteins reported in Africa are shown in Table 5. The fermentation process during production has been described as alkaline fermentation due to microbial enzymatic changes that involve hydrolysis of proteins to polypeptides, peptides, amino acids and ammonia, thereby bringing about an increase

Table 5: Examples of Mixed Lactic, Acetic Acid and Alcoholic-Fermented Vegetable Starch Beverages in Africa

Fermented food product	Country	Vegetal substrate	Microorganisms	Reference
Tella	Ethiopia	Sorghum	Yeasts and LAB	Faparusi et al. (1973)
Burukutu	Ethiopia Nigeria Ghana	Guinea corn and cassava	*S. cerevisiae* yeasts, *Lb. plantarum* and *Lb. fermentum*	Faparusi et al. (1973)
Pito	Nigeria Ghana	Guinea corn and maize	*Leuconostoc* spp., *Lactobacillus* spp., *Saccharomyces* spp., *Candida* spp. and *Geotricum candidium*	Ekundayo (1977)
Kaffir beer	South Africa	Kaffir corn or maize	*S. cerevisiae*, *Lactobacillus* spp., *Acetobacter* spp.	Hesseltine (1979), Odunfa and Oyewole (1998)
Busaa	East Africa	Maize	*S. cerevisiae*, *C. krusei*, *Lb. plantarum*, *Lb. helveticus*, *Lb. salivarius*, *Lb. brevis*, *W. viridescens*, *Pd. damnosus*, *Pd. parvulus*.	Nout (1980), Harkishor (1977)
Malawa beer	Uganda	Maize	Unknown	—
Zambian opaque beer	Zambia	Maize	Unknown	—
Merissa	Sudan	Sorghum	LAB, yeasts	Dirar (1978)
Sekete	Nigeria (south)	Maize	*Ac. aceti*, *Ac. pasteurianus*, *Lb. brevis*, *Lb. buchneri*, *Lb. plantarum*, *Lactobacillus* spp., *S. cerevisiae*, *Saccharomyces* spp., *Flavobacterium* spp., *Micrococcus varians*, *B. licheniformis*	Sanni et al. (1999)
Bouza	Egypt	Wheat and maize	Unknown	
Kishk	Egypt	Wheat and milk	*Lactobacillus*, yeasts and *B. subtilis*	Morcos et al. (1973)
Tchoukoutou	Benin	Sorghum	Yeasts and LAB	Greppi et al. (2013)

in the pH from 6.8 to 8.0. Fermented vegetable proteins are found to be very rich in polyglutamic acid as a result of *Bacillus* metabolism, with compounds such as 3-hydroxybutanone (acetoin) and derivatives [butanedione (diacetyl) and 2, 3-butanediol], acids (acetic, propanoic, 2-methylpropanoic, 2-methylbutanoic and 3-methylbutanoic), as well as pyrazine also being produced.

The microbial ecology of microbes predominantly responsible for the important biochemical changes during traditional fermentation of vegetable proteins was shown to involve diverse bacterial species. Starter cultures are not used and natural fermentation is dominated by different bacteria with enzymatic activities, including *B. subtilis* group bacteria, such as *B. subtilis sensu stricto*, *B. licheniformis*, *B. amyloliquefaciens*, and *B. pumilus*. Studies showed high proteinase and amylase activities from the onset of fermentation up to 48 hours. Different species of LAB were isolated during fermentation of vegetable proteins for condiment production in Africa. Ouoba et al., (2010) reported the diversity of LAB in alkaline fermented foods in Africa to include *Ec. faecium*, *Ec. hirae* and *Pd. acidilactici* from *bikalga* and *soumbala*. Oguntoyinbo et al. (2007) isolated *Enterococcus* spp. from *okpehe* which exhibited a cheese-like aroma develop during model fermentation, demonstrating that these bacteria did not play a significant role in the fermentation but rather affected the product characteristics in a negative way. A recent study could, on the other hand, show that LAB plays a role in the flavor development during fermentation of vegetable proteins of other legumes, as shown in Table 4. An *in vitro* determination of volatile compound development during starter culture-controlled fermentation of *Cucurbitaceae* cotyledons showed that a mixed culture of *Lb. plantarum*, *Torulaspora delbrueckii* and *Pd. acidilactici* contributes to development of volatile compounds, such as esters and low concentrations of aldehydes and ketones during fermentation (Kamda et al. 2015).

4.4. Alcoholic and Lactic Fermented Starchy-Vegetables

Tubers and malted cereals are starchy vegetables fermented in Africa to produce a variety of alcoholic beverages by involving mixed cultures of LAB. Examples of such alcoholic beverages are shown in Table 5. The production process involves malting of cereals, grinding, submerging of grits in water for a fermentation period of two to five days, before filtration, in the case of *pito* brewing in Ghana or Nigeria, or without filtration, for example, in production of *burukutu* in Nigeria (Blandino et al. 2003, Faparusi et al. 1973). Traditional methods of production of some of these foods are shown in Figure 5. *Pito* and *dolo* are important alcohol-fermented malted cereal beverages popularly consumed in Nigeria, Ghana, Burkina Faso, Mali and Ivory Cost (Demuyakor and Ohta 1991, Sawadogo-Lingani et al. 2008, Onaghise and Izuagbe 1989). Also, *burukutu* consisting of fermented malted sorghum, is consumed in northern

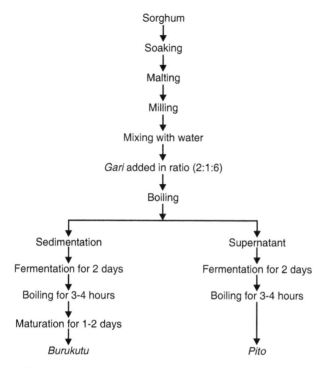

Fig. 5: Mixed Lactic, Acetic Acid and Alcohol Fermented Beverages in Africa

Nigeria (Sawadogo-Lingani et al. 2007, Faparusi 1970, Faparusi et al. 1973), and *sekete* in Nigeria (Adegoke et al. 1995, Fapohunda 1988, Sanni 1989b). In Ethiopia, *tella* is an important beer-like product made by fermentation of cereals. The production process includes addition of *gesho*, which is an extract from the leaves of *Rhamnus prinioides*. This extract imparts the characteristic bitter taste to beer and has been shown to possess medicinal properties similar to that of hops (Berhanu 2014). Microorganisms involved in fermentation include *Lb. fermentum, Lb. plantarum, Ln. mesenteroides* and the yeasts *S. cerevisiae* and *Schizosaccharomyces pombe* (Kolawole et al. 2007, Sawadogo-Lingani et al. 2007). The production process does not involve the use of starter cultures and the process is often uncontrolled. Acetic acid bacteria also proliferate, leading to production of vinegar. Sanni et al. (1999) reported the involvement of *Acetobacter aceti, Ac. hansenii* and *Ac. pasteurianus*, as well as other bacteria, such as *Alcaligenes, Flavobacterium* and *Lb. brevis*. These bacteria are presumed to contribute to the vinegary flavor and off-odors that become pronounced during deterioration of the beverages. Biogenic amine in traditional West African alcoholic beverages, the presence of zearalenone in cereals used in brewing and participation of spoilage yeast strains and pathogenic bacteria like *E. coli, St. aureus, B. subtilis, Streptococcus* spp., *R. stolonifer, A. niger, A. flavus*, and *Mucor* spp.,

St. aureus, B. subtillis and *Proteus* spp. has been reported (Kolawole et al. 2007, Lasekan and Lasekan 2000, Okoye 1986). This again points to the need for controlled fermentation with added starter cultures, which could inhibit the growth of contaminants and help to stabilize the product with respect to quality attributes. The major biochemical changes occurring during the fermentation process involve metabolic activities of yeast, leading to production of alcohol in the fermenting matrix. Studies on the nutritional composition of traditional fermented alcoholic beverages in Africa showed a decrease in ascorbic acid content but a concomitant increase in amino acids, such as leucine, lysine, valine, asparagine, glutamic acid, methionine and phenylalanine (Sanni and Oso 1988). Industrial production of sorghum beer in South Africa is one example of African fermented vegetable starch that has been industrially scaled-up and for which the process involves an optimized, well-studied and well-defined starter culture. The production and consistency of the product quality is thereby sustained.

5. Fermented Fruits

5.1. Alcoholic Beverages from Fruits Involving Lactic Acid Bacteria in Fermentation

Different fruits are grown in Africa and harvested annually in different regions of the continent. Fruits used in Africa include banana, papaya, *marula*, mango, tomato, sausage tree-fruit *Kigelia africana* and *Ziziphus mauritiana* (*masau*). Because of low pH and high acidity of fruits, microbial deterioration is very slow and usually only osmophilic microbes are responsible for the major biochemical changes. Fruits are processed into different products that include juices, pickles, alcoholic beverages and vinegar. A typical method of processing and fermentation of African fruit during *muratina* production is shown in Figure 6. Other examples of fermented fruits in Africa are shown in Table 6. In Nigeria, information is available on *agadagidi*, an effervescent drink produced from ripe plantain (*Musa paradisiaca*) pulp. It is a popular drink in south western Nigeria during ceremonies (Sanni 1989a, Sanni and Oso 1988). Similarly, *uruaga* is a fermented used banana in Uganda, while cashew and cocoa wine are popular in Nigeria.

In Zimbabwe, wild fruits, such as *masau*, are usually processed into porridge, traditional cakes, *mahewu* and jam (Nyanga et al. 2007). Moreover, they are fermented to produce alcoholic beverages such as *kachasu*. Nyanga et al. (2008) reported that *masau* is rich in citric acid, tartaric, malic, succinic and oxalic acids, minerals, fiber, crude protein and vitamin C (Nyanga et al. 2013a). *Lb. agilis, Lb. plantarum, W. minor, W. divergens, W. confusa, Lb. hilgardii, Lb. fermentum* and *Streptococcus* spp. were isolated from *masau* fruit products and were identified as bacteria that could be developed as starter cultures for fermentation of fruit products (Nyanga et al. 2007).

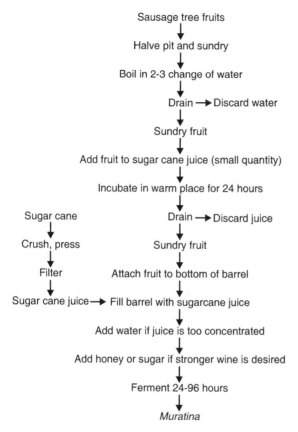

Fig. 6: Flow Diagram of Fermented Sausage Tree Fruit for *muratina* Production.
Adapted from Harkishor, 1977

Recently, a fortified lactic acid-fermented probiotic dairy with a 14 per cent (wt/vol) concentrated baobab fruit pulp, *mutandabota*, was developed in Zimbawe. *Lb. rhamnosus* (yoba) was used as starter culture for the fermented dairy drink, leading to a product with pH of 3.5, rich in protein and vitamin C, as well as possessing the potential for improvement of intestinal health (Mpofu et al. 2014a, b). The need for development of lactic acid-fermented beverages that could support a healthy life of human population and contribute to the dietary intake based on the knowledge of fermented foods has been suggested (Franz et al. 2014).

The fermented juice from palm sap of both *Rafia guinensis* and *Borassus akeassii*, popularly known as palm wine, is consumed widely in many African countries. During the fermentation process, *S. cerevisiae* ferments the glucose and many other plant-derived carbohydrates, such as sucrose, maltose and raffinose to produce alcohol. Studies show that other yeasts, such as *Sch. pombe* are involved in the traditional fermentation process. Apart from yeasts, *Lactobacillus* and acetic acid bacteria are described to

Table 6: Examples of Mixed Lactic, Acetic Acid and Alcoholic Fermented Fruit Beverages in Africa

Fermented food product	Country	Fruit and vegetable	Fermentation	Microorganisms	Reference
Agadagidi	Nigeria	Plantain	Alcoholic	*Saccharomyces* spp., *Leuconostoc* spp., *Streptococcus* spp., *Bacillus* spp., *Micrococcus* spp.	Sanni (1989a)
Cashew wine	Nigeria	Cashew	Alcoholic	Unknown	—
Cocoa wine	Nigeria	Cocoa	Alcoholic	Unknown	—
Palm wine *Emu* or *oguro*	Africa	Palm sap	Lactic, later alcoholic and acetic acid	*Lb. plantarum, Ln. mesenteroides, Fructobacillus durionis, Str. mitis.* Acetic acid bacteria. *S. cerevisiae, Arthroascus* spp., *Issatchenkia* spp., *Candida* spp., *Trichosporon* spp., *Hanseniaspora* spp., *Kodamaea* spp., *Schizosaccharomyces* spp., *Trigonopsis* spp,. *Galactomyces* spp.	Ehrmann et al. (2009), Faparusi and Bassir (1972), Ouoba et al. (2012)
Uruaga	Kenya	Banana	Alcoholic and lactic	Unknown	—
Ulansi	East and South Africa	Bamboo	Alcholic and lactic	Unknown	—
Muratina	Kenya	*Sausage tree fruit (Kigelia africana)*	Alcholic and lactic	Uknown	—

play a role in fermentation and these were isolated at the initial and later stages of the fermentation respectively (Amoa-Awua et al. 2007, Ouoba et al. 2012).

Fruit processing into wine is developed on an industrial scale in South Africa. Grapes are commonly used for wine production and LAB play an important role during malolactic fermentation. This is a decarboxylation process by which malic acid, which is naturally present in grapes, is converted to lactic acid with concurrent liberation of carbon dioxide. Fermentation plays important role in aroma development of specific wines. LAB, such as *Oenococcus oenii* and various species of *Lactobacillus* and *Pediococcus*, are reported to occur in wine or to play a role in malolactic fermentation during South African wine production (Du Toit et al. 2011, Miller et al. 2011). Recently, a bacteriocin-producing *Ec. faecium* was isolated from South African wine (Ndlovu et al. 2015), but whether such bacteria play a beneficial or detrimental role is currently not known. Co-metabolic activities of LAB with *S. cerevisiae* in a transcriptomic study show that LAB metabolic actives have reciprocal responses, justifying yeast development of oenological aroma during fermentation (Rossouw et al. 2012).

In Africa, different fruits are processed into vinegar on a very small scale. Fruit vinegars are made from fruit wines that are processed from fruits, such as plum, mango, apple cider, marula, honey, coconut and grapefruit (Ndoye et al. 2007).

5.2. Effect of Fermentation on Nutrients Bioavailability and Detoxification

An important outcome of fermentation is the extensive hydrolysis that impacts the substrates and contributes the preservative influence. Fermentation thus has reciprocal effects on decomposition of complex molecules. Protease activities during alkaline fermentation of vegetable proteins, amylase and pectinase in starchy vegetables are important in these macromolecule degradative processes, resulting in the breakdown of proteins, carbohydrates and oligosaccharides and to the release of important compounds essential to human nutritional requirements (Motarjemi and Nout 1996). Processing, that includes microbial fermentation and germination as well as the endogenous enzymes produced during processing, contributes significantly to the bio-availability and enrichment of fermented products. Microbial phytase activities also contributes to the reduction of phytate composition in cereals (Kayode et al. 2007). Compounds, such as flavonoids, alkylresorcinols, glucosinolates and saponins are present in the many fruits and vegetables that are consumed in Africa and may contribute to the consumer's health.

Yeast activity in fermentation may also increase the vitamin content of vegetables and fruits, such as the availability of riboflavin, vitamin B_{12} and niacin. Riboflavin and niacin concentrations increase in alcohol-fermented vegetable starches in products, such as sorghum beer, a popular drink in South Africa, which significantly reduce incidences of pellagra (Steinkraus 2002). Palm wine is also a very rich source of ascorbic acid, thiamine and pyridoxine as well as vitamin B_{12} and other B vitamins (Steinkraus 1997). Also, fermented foods are a rich source of folate, which is present in various green leafy vegetables, cereals and legumes and are believed to prevent heart disease, cancer and neuropsychiatric disorders (Brouwer et al. 1999).

From a safety point of view, mycotoxicosis may be a problem in many African countries due to contamination of poorly processed fruits, legume and cereals (Roige et al. 2009). In some cases, fermentation reduces aflatoxin B_1 concentration up to 30 per cent in fermented cereal gruels (Dogi et al. 2013). Processing and specific LAB with the potential to inhibit activity of *A. parasiticus* and *Fusarium germineraum*, as well as reduce the composition of aflatoxin B_1 and zearalenone are found to be important for fermentation of cereals (Adegoke et al. 1994, Fandohan et al. 2005). The multiple problems that still occur on the African continent include problems of infrastructure, water supply, sanitation and hygiene during processing. These, however, do compromise the safety and quality of many traditional lactic acid-fermented foods. Home and cottage sized, small-scale food processing endeavors, using crude techniques and rudimentary utensils, are mainly adopted and these are relatively uncontrolled processes, thereby exposing many of these foods to inconsistent quality or to different pathogenic microbes (Oguntoyinbo 2014).

6. Recommendations

As mentioned above, starter cultures are generally not used for the production of lactic-acid fermented African fruits and vegetables. As a consequence, different consortia of microbes participate during traditional fermentations in a relatively uncontrolled fermentation. This leads to products with inconsistent quality and posing safety concerns. Studies emphasize the use of adequately characterized microbial strains as starter cultures in the production of lactic acid-fermented foods in Africa. This guarantees a predictable process and facilitates a desired product consistency and safety. The use of starter culture could be combined with back-slopping techniques, especially in a household or on a small-enterprise scale, where the known microbial population of LAB in a successful batch produced with starters could be used for subsequent batches. Some of the challenges that starter cultures could face in African food processing has been previously identified (Holzapfel 2002). However, a pilot starter-culture fermentation study in Africa during *gari* production in Benin showed rapid acidity and effective pathogen control using defined starter cultures strains (Huch Nee Kostinek et al. 2008). Other studies showed that bacteriocins produced by LAB is an important characteristic of starter cultures targeted for use by small-scale food processor to facilitate pathogen control (Galvez et al. 2007, Oguntoyinbo et al. 2007). Recent review of the use of microbial strains in the development of probiotic foods as functional foods in Africa indicated that certain traditional fermented fruits and vegetables could be optimized to become functional foods, which would be in line with the WHO recommendation of the need to use traditional foods for nutritional and health improvement (Franz et al. 2014). However, some of the challenges associated with the use of starter cultures include accumulation of biogenic amines formed by decarboxylation of amino acids or by amination and transamination of aldehydes and ketone, and have toxicological effect on humans. Examples include tyramine, histamine, putrescine, and cadaverine (Bover-Cid and Holzapfel 1999, Simon-Sarkadi and Holzapfel 1995). Biogenic amine-producing strains were identified among enterococci isolated from *hussuwa*. Studies showed that the occurrence in *kenkey* was a low total of <60 ppm but increased in cadaverine and tyramine <200–500ppm as well as putrescine was reported when white and red cowpea were added to *kenkey* as protein-enrichment. This indicates that potential starter cultures should first be carefully screened for the absence of biogenic amine activity, for technologically useful traits such as reduction of α-galactoside sugars and for production of bacteriocin(s), α-amylase activity and fast acidification capacity. The most urgent need, however, is for the local industry, governmental agencies and research institutions to cooperate in an effort to industrialize fermentation technology for processing African fruits and vegetables into traditional products.

Acknowledgments

The authors thank the Georg Forster Fellowship for Experienced Researcher of Alexander von Humboldt Stiftung, Germany, for funding research visit of FAO. The responsibility for the article content lies solely with the authors.

References

Abriouel, H., N. Ben Omar, R.L. Lopez, M. Martinez-Canamero, S. Keleke and A. Galvez. 2006. Culture-independent analysis of the microbial composition of the African traditional fermented foods *poto poto* and *degue* by using three different DNA extraction methods. *International Journal of Food Microbiology* 111(3): 228-233.

Abukutsa-Onyango, M.O. 2007. Seed production and support systems for African leafy vegetables in three communities in western Kenya. *African Journal of Food, Agriculture, Nutrition and Development* 7 (3).

Achi O.K. and P.I. Akubor. 2000. Microbiological characterization of yam fermentation for 'Elubo' (yam flour) production. *World Journal of Microbiology and Biotechnology* 16(1): 3-7.

Adegoke, G.O., R.N. Nwaigwe and G.B. Oguntimein. 1995. Microbiological and biochemical changes during the production of *sekete* — A fermented beverage made from maize. *Journal of Food Science and Technology* 32: 516-518.

Adegoke, G.O., E.J. Otumu and A.O. Akanni. 1994. Influence of grain quality, heat and processing time on the reduction of aflatoxin B$_1$ levels in 'tuwo' and 'ogi': two cereal-based products. *Plant Foods for Human Nutrition* 45(2): 113-117.

Adewumi, G.A., F.A. Oguntoyinbo, S. Keisam, W. Romi and K. Jeyaram. 2012. Combination of culture-independent and culture-dependent molecular methods for the determination of bacterial community of *iru*, a fermented *Parkia biglobosa* seeds. *Frontiers in Microbiology* 3: 436.

Amoa-Awua, W.K., F.E. Appoh and M. Jakobsen. 1996. Lactic acid fermentation of cassava dough into agbelima. *International Journal of Food Microbiology* 31(1-3): 87-98.

Amoa-Awua, W.K., E. Sampson and K. Tano-Debrah. 2007. Growth of yeasts, lactic and acetic acid bacteria in palm wine during tapping and fermentation from felled oil palm (*Elaeis guineensis*) in Ghana. *Journal of Applied Microbiology* 102(2): 599-606.

Axtell, J.D., A.W. Kirleis, M.M. Hassen, N. D'Croz Mason, E.T. Mertz and L. Munck. 1981. Digestibility of sorghum proteins. *Proceedings of the National Academy of Sciences, USA* 78(3): 1333-1335.

Ayua, E. and J. Omware. 2013. Assessment of processing methods and preservation of African leafy vegetables in Siaya county, Kenya. *Global Journal of Biology Agriculture and Health Sciences* 2: 46-48.

Bankole, S.A. and O.O. Mabekoje. 2004. Occurrence of aflatoxins and fumonisins in pre-harvest maize from south-western Nigeria. *Food Additives and Contaminants* 21(3): 251-255.

Bankole, S., M. Schollenberger and W. Drochner. 2006. Mycotoxins in food systems in Sub-Saharan Africa: A review. *Mycotoxin Research* 22(3): 163-169.

Battcock, M. and S. Azam-Ali. 1998. Fermented fruits and vegetables: A global perspective. *FAO Agricultural Services Bulletin No. 134*. M-17 ISBN 92-5-104226-8.

Berhanu, A. 2014. Microbial profile of *Tella* and the role of *gesho* (*Rhamnus prinoides*) as bittering and antimicrobial agent in traditional *Tella* (Beer) production. *International Food Research Journal* 21(1): 357-365.

Blandino, A., M.E. Al-Aseeria, S.S. Pandiella, D. Cantero and C. Webb. 2003. Cereal-based fermented foods and beverages. *Food Research International* 36: 527-543.

Bover-Cid, S. and W.H. Holzapfel. 1999. Improved screening procedure for biogenic amine production by lactic acid bacteria. *International Journal of Food Microbiology* 53(1): 33-41.

Brauman, A., S. Keleke, M. Malonga, E. Miambi and F. Ampe. 1996. Microbiological and biochemical characterization of cassava retting, a traditional lactic acid fermentation for foo-foo (cassava flour) production. *Applied and Environmental Microbiology* 62(8): 2854-2858.

Brouwer, I.A., M. van Dusseldorp, C.E. West, S. Meyboom, C.M. Thomas, M. Duran, K.H. van het Hof, T.K. Eskes, J.G. Hautvast and R.P. Steegers-Theunissen. 1999. Dietary folate from vegetables and citrus fruit decreases plasma homocysteine concentrations in humans in a dietary controlled trial. *Journal of Nutrition* 129(6): 1135-1139.

Calderon Santoyo, M., G. Loiseau, R. Rodriguez Sanoja and J.P. Guyot. 2003. Study of starch fermentation at low pH by *Lactobacillus fermentum* Ogi E1 reveals uncoupling between growth and alpha-amylase production at pH 4.0. *International Journal of Food Microbiology* 80(1): 77-87.

Capozzi, V., P. Russo, M.T. Duenas, P. Lopez and G. Spano. 2012. Lactic acid bacteria producing B-group vitamins: A great potential for functional cereals products. *Applied Microbiology and Biotechnology* 96(6): 1383-1394.

Collins, J.L. and G.G. Saunders. 1976. Changes in trypsin inhibitory activity in some soybean varieties during maturation and germination. *Journal of Food Science* 41: 168-172.

De Angelis, M., F. Bottacini, B. Fosso, P. Kelleher, M. Calasso, R. Di Cagno, M. Ventura, E. Picardi, D. van Sinderen and M. Gobbetti. 2014. *Lactobacillus rossiae*, a vitamin B_{12} producer, represents a metabolically versatile species within the Genus *Lactobacillus*. *PLoS One*, 9(9): e107232.

Demuyakor, B. and Y. Ohta. 1991. Characteristics of *pito* yeasts from Ghana. *Food Microbiology* 8: 183-193.

Deshpande, S.S., D.K. Salunkhe, O.B. Oyewole, S. Azam-Ali, M. Battcock and R. Bressani. 2000. Fermented grain legumes, seeds and nuts: A global perspective. FAO *Agricultural Services. Bulletin 142*. Food and Agricultural Organization of the United Nation. Rome.

Dirar, H.A. 1978. A microbiological study of Sudanese *merissa* brewing. *Journal of Food Science* 43: 1683-1686.

Dirar, H.A., D.B. Harper and M.A. Collins. 1985. Biochemical and microbiological studies on *kawal*, a meat substitute derived by fermentation of *Cassia obtusifolia* leaves. *Journal of the Science of Food and Agriculture* 36: 881-892.

Diwani, T. and G. Janssens. 2001. Effects of Harvesting and Deflowering on Leafy Vegetables (*Amaranthus and Solanum*) under Drought Stress. University of Bonn, Bonn, Germany. Institut für Obst-und Gemüsebau.

Dogi C.A., A. Fochesato, R. Armando, B. Pribull, M.M. de Souza, I. da Silva Coelho, D, Araujo de Melo, A. Dalcero and L. Cavaglieri. 2013. Selection of lactic acid bacteria to promote an efficient silage fermentation capable of inhibiting the activity of *Aspergillus parasiticus* and *Fusarium gramineraum* and mycotoxin production. *Journal of Applied Microbiology* 114: 1650-1660.

Du Toit, M., L. Engelbrecht, E. Lerm and S. Krieger-Weber. 2011. *Lactobacillus*: The next generation of malolactic fermentation starter cultures — An overview. *Food and Bioprocess Technology* 4: 876-906.

Efiuvwevwere, B.J. and O. Akona. 1995. The microbiology of 'kunun-zaki', a cereal beverage from northern Nigeria, during the fermentation (production) process. *World Journal of Microbiology and Biotechnology* 11(5): 491-493.

Ehrmann, M.A., S. Freiding and R.F. Vogel. 2009. *Leuconostoc palmae* sp. nov., a novel lactic acid bacterium isolated from palm wine. International *Journal of Systematic and Evolutionary Microbiology* 59(Pt 5): 943-947.

Ekundayo, J.A. 1977. Nigerian *pito*. *In*: The proceding and symposium on indigenous fermented foods, November 21-27, 1977, GIAMI, Bangkok, Thailand.

Fandohan, P., D. Zoumenou, D.J. Hounhouigan, W.F. Marasas, M.J. Wingfield and K. Hell. 2005. Fate of aflatoxins and fumonisins during the processing of maize into food products in Benin. *International Journal of Food Microbiology* 98(3): 249-259.

Faparusi, S.I. 1970. Sugar changes during preparation of *burukutu* beer. *Journal of the Science of Food and Agriculture* 21: 79.

Faparusi, S.I. and O. Bassir. 1972. Effect of extracts of the bark of *Saccoglottis gabonensis* on the microflora of palm wine. *Applied Microbiology* 24(6): 853-856.

Faparusi, S.I., M.O. Olofinbo and J.A. Ekundayo. 1973. Microbiology of *burukutu* beer. *Zeitschrift fuer Allgemeine Mikrobiologie* 13: 563-568.

Fapohunda, S.O. 1988. Amino-acid-composition and mycology of *sekete*, a fermented product from *Zea-mays*-L. *Die Nahrung* 32: 571-575.

Fischer, M.M., I.M. Egli, I. Aeberli, R.F. Hurrell and L. Meile. 2014. Phytic acid degrading lactic acid bacteria in tef-injera fermentation. *International Journal of Food Microbiology* 190: 54-60.

Fleming S.E. 1981. A study of relationships between flatus potential and carbohydrate distribution in legume seeds. *Journal of Food Science* 46: 794-798.

Franz, C.M., M. Huch, J.M. Mathara, H. Abriouel, N. Benomar, G. Reid, A. Galvez and W.H. Holzapfel. 2014. African fermented foods and probiotics. *International Journal of Food Microbiology* 190: 84-96.

Galvez, A., H. Abriouel, R.L. Lopez and N. Ben Omar. 2007. Bacteriocin-based strategies for food bio-preservation. *International Journal of Food Microbiology* 120(1-2): 51-70.

Greppi A., K. Rantsiou, W. Padonou, J. Hounhouigan, L. Jespersen, M. Jakobsen and L. Cocolin. 2013. Determination of yeast diversity in ogi, mawè, gowé and tchoukoutou by using culture-dependent and independent methods. *International Journal of Food Microbiology* 165: 84-88.

Gustavsson, J., C. Cederberg, U. Sonesson, R. van Otterdijk and A. Meybeck. 2011. *Global Food Losses and Food Waste*: *Extent*, *Causes and Prevention*. Study conducted for the International Congress SAVE FOOD! At Interpack 2011, Düsseldorf, Germany. Food and Agriculture Organization of the United Nations, Rome, Italy.

Haard, N.F., S.A. Odunfa, L. Cherl-Ho, R. Quintero-Ramírez, A. Lorence-Quiñones and C. Wacher- Radarte. 1999. Fermented cereals: A global perspective. *FAO Agricultural Services Bulletin* No. 138. M-17 ISBN 92-5-104296-9.

Halász, A., Á. Baráth and W.H. Holzapfel. 1999. The influence of starter culture selection on sauerkraut fermentation. *Zeitschrift fuer Lebensmittel-Untersuchung und-Forschung* A 208: 434-438.

Halm, M., A. Lillie, A.K. Sorensen and M. Jakobsen. 1993. Microbiological and aromatic characteristics of fermented maize doughs for *kenkey* production in Ghana. International Journal of Food Microbiology 19(2): 135-143.

Harkishor, K.M. 1977. *Kenyan Sugarcane Wine — Muratina Symposium on Indigenous Fermented Foods*. Bangkok Thailand.

Hellstrom, A.M., A. Almgren, N.G. Carlsson, U. Svanberg and T.A. Andlid. 2012. Degradation of phytate by *Pichia kudriavzevii* TY13 and *Hanseniaspora guilliermondii* TY14 in Tanzanian togwa. *International Journal of Food Microbiology* 153(1-2): 73-77.

Hesseltine, C.W. 1979. Some important fermented foods of Mid-Asia, the Middle East and Africa. *Journal of the American Oil Chemists' Society* 56: 367-37.

Holzapfel, W.H. 2002. Appropriate starter culture technologies for small-scale fermentation in developing countries. *International Journal of Food Microbiology* 75(3): 197-212.

Huang, J., C. Pray and S. Rozelle. 2002. Enhancing the crops to feed the poor. *Nature*, 418(6898): 678-684.

Huch Nee Kostinek, M., A. Hanak, I. Specht, C.M. Dortu, P. Thonart, S. Mbugua, W.H. Holzapfel, C. Hertel and C.M. Franz. 2008. Use of *Lactobacillus* strains to start cassava fermentations for *Gari* production. *International Journal of Food Microbiology* 128(2): 258-267.

Humblot, C. and J.P. Guyot. 2009. Pyro-sequencing of tagged 16S rRNA gene amplicons for rapid deciphering of the microbiomes of fermented foods such as pearl millet slurries. *Applied and Environmental Microbiology* 75(13): 4354-4361.

Humblot, C., R. Perez-Pulido, D. Akaki, G. Loiseau and J.P. Guyot. 2012. Prevalence and fate of *Bacillus cereus* in African traditional cereal-based foods used as infant foods. *Journal of Food Protection* 75(9): 1642-1645.

Humblot, C., W. Turpin, F. Chevalier, C. Picq, I. Rochette and J.P. Guyot. 2014. Determination of expression and activity of genes involved in starch metabolism in *Lactobacillus plantarum* A6 during fermentation of a cereal-based gruel. *International Journal of Food Microbiology* 185: 103-111.

Kamda, A.G., C.L. Ramos, E. Fokou, W.F. Duarte, A. Mercy, K. Germain, D.R. Dias and R.F. Schwan. 2015. *In vitro* determination of volatile compound development during starter culture-controlled fermentation of *Cucurbitaceae* cotyledons. *International Journal of Food Microbiology* 192: 58-65.

Kawamura, S. 1954. Studies on soybean carbohydrate. IV. Determination of oligosaccharides in soybeans. *Nippon Nogei Kogaku Kaishi* 28: 851-852.

Kayode, A.P.P., J.D. Hounhouigan and M.J.R. Nout. 2007. Impact of brewing process operations on phytate, phenolic compounds and in vitro solubility of iron and zinc in opaque sorghum beer. *LWT Food Science and Technology* 40: 834-841.

Kimaryo, V.M., G.A. Massawe, N.A. Olasupo and W.H. Holzapfel. 2000. The use of a starter culture in the fermentation of cassava for the production of 'kivunde', a traditional Tanzanian food product. *International Journal of Food Microbiology* 56(2-3): 179-190.

Kobawila, S.C., D. Louembe, S. Keleke, J. Hounhouigan and C. Gamba. 2005. Reduction of the cyanide content during fermentation of cassava roots and leaves to produce *bikedi* and *ntoba mbadi*, two food products from Congo. *African Journal of Biotechnology* 4(7): 689-696.

Kolawole, O.M., R.M.O. Kayode and B. Akinduyo. 2007. Proximate and microbial analyses of *burukutu* and *pito* produced in Ilorin, Nigeria. *African Journal of Biotechnology* 6: 587-590.

Lasekan, O.O. and W.O. Lasekan. 2000. Biogenic amines in traditional alcoholic beverages produced in Nigeria. *Food Chemistry* 69: 267-271.

Lei, V. and M. Jakobsen. 2004. Microbiological characterization and probiotic potential of *koko* and *koko* sour water, African spontaneously fermented millet porridge and drink. *Journal of Applied Microbiology* 96(2): 384-397.

Maktouf, S., A. Kamoun, C. Moulis, M. Remaud-Simeon, D. Ghribi and S.E. Chaabouni. 2013. A new raw-starch-digesting alpha-amylase: Production under solid-state fermentation on crude millet and biochemical characterization. *Journal of Microbiology and Biotechnology* 23(4): 489-498.

Mbugua, S.K. and J. Njenga. 1992. The anti-microbial activity of fermented uji. *Ecology of Food and Nutrition* 28(3): 191-198.

Mbugua, S.K. 1985. Microbial growth during spontaneous uji fermentation and its influence on the end product. *East African Agricultural and Forestry Journal* 50: 101-110.

Miller, B.J., C.M. Franz, G.S. Cho and M. du Toit. 2011. Expression of the malolactic enzyme gene (*mle*) from *Lactobacillus plantarum* under winemaking conditions. *Current Microbiology* 62(6): 1682-1688.

Mohammed, S.I., L.R. Steenson and A.W. Kirleis. 1991. Isolation and characterization of microorganisms associated with the traditional sorghum fermentation for production of Sudanese *kisra*. *Applied and Environmental Microbiology* 57(9): 2529-2533.

Morcos, S.R., S.M. Hegazi and S.T. El-Damhoughy. 1973. Fermented foods of common use in Egypt: The nutritive value of kishk. *Journal of the Science of Food and Agriculture* 24: 1153-1156.

Motarjemi, Y. and M.J. Nout. 1996. Food fermentation: A safety and nutritional assessment. Joint FAO/WHO Workshop on Assessment of Fermentation as a Household Technology for Improving Food Safety. *Bulletin of the World Health Organization* 74(6): 553-559.

Mpofu, A., A.R. Linnemann, M.J. Nout, M.H. Zwietering and E.J. Smid. 2014a. Mutandabota, a food product from Zimbabwe: Processing, composition, and socioeconomic aspects. *Ecology of Food and Nutrition* 53(1): 24-41.

Mpofu, A., A.R. Linnemann, W. Sybesma, R. Kort, M.J. Nout and E.J. Smid. 2014b. Development of a locally sustainable functional food based on *mutandabota*, a traditional food in southern Africa. *Journal of Dairy Science* 97(5): 2591-2599.

Muchoki, C., J.K. Imungi and P.O. Lamuka. 2007. Changes in ß-carotene, ascorbic acid and sensory properties in fermented, solar-dried and stored cowpea leaf vegetables *In: African Journal of Food, Agriculture, Nutrition and Development* 7(3).

Mugula, J.K., J.A. Narvhus and T. Sørhaug. 2003a. Use of starter cultures of lactic acid bacteria and yeasts in the preparation of *togwa*, a Tanzanian fermented food. *International Journal of Food Microbiology* 83(3): 307-318.

Mugula, J.K., J.A. Narvhus and T. Sørhaug 2003b Microbiological and fermentation characteristics of togwa, a Tanzanian fermented food. *International Journal of Food Microbiology* 80(3): 187-99.

Mukisa, I.M., D. Porcellato, Y.B. Byaruhanga, C.M. Muyanja, K. Rudi, T. Langsrud and J.A. Narvhus. 2012. The dominant microbial community associated with fermentation of *Obushera* (sorghum and millet beverages) determined by culture-dependent and culture-independent methods. *International Journal of Food Microbiology* 160(1): 1-10.

Ndlovu, B., H. Schoeman, C.M. Franz and M. Du Toit. 2015. Screening, identification and characterization of bacteriocins produced by wine-isolated LAB strains. *Journal of Applied Microbiology* 118: 1007-1022.

Ndoye, B., S. Lebecque, J. Destain, A.T. Guiro and P. Thonart. 2007. A new pilot plant scale acetifier designed for vinegar production in Sub-Saharan Africa. *Process Biochemistry* 39: 916-923.

Nigatu, A. 2000. Evaluation of numerical analyses of RAPD and API 50 CH patterns to differentiate *Lactobacillus plantarum, Lact. fermentum, Lact. rhamnosus, Lact. sake, Lact. parabuchneri, Lact. gallinarum, Lact. casei, Weissella* minor and related taxa isolated from kocho and tef. *Journal of Applied Microbiology* 89(6): 969-978.

Nout, M.J. 2009. Rich nutrition from the poorest — Cereal fermentations in Africa and Asia. *Food Microbiology* 26(7): 685-692.

Nout, M.J. and P.K. Sarkar. 1999. Lactic acid food fermentation in tropical climates. *Antonie Van Leeuwenhoek* 76(1-4): 395-401.

Nout, M.J., F.M. Rombouts and A. Havelaar. 1989. Effect of accelerated natural lactic fermentation of infant food ingredients on some pathogenic microorganisms. *International Journal of Food Microbiology* 8(4): 351-361.

Nout, M.J.R. 1980. Microbiological aspects of the traditional manufacture of busaa, a Kenyan opaque maize beer. *Chemie Mikrobiologie Technologie der Lebensmittel* 6: 137-142.

Nyanga, L.K., M.J.R. Nout, T.H. Gadaga, T. Boekhout and M.H. Zwietering. 2008. Traditional processing of *Masau* fruits (*Ziziphus mauritiana*) in Zimbabwe. *Ecology of Food and Nutrition* 47: 95-107.

Nyanga, L.K., T.H. Gadaga, M.J.R. Nout, E.J. Smid, T. Boekhout and M.H. Zwietering. 2013a. Nutritive value of *masau* (*Ziziphus mauritiana*) fruits from Zambezi Valley in Zimbabwe. *Food Chemistry* 138(1): 168-172.

Nyanga, L.K., M.J.R. Nout, T.H. Gadaga, B. Theelen, T. Boekhout and M.H. Zwietering. 2007. Yeasts and lactic acid bacteria microbiota from *masau* (*Ziziphus mauritiana*) fruits and their fermented fruit pulp in Zimbabwe. *International Journal of Food Microbiology* 120(1-2): 159-166.

Obadina, A.O., O.B. Oyewole and M.O. Odubay. 2007. Effect of storage on the safety and quality of *fufu* flour. *Journal of Food Safety* 27: 148-156.

Obilie, E.M., K. Tano-Debrah and W.K. Amoa-Awua. 2003. Microbial modification of the texture of grated cassava during fermentation into *akyeke*. *International Journal of Food Microbiology* 89(2-3): 275-280.

Obilie, E.M., K. Tano-Debrah and W.K. Amoa-Awua. 2004. Souring and breakdown of cyanogenic glucosides during the processing of cassava into *akyeke*. *International Journal of Food Microbiology* 93(1): 115-121.

Odunfa, S.A. 1985. African Fermented Foods. pp. 155-191. *In*: B.J.B. Wood (Ed.). *Microbiology of Fermented Foods*. Vol. 2. Elsevier Applied Science Publisher, New York.

Odunfa, S.A. 1986. Microbiological assay of Vitamin B and biotin in some Nigerian fermented vegetable proteins. *Food Chemistry* 19: 129-136.

Odunfa, S.A. 1988. African fermented foods: From art to science. Presentation at IFS/UNU Workshop of Development of indigenous fermented foods and food technology in Africa, Douala, 1985. *MIRCEN Journal of Applied Microbiology and Biotechnology* 4: 259-273.

Odunfa, S.A. and O.B. Oyewole. 1998. African Fermented Foods. Pp. 713-752. *In*: B.J.B. Wood (Ed.). *Microbiology of Fermented Foods*. Second edn. Blackie Academic and Professional, London.

Odunfa, S.A., O.D. Teniola and J. Nordstrom. 2001. Evaluation of lysine and methionine production by some *Lactobacilli* and yeasts from *ogi*. *International Journal of Food Microbiology* 63: 159-163.

OECD-FAO 2011. *OECD-FAO Agricultural Outlook* 2011-2020. Paris: OECD/FAO.

Oguntoyinbo F.A. 2014. Safety challenges associated with traditional foods of West Africa. *Food Reviews International* 30: 338-358.

Oguntoyinbo, F.A. and E.R. Dodd. 2010. Bacterial dynamics during solid substrate fermentation of cassava for gari production. *Food Control* 21: 306-312.

Oguntoyinbo, F.A. and A. Narbad. 2012. Molecular characterization of lactic acid bacteria and *in situ* amylase expression during traditional fermentation of cereal foods. *Food Microbiology* 31(2): 254-262.

Oguntoyinbo, F.A., M. Huch, G.S. Cho, U. Schillinger, W.H. Holzapfel, A.I. Sanni and C.M. Franz. 2010. Diversity of *Bacillus* species isolated from *okpehe*, a traditional fermented soup condiment from Nigeria. *Journal of Food Protection* 73(5): 870-878.

Oguntoyinbo, F.A., A.I. Sanni, C.M. Franz and W.H. Holzapfel. 2007. *In vitro* fermentation studies for selection and evaluation of *Bacillus* strains as starter cultures for the production of *okpehe*, a traditional African fermented condiment. *International Journal of Food Microbiology* 113(2): 208-218.

Oguntoyinbo, F.A., P. Tourlomousis, M.J. Gasson and A. Narbad. 2011. Analysis of bacterial communities of traditional fermented West African cereal foods using culture-independent methods. *International Journal of Food Microbiology* 145(1): 205-210.

Okoye, Z.S.C. 1986. Zearalenone in native cereal beer brewed in Jos metropolis of Nigeria. *Journal of Food Safety* 7: 233-239.

Olsen, A., M. Halm and M. Jakobsen. 1995. The antimicrobial activity of lactic acid bacteria from fermented maize (*kenkey*) and their interactions during fermentation. *Journal of Applied Bacteriology* 79(5): 506-512.

Onaghise, E.O. and Y.S. Izuagbe. 1989. Improved brewing and preservation of *pito*, a Nigerian alcoholic beverage from maize. *Acta Biotechnologica* 9: 137-142.

Ouoba, L.I., C. Kando, C. Parkouda, H. Sawadogo-Lingani, B. Diawara and J.P. Sutherland. 2012. The microbiology of *Bandji*, palm wine of *Borassus akeassii* from Burkina Faso: Identification and genotypic diversity of yeasts, lactic acid and acetic acid bacteria. *Journal of Applied Microbiology* 113(6): 1428-1441.

Ouoba, L.I., C.A. Nyanga-Koumou, C. Parkouda, H. Sawadogo, S.C. Kobawila, S. Keleke, B. Diawara, D. Louembe and J.P. Sutherland. 2010. Genotypic diversity of lactic acid bacteria isolated from African traditional alkaline-fermented foods. *Journal of Applied Microbiology* 108(6): 2019-2029.

Ouoba, L.I., C. Parkouda, B. Diawara, C. Scotti and A.H. Varnam, 2008. Identification of Bacillus spp. from *Bikalga*, fermented seeds of *Hibiscus sabdariffa*: phenotypic and genotypic characterization. *Journal of Applied Microbiology* 104(1): 122-131.

Oyewole O.B. 1997. Lactic fermented foods in Africa and their benefits. Food Control 8: 289-297.

Padonou, W.S., D.S. Nielsen, J.D. Hounhouigan, L. Thorsen, M.C. Nago and M. Jakobsen. 2009. The microbiota of *Lafun*, an African traditional cassava food product. *International Journal of Food Microbiology* 133(1-2): 22-30.

Pariza, M.W. 1996. Toxic substances. Pp. 825-840. *In*: O.R. Fennema (Ed.). *Food Chemistry*, 3rd edn. Marcel Dekker. New York.

Parkouda, C., D.S. Nielsen, P. Azokpota, L.I. Ouoba, W.K. Amoa-Awua, L. Thorsen, J.D. Hounhouigan, J.S. Jensen, K. Tano-Debrah, B. Diawara and M. Jakobsen. 2009. The microbiology of alkaline-fermentation of indigenous seeds used as food condiments in Africa and Asia. *Critical Reviews in Microbiology* 35(2): 139-156.

Parkouda, C., L. Thorsen, C.S. Compaore, D.S. Nielsen, K. Tano-Debrah J.S. Jensen, B., Diawara and M. Jakobsen. 2010. Microorganisms associated with Maari, a Baobab seed-fermented product. *International Journal of Food Microbiology* 142(3): 292-301.

Rackis, J.J., D.J. Sessa, F.R. Steggerda, J. Shimizu, J. Anderson and S.L. Pearl. 1970. Soybean factors relating to gas production by intestinal bacteria. *Journal of Food Science* 35: 634-639.

Rautenbach, F., M. Faber, S. Laurie and R. Laurie. 2010. Antioxidant capacity and antioxidant content in roots of 4 sweetpotato varieties. *Journal of Food Science* 75(5): C400-405.

Roige, M.B., S.M. Aranguren, M.B. Riccio, S. Pereyra, A.L. Soraci and M.O. Tapia. 2009. Mycobiota and mycotoxins in fermented feed, wheat grains and corn grains in Southeastern Buenos Aires province, Argentina. *Revista Iberoamericana de Micologia*, 26(4): 233-237.

Rosenzweig, C.A. and M.L. Parry. 1994. Potential impact of climate change on world food supply. *Nature* 367: 133-137.

Rossouw, D., M. Du Toit, and F.F. Bauer. 2012. The impact of co-inoculation with *Oenococcus oeni* on the trancriptome of *Saccharomyces cerevisiae* and on the flavour-active metabolite profiles during fermentation in synthetic must. *Food Microbiology* 29(1): 121-131.

Sanni, A.I. and F.A. Oguntoyinbo. 2014. *Maari, Ntoba mbodi* and *Owoh*. Pp. 140-143. *In*: M.J.R. Nout and P.K. Sarkar (Eds.). *Handbook of Indigenous Foods Involving Alkaline Fermentation*. CRC Press/Taylor and Francis Group, Boca Raton.

Sanni, A.I. 1989a. Some environmental and nutritional factors affecting growth of associated microorganisms of *agadagidi*. *Journal of Basic Microbiology* 29(9): 617-622.

Sanni, A.I. 1989b. Chemical studies on *sekete* beer. *Food Chemistry* 33: 187-191.

Sanni, A.I. 1993. The need for process optimization of African fermented foods and beverages. *International Journal of Food Microbiology* 18(2): 85-95.

Sanni, A.I. and B.A. Oso. 1988. Nutritional studies on *agadagidi*. *Nahrung* 32(2): 169-172.

Sanni, A.I., A.A. Onilude, I.F. Fadahunsi and R.O. Afolabi. 1999. Microbial deterioration of traditional alcoholic beverages in Nigeria. *Food Research International* 32: 163-167.

Sawadogo-Lingani, H., V. Lei, B. Diawara, D.S. Nielsen, P.L. Moller, A.S. Traore and M. Jakobsen. 2007. The biodiversity of predominant lactic acid bacteria in dolo and pito wort for the production of sorghum beer. *Journal of Applied Microbiology* 10: 765-777.

Sawadogo-Lingani, H., B. Diawara, A.S. Traore and M. Jakobsen. 2008. Technological properties of *Lactobacillus fermentum* involved in the processing of dolo and pito, West African sorghum beers, for the selection of starter cultures. *Journal of Applied Microbiology* 104: 873-882.

Schoustra, S.E., C. Kasase, C. Toarta, R. Kassen and A.J. Poulain. 2013. Microbial community structure of three traditional zambian fermented products: *mabisi, chibwantu* and *munkoyo*. *PLoS One*, 8(5): e63948.

Sekwati-Monang, B., R. Valcheva and M.G. Ganzle. 2012. Microbial ecology of sorghum sourdoughs: Effect of substrate supply and phenolic compounds on composition of fermentation microbiota. *International Journal of Food Microbiology* 159(3): 240-246.

Shiundu, K.M. and R.K. Oniang'o. 2007. Marketing African leafy vegetables: Challenges and opportunities in the Kenyan context. *African Journal of Food, Agriculture, Nutrition and Development* 7 (4).

Simon-Sarkadi, L. and W.H. Holzapfel. 1995. Biogenic amines and microbial quality of sprouts. *Zeitschrift fuer Lebensmittel-Untersuchung und-Forschung*, 200(4): 261-265.

Smith, I.F. and P. Eyzaguirre. 2007. African leafy vegetables: Their role in the World Health Organization's global fruit and vegetable initiative. *African Journal of Food, Agriculture, Nutrition and Development*, 7, Nos. 3 & 4: 1-8.

Songre-Ouattara, L.T., C. Mouquet-Rivier, C. Icard-Verniere, C. Humblot, B. Diawara and J.P. Guyot. 2008. Enzyme activities of lactic acid bacteria from a pearl millet fermented gruel (*ben-saalga*) of functional interest in nutrition. *International Journal of Food Microbiology* 128(2): 395-400.

Stanton, R.W. 1985. Food Fermentation in the Tropics. Pp. 696-712. *In*: B.J.B. Wood (Ed.). *Microbiology of Fermented Foods*. Elsevier Applied Science Publishers, UK.

Steinkraus, K.H. 1995. *Classification of Household Fermentation Techniques.* Background paper for WHO/FAO Workshop on assessment of fermentation as household technology for improving food safety. Dec. 11-15, 1995. Dept. of Health. Pretoria, South Africa.

Steinkraus, K.H. 2002. Fermentations in world food processing. *Comprehensive Reviews in Food Science and Food Safety* 1: 23-32.

Steinkraus, K.H. 1997. Classification of fermented foods: Worldwide review of household fermentation techniques. *Food Control* 8: 311-317.

Steinkraus, K.H. 1985. Potential of African indigenous fermented foods. *In*: *Development of Indigenous Fermented Foods and Food Technology in Africa*. Stockholm, Sweden: International Foundation for Science (IFS). pp. 34-70.

Steinkraus, K.H. 1996. Introduction to indigenous fermented foods. Pp. 1-5. *In*: K.H. Steinkraus (Ed.). *Handbook of Indigenous Fermented Food*, second edn. Marcel Dekker. New York.

Suliman, H.B., A.M. Shommein and S.A. Shaddad. 1987. The pathological and biochemical effects of feeding fermented leaves of *Cassia obtusifolia 'kawal'* to broiler chicks. *Avian Pathology* 16(1): 43-49.

Willcox, J.K., G.L. Catignani and S. Lazarus. 2003. Tomatoes and cardiovascular health. *Critical Reviews in Food Science and Nutrition* 43(1): 1-18.

Willcox, J. K., S.L Ash, G.L Catignani. 2004 Antioxidants and prevention of chronic disease. *Critical Reviews in Food Science and Nutrition* 44(4): 275-95.

Yousif, N.M., P. Dawyndt, H. Abriouel, A. Wijaya, U. Schillinger, M. Vancanneyt, J. Swings, H.A. Dirar, W.H. Holzapfel and C.M. Franz. 2005. Molecular characterization, technological properties and safety aspects of enterococci from 'hussuwa', an African fermented sorghum product. *Journal of Applied Microbiology* 98(1): 216-228.

Yousif, N.M., M. Huch, T. Schuster, G.S. Cho, H.A. Dirar, W.H. Holzapfel and C.M. Franz. 2010. Diversity of lactic acid bacteria from *hussuwa*, a traditional African fermented sorghum food. *Food Microbiology* 27(6): 757-768.

12

Lactic Acid Fermentation of Smoothies and Juices

Raffaella Di Cagno[1,*], Pasquale Filannino[2], and Marco Gobbetti[3]

1. Lactic Acid Fermentation of Vegetable- and Fruit-based Beverages

Plant-based beverages produced by controlled fermentation of lactic acid bacteria are the new products meeting to the consumer's demand for minimally processed foods characterized by high nutritional value, rich flavor and enhanced shelf-life. In these products high proportions of protective substances contained in the raw material are preserved. Moreover, during fermentation, lactic acid bacteria produce additional health-promoting components and allow a better preservation (Di Cagno et al. 2013). Lactic acid bacteria and, more in general, the lactic acid fermentation is considered as one of the most suitable tools to exploit the biogenic/functional potential of plant matrices and to enrich them with bioactive compounds (Gobbetti et al. 2010). Indeed, fermentation by select lactic acid bacteria was largely used to enhance the anti-microbial, antioxidant and immune-modulatory features of several vegetables and fruits as well as of medicinal plants, like *Echinacea* spp. (Rizzello et al. 2013) as a result of the ingestion of microbial metabolites (biogenic effect). Under optimal processing conditions, microbes contribute to plant functionality through their enzyme portfolio, which promotes the synthesis of various metabolites and/or the release of functional compounds that are cryptic in the raw matrix. Recently, lactic acid bacteria were used to synthesize γ-amino butyric acid (GABA) from grape juice (Di Cagno et al. 2010a). Besides, health benefits from fermentation of plant materials are usually direct, through interaction of ingested live microorganisms with the host (probiotic effect).

[1] Department of Soil Plant and Food Sciences, University of Bari Aldo Moro, 70126 Bari, Italy
[2] E-mail: pasquale.filannino1@uniba.it
[3] E-mail: marco.gobbetti@uniba.it
* Corresponding author: E-mail: raffaella.dicagno@uniba.it

The technological options for production of fermented vegetable juices are mainly three: (i) spontaneous fermentation by autochthonous lactic acid bacteria; (ii) fermentation by starter cultures added into raw vegetables; and (iii) fermentation of mild heat-treated vegetables by starter cultures (Di Cagno et al. 2013). Lactic acid bacteria microbiota of spontaneously-fermented vegetable juices is mainly represented by the genera *Lactobacillus*, *Leuconostoc* and *Pediococcus* while the starter cultures most widely used were *Lactobacillus plantarum*. Overall, spontaneously-fermented vegetable juices correspond to the traditional fermented beverages, which represent a cultural heritage from the tradition of rural communities.

Shalgam is a traditional fermented, red colored, cloudy and sour soft beverage from Turkey, which is produced by lactic acid fermentation of a mixture of turnips, black carrot, bulgur flour, salt and water. It is widely consumed in cities like Adana, Hatay and Icel (the Mediterranean region of Turkey). In recent years, it has become popular in metropolises too, such as Istanbul, Ankara and Izmir (Tangüler and Erten 2012). Shalgam juice is a nutritional beverage due to its high mineral, vitamin, amino acid and polyphenol contents (Erten et al. 2008). This vegetable mixture contains a high level of anthocyanin as cyanidin-3-glycoside (88.3–114.1 mg/L) and is extracted from black carrot to shalgam juice during processing (Kammerer et al. 2004). The microbiota of shalgam juice is mainly composed of lactic acid bacteria (within the range of 10^6–10^8 cfu/mL) that belong to the genus *Lactobacillus* (89.63 per cent) and followed by *Leuconostoc* (9.63 per cent) and *Pediococcus* (0.74 per cent). Among the lactic acid bacteria, *Lb. plantarum*, *Lb. brevis* and *Lb. paracasei* subsp. *paracasei* were predominant (Tangüler and Erten 2012). Another example of traditional fermented non-alcoholic beverages from Turkey is Hardaliye, which is obtained from red grape or grape juice, crushed black mustard seeds and cherry leaf. Washed red grapes and mustard seeds are pressed separately till the crusts rupture. The ruptured crust of grapes gives the dark color to the final product, depending on the grape varieties. Pressed grapes and cherry leaves are placed in a barrel and 0.2 per cent pressed mustard seeds and/or 0.1 per cent of benzoic acid are added. The barrels are closed and incubated at room temperature for five to 10 days. After incubation, the mixture is filtered and kept to cool (Altay et al. 2013). Hardaliye is an example of a spontaneous fermented juice where *Lb. paracasei* subsp. *paracasei* and *Lb. casei* subsp. *pseudoplantarum* mainly dominate the lactic acid bacteria microbiota, which may be made up of several different species, such as *Lb. pontis*, *Lb. brevis*, *Lb. acetotolerans*, *Lb. sanfranciscensis* and *Lb. vaccinostercus*. The cell number of lactic acid bacteria was reported within the range of 10^2–10^4 cfu/mL.

Overall, it was seen that fermented vegetable juices with optimal characteristics are achieved by selecting lactic acid bacteria strains suitable for fermentation of individual raw materials, according to certain factors, such as the specific dependence on the supply of nutrients for growth and the chemical and physical environment (Figure 1). Consequently, selection of starter cultures within the autochthonous microbiota of vegetables and fruits is recommended since autochthonous cultures ensure a prolonged shelf-life and targeted nutritional, rheology and sensory properties (Di Cagno et al. 2013).

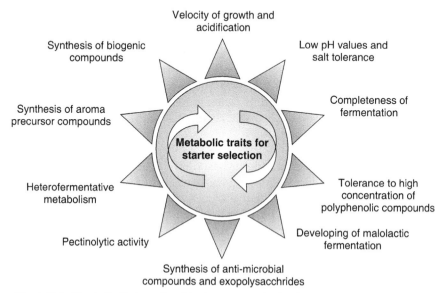

Fig. 1: Main Metabolic Traits to Consider for Starter Selection in Fermentation of Vegetable Products

2. Innovative Plant-based Fermented Products

Nowadays there is a strong tendency towards consumption of fresh-like, high nutritional value, health-promoting and rich flavor ready-to-drink foods and beverages that are minimally processed. Juices directly obtained from fruits (not from concentrates), and distributed through the refrigerated chain with a relatively short shelf-life are a good example of this (Esteve et al. 2005). The consumer demand for non-dairy beverages with high functionality is growing as a consequence of the ongiong trend towards trend of vegetarianism and the increasing prevalence of lactose intolerance. Innovation in food biotechnology is essential to improve the nutritional features while enhancing the hedonistic aspects. Smoothies, originally consisting of pure fresh fruits and vegetables, represent an example of an innovative, vegetable-based fermented beverage.

2.1. Smoothies

Smoothies were first introduced in the 1960 in the United States. In 1980, United States produced the first 'smoothie bar' in areas dedicated to the exclusive sale of these products, which re-emerged in 2000 (Titus 2008) as part of a trend towards health-promoting nutrition (Robinson and Ogawa 1999, McCorquodale et al. 2006). Consumption of fruits and vegetables is associated with a decreased risk of diet-related diseases, such as obesity, diabetes, stroke and some types of cancer (Dauchet et al. 2007). In spite of this knowledge, guidelines and programs to promote a healthy lifestyle, intake higher fruits and vegetables is far from promising, in particular in children who consume far below the recommendations, that is, less than five fruits and vegetable servings per day. Therefore, the increase in fruit and vegetable intake in the form of smoothies or juices has become very popular, especially in the young age groups and might thus be a source of health-promoting polyphenols with antioxidant and anti-inflammatory activities. Consequently, in recent years, smoothies have rapidly increased in popularity (e.g. product growth rising 2.39 times from 2002 to 2007 according to food merchandisers) (Lal 2007) with a loyal consumer base developed.

The manufacture of smoothies is based on the use of a mixture of fruits and vegetables, often after removing the seeds and peel, which are mainly processed to pulp or puree (Qian 2006). In most cases, mixtures of fruits and vegetables are selected based on color, flavor, drinkable texture and, especially, to guarantee high concentration of nutrients with low energy content (Watzl 2008). As a consequence, smoothies depend on the supply of fruits and vegetables, especially to people who cannot consume fresh fruits and vegetables mainly due to market availability and/or convenience (Watzl 2008). Depending on the manufacture and composition, one smoothie is enough to replace the nutritional value of at least one portion of fruits or vegetables.

Recently (Di Cagno et al. 2008, 2009, 2010b, 2011a), it was shown that the use of selected autochthonous lactic acid bacteria starters guaranteed a prolonged shelf-life of fermented vegetables and fruits, which also maintained agreeable nutritional, rheological and sensorial properties. To get desirable properties from fermented fruits, lactic acid bacteria have to be adapted to the intrinsic characteristics of the raw materials. Spontaneous fermentation results from competitive activities between a variety of autochthonous and contaminating microorganisms, leading to a high risk of failure. Both from the hygiene and safety points of view, the use of starter cultures is recommended, as it leads to rapid inhibition of spoilage and pathogenic bacteria, and to a processed fruit with consistent sensory and nutritional quality. A novel protocol for the manufacture of fermented red and green smoothies was set up (Di Cagno et al. 2011b) (Figure 2). White grape juice and *Aloe vera* extract were mixed with red (cherries, blackberries prunes and tomatoes) or green (fennels, spinach, papaya and kiwi) fruits and vegetables and fermented by mixed autochthonous starters. The combination of vegetables and fruits in the smoothie formula is considered difficult, especially due to its very intense flavor and taste (Max-Rubner-Institut 2008). However, the choice of the ingredients was

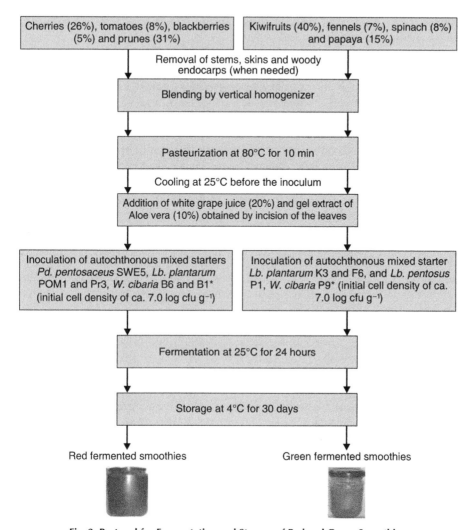

Fig. 2: Protocol for Fermentation and Storage of Red and Green Smoothies

not binding but, based on preliminarily assays, it was promising to ensure optimal viscosity, mouthfeeling, color, health-promoting components and sensory acceptance. In fact, white grape juice and *A. vera* extract were used as the common ingredients for red and green smoothies since they supplied polyphenolic compounds and biologically active glucomannans (Rodrìguez et al. 2010), while conferring fluidity on the preparation. Red fruits and vegetables used in the RS formula were mainly rich in polyphenolic compounds (cherries, blackberries and prunes), lycopene (tomatoes), soluble fibers and antioxidant activity (especially, blackberries). Green fruits and vegetables used in the green smoothie formula were mainly rich in ascorbic acid and soluble fibers (kiwifruits), polyphenolic compounds (kiwifruits and papaya), inuline (spinach) and contributed to sweetness (fennels). *Weissella cibaria, Lb. plantarum, Lactobacillus* sp. and *Lb. pentosus* were variously identified in blackberries, prunes, kiwifruits, papaya and fennels by partial 16S rRNA gene sequence. Representative isolates from each plant species were screened, based on the kinetics of growth on fruit juices. A protocol for processing and storage of red and green smoothies was set up, which included fermentation by select lactic acid bacteria starters and exo-polysaccharide-producing strains. Starters grew and remained viable at 10^9 cfu/mL for 30 days of storage at 4°C. No contaminating *Enterobacteriaceae* and yeast were found throughout storage. Lactic acid fermentation by select starters positively affected the content of antioxidant compounds (polyphenolic compounds and, especially, ascorbic acid) and enhanced the sensory attributes of the smoothies (Figure 3). Viscosity and, especially, color difference and browning index of red and green started samples of smoothies were clearly preferable to unstarted samples or commercial smoothies. Probably the higher value of viscosity was related to the synthesis of exo-polysaccharides (EPS) by select strains of *W. cibaria*. Synthesis of EPS by autochthonous lactic acid bacteria was seen during fermentation of tomato juice and other fruits (Di Cagno et al. 2009). However, a statistical correlation was found between the level of ascorbic acid and free radical scavenging activity. As shown by a first-order equation, the rate of degradation of ascorbic acid through storage was found to be higher in the unstarted as against the started red and green smoothies.

A novel protocol for the manufacture of cherry puree added (10 per cent, v/v) of stem infusion was also set up at the pilot plant (Di Cagno et al. 2011a). The addition of stem infusion to cherry puree enlarged the

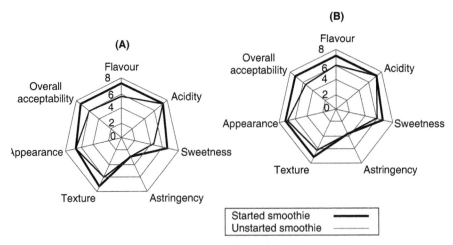

Fig. 3: Sensory Analysis of Red (A) and Green (B) Smoothies Fermented for 24 hours at 25°C by the Mixed Starters of Lactic Acid Bacteria except for the use of starters, red and green smoothies were used as control (unstarted).

profile of phenolic compounds, mainly quercetin. The addition of cheery stems contributed to the health-promoting activity attributed to cherry and, more in general, it contained flavonoids and potassium ions. Despite the hostile environment (e.g. pH 3.9, presence of phenolic compounds, high concentration of carbohydrates), the selected autochthonous *Pd. pentosaceus* SWE5 and *Lb. plantarum* FP3 starters grew well in the puree (ca. 10^9 cfu/g) and both remained viable for 60 days of storage at cell numbers exceeding those of potential probiotic beverages.

2.2. *Plant-based Fermented Juices* (*Lactojuices*)

One of the most attractive utilities is the development of lactic acid-fermented fruit and vegetable juices, also named 'lactojuice'. Since the latter have flavor that is pleasing to all age groups, and are perceived as healthy and refreshing beverages, they promise a considerable market value. Fruits and vegetables tested in novel lactojuices include watermelon, sapodilla, carrot, beetroot, pepper, parsley, lettuce, lemon, cabbage, spinach, tomato, pomegranate, blackcurrant, orange, grapes and sweet potatoes. Fermentation of cabbage into sauerkraut juice is mostly obtained by spontaneous microbiota although, as with other fermented vegetables, the use of starters was required to obtain a uniform product. Besides the choice of the starter culture, to improve the quality and sensory properties of sauerkraut juice, the addition of other ingredients and the mixture with other juices, was also considered. Carrot juice is one of the most common vegetable juices that can be strongly improved by lactic acid bacteria fermentation. Fermented carrot juice is microbiologically stable, has good sensory properties and potentially high nutritional value. The use of select starter cultures for carrot juice may improve the juice yield, thanks to the activity of pectinolytic enzymes, iron solubility and availability of more minerals. The demonstration of suitability of tomatoes as the raw material for the manufacture of lactic acid fermented juice has opened new perspectives for this product. Commercially-available tomato juices are usually subjected to thermal processing, which induces undesirable changes in color, flavor and nutritional value. On the contrary, processing of tomatoes into fermented juice influences the nutritional and sensory properties. Fermentation by singly used ten autochthonous strains belonging to *Lb. plantarum*, *W. cibaria/confusa*, *Lb. brevis*, *Pd. pentosaceous*, *Lactobacillus* sp. and *Enterococcus faecium/faecalis* positively affected the health-promoting and sensory properties of tomato juice via a protocol which included fermentation at 25°C for 17 hours and further storage at 4°C for 40 days (Di Cagno et al. 2009), leading to improved viscosity, color, and ascorbic acid, glutathione and total antioxidant contents. The use of autochthonous strains conferred the highest viscosity to the juice, especially when the EPS producer, *W. cibaria/confusa*, was used. A large number of volatile components characterized the tomato juices fermented with autochthonous strains as they had a positive log odour value in tomato juice, such as 3-methyl-3-butan-1-ol, 2-3 butanedione, 3-hydroxy-2-butanone, dimethyl sulphide and terpens. Pomegranate juices fermented with selected strains of *Lb. plantarum* have, in general, improved the health-promoting and sensory properties (Filannino et al. 2013). *Lb. plantarum* strains consumed glucose, fructose, malic acid and branched chain and aromatic amino acids. Compared to unstarted pomegranate juice, the color and browning indexes of fermented pomegranate juice were preferable. Fermentation increased the concentration of ellagic acid and enhanced the anti-microbial activity. Besides higher phenolic concentration and antioxidant activity, a positive immunomodulation activity distinguished fermented pomegranate juices from unfermented ones; in particular, fermented pomegranate juices scavenged the reactive oxygen species generated by H_2O_2 (hydrogen peroxide) and modulated the synthesis of immune-mediators from peripheral blood mononuclear cells (PBMC).

Red beets were also evaluated as a potential substrate for the production of probiotic beet juice by four species of lactic acid bacteria (*Lb. acidophilus*, *Lb. casei*, *Lb. delbrueckii*, and *Lb. plantarum*). Although the probiotication of several fruit and vegetable juices with lactic acid bacteria and bifidobacteria seems promising, some cultures may negatively affect the sensory properties of the juices or show sensitivity towards low pH of the raw matrices and post-fermented products, fermentation metabolites, processing temperature and gastrointestinal tract conditions. The addition of yeast autolysate (e.g. spent Brewer's yeast) in juices may increase the number of lactic acid bacteria during fermentation, reduce the time of fermentation and enrich the juices with amino acids, vitamins, minerals and antioxidants. Thus a proper selection of fruit matrices and probiotic strains, and if required the addition of other ingredients, are decisive steps to produce a palatable beverage that can be consumed regularly, providing functional benefits.

2.3. Vegetable Milks

The so-called vegetable milks are mainly obtained from soy, cereals and nuts; moreover, other minor raw materials may be hemp, sunflower, legumes (e.g. lupin and peanut seeds) and tubers (e.g. tigernuts). Vegetable milks are commonly used as raw materials to develop yogurt-like product in a process that involves the following main steps: (i) milk conditioning to the optimal growth temperature of the starters, (ii) inoculation and incubation procedures, (iii) and cooling to 4°C (Bernat et al. 2014a). Usually, the viscoelastic properties of mixed gels are altered by changing the level and type of ingredients, and by varying the level of acidity of the food matrices, as determined by lactic acid fermentation. Protein coagulation and water separation, also termed wheying-off, is the common defect that may occur during storage of yogurt-like products. Prevention of this defect is achieved by increasing the content of total solids or by adding stabilizers (e.g. xanthan gum, modified starches, pectin and cellulose derivatives). *In situ* production of oligosaccharides and EPS by lactic acid bacteria promote physical stability and positively affect the texture, since these compounds act as emulsifiers or stabilizers and increase viscosity and mouthfeel (Coda et al. 2012). Prebiotic compounds (e.g. starch, fibers) are sometimes added with technological benefits (for instance, viscosity increases in the food matrices) and a beneficial effect on survival of probiotics during food processing and storage, and within the human gastrointestinal tract, since prebiotics act as protective barriers (Bernat et al. 2014a).

Soy milk is the most popular vegetable milk, and lactic acid fermentation of soy milk for making yogurt is carried out at industrial level. The low concentration of soluble carbohydrates of soy milk may limit the growth and acidification of lactic acid bacteria, and make indispensible the addition of glucose and fructose (Coda et al. 2012). The potential of soy-based beverages as a functional food is linked to the high nutritional value (e.g. low cholesterol and low saturated fats, high quality protein and amino acids, dietary fiber, magnesium, phosphorus, vitamin K, riboflavin, thiamine, folic acid, isoflavones and other flavonoids) and to their role in the prevention of some chronic degenerative diseases (Bernat et al. 2014a). Fermentation may increase the efficacy of soy products by enhancing the bioactivity and bioavailability of soy constituents. *In vitro* fermentation by probiotics increases the bioavailability of isoflavones by removing glycosyl moieties and transforming them into aglycone structures. Soy isoflavone aglycones are absorbed more easily by the intestine. Soy milk fermentation by *Lb. plantarum* enhances the beneficial effects of soy on lipid metabolism (Kim et al. 2014). Over the last decades, even peanut milk has been successfully converted into low-cost edible products with high nutritional value. Peanut milk, containing added sucrose and fruit flavorings was fermented by *Lb. bulgaricus* and *Lb. acidophilus*, resulting in a product with an acceptable custard-like texture and favorably competitive with flavored buttermilk (Beuchat and Nail 1978). Other minor raw materials, such as tubers, are exploited for making yogurt-like products (Bernat et al. 2014a). Fermentation of tigernuts produces varieties of milk products, such as natural tigernut milk, pasteurized tigernut milk, sterilized tigernut milk, ultra-high temperature tigernut milk, concentrated and condensed tigernut milk. A fermented tigernut milk was produced through pasteurization at 90°C for 15 min. and fermentation at 45°C for 18 hours, using mixed autochthonous cultures of *Lb. plantarum, Lactococcus lactis* and *Lc. thermophilus*. Besides being a good supplier of protein and moderate fat, fermented tigernut milk was positively evaluated for sensory acceptability (Wakil et al. 2014).

Nuts-derived milk represents a suitable raw substrate to develop healthy yogurt-like beverages, since it is a good source of phytochemicals, dietary fibers and carbohydrates with low glycemic index, mono- and polyunsaturated fatty acids, proteins, vitamins and minerals (Bernat et al. 2014a). Probiotic hazelnut milk was successfully obtained after the addition of glucose and inulin through fermentation with *Lb. rhamnosus* and *Streptococcus thermophilus* for 3.5 hours. Starters viability was maintained both throughout the 28 storage days (above the level recommended as being the minimum to ensure health benefits) and also after a simulated digestion (Bernat et al. 2014b). Almond milk has high antioxidant activity, owing to the α-tocopherol and phenolic constituents, a low ratio of Na/K (sodium/potassium) and a balanced ratio of Ca/P (calcium/phosphorus). Fermentation of almond milk with *Bifidobacterium longum* and *Lb. rhamnosus* had a positive immunomodulatory effect on macrophages and did not impair the negative effects on the energetic metabolism of intestinal epithelial cells (Bernat et al. 2015).

3. Comprehensive Understanding of Physiology and Biochemistry of Lactic Acid Bacteria During Fermentation of Fruit and Vegetable Juices

Understanding of physiology and biochemistry of lactic acid bacteria may provide insight into the effect of metabolic activity on flavor, texture, shelf-life and nutritional properties of fermented vegetable and fruit juices. Even the ability to release biogenic compounds by microorganisms depends on specific metabolic traits. The contribution of 'omic' technologies should permit the full exploitation of microbial cells, through a combination of interdisciplinary approaches such as genomics, transcriptomics, proteomics and metabolomics. Genomic-based studies provide useful insight into the mechanisms followed by lactic acid bacteria to sense and adapt to the plant environment. Proteomics plays a pivotal role in linking genome and transcriptome to potential biological functions. Metabolomics allows simultaneous identification and quantitative analysis of intracellular metabolites that are produced or modified by the bacterial metabolism and the possibility to predict, among others, the sensory and nutritional qualities of the fermented final product.

There is enough evidence that chemical composition of plant matrices markedly affects the functional features of lactic acid bacteria. Raw vegetables and fruits possess intrinsic chemical and physical parameters that make them particularly complex and harsh matrices, with numerous factors influencing microbial growth and metabolite production (Filannino et al. 2014). Results observed with functional starters under optimal laboratory conditions are not necessarily obtained in the actual food matrix. For instance, some lactic acid bacteria produce bacteriocins that are active towards food spoilage organisms and/ or pathogenic bacteria when cultivated in complex media (*in vitro*), but fail to show inhibitory activity when grown in a food environment (*in situ*), as synthesis of bacteriocins by lactic acid bacteria is controlled by sophisticated molecular regulatory systems related to detection of specific environmental factors (e.g. physical and chemical characteristics of plant materials), cell density and survival under a variety of conditions (e.g. storage conditions). Therefore, in order to select bacteriocin-producing strains for juice fermentation, the physical and chemical characteristics of laboratory media should mimic those of the natural food ecosystems. A study on quorum-sensing regulation of constitutive plantaricin by *Lb. plantarum* showed that bacteriocin synthesis was constitutive under solid conditions of growth on vegetable and fruit agar plates but in contrast, it depended on the size of the inoculum when *Lb. plantarum* was grown in carrot juice (Rizzello et al. 2014). *In situ* production of EPS is another feature, which may be affected by the food matrix and the applied process technology. Also the metabolism of phenolic compounds is strongly dependent on the composition and intrinsic factors of fruits and vegetables; thus information derived from laboratory media is not fully transferable to food fermentation. During cherry juice and broccoli puree fermentations, phenolic acid metabolism was indicated as a possible stress-response mechanism for *lactobacilli* (Filannino et al. 2015). The use of model systems may help to get a better understanding of interaction between microbial systems and the wide variety of plant environments. Modeling focuses on cell growth, but growth or non-growth related models are also applied to describe the changes of other biochemical compounds (e.g. primary or secondary metabolites concentrations, volatile production) and physical properties (e.g. rheological and textural properties) in these food systems. In the last decade there has been an increasing interest in mathematical modeling aiming to relate the biochemical properties (response variables) to environmental factors (controlling factors). The high number of components playing a role makes it difficult to develop mathematical models applicable to all plant matrices and bacterial strains. In fact, diversity of plant environments and bacterial enzymatic activities makes microbial adaptation to vegetable and fruit ecosystems markedly heterogeneous. Lactic acid bacteria may encounter a multitude of stresses in plant fermentation, such as acidic pH, various temperatures, oxidative stress, osmotic stress, competing microorganisms, nutrient limitations and anti-microbial compounds. Adaptation depends on the microbial ability to adopt specific metabolic pathways (e.g. use of non-conventional carbon sources or pathways aimed at homeostasis of the intracellular pH), and the capacity to share metabolic energy between biosynthesis (e.g. use of alternative substrates) and maintenance (e.g. cell membrane modification, defense responses, reactive oxygen detoxification and ion transport) (Filannino et al. 2014). Recently, *Arabidopsis thaliana* leaf tissue lysate was used as a model fermentation system for elucidating the role of inherent features of plant tissues in driving the lactic acid bacteria metabolism (Golomb and Marco 2015). A combined approach of genomics, transcriptomics and metabolomics showed that a plant-derived strain of

Lc. lactis has a competitive edge compared to non plant-associated strains (e.g. capacity to consume a broader array of plant carbohydrates, increased biosynthetic capacity for non-ribosomal peptides and polyketides, fewer amino acid auxotrophies). Furthermore, several genes related to the general stress response were down-regulated, supporting the conclusion that autochthonous lactic acid bacteria from the plant environment are better suited for plant-based fermentation than allochthonous strains (Golomb and Marco 2015).

A panel of various metabolomic approaches and multidimensional statistical analyses was recently used to investigate the metabolic responses of six *Lb. plantarum* strains under plant-like conditions (Filannino et al. 2014). *Lb. plantarum* has a relatively simple carbon metabolism mainly devoted to lactic acid synthesis, but one of its striking features is its enormous flexibility with respect to catabolic substrates. Fruit (cherry and pineapple) and vegetable (carrot and tomato) juices were chosen as model systems for the study as representatives of diverse ecosystems. As expected, fruit juices were most stressful for microbial growth. Growth was negatively correlated with the initial concentrations of malic acid and carbohydrates (glucose and fructose), and positively correlated with the initial pH. Decrease of external and intracellular pH, alteration of cell membrane permeability, and/or reduction of proton motive force are the main side-effects of malic acid. The consumption of malic acid was noticeable in all juices. Decarboxylation of malic acid provides energy advantages due to the increased intracellular pH and the synthesis of reducing power. Despite the increase in the organic acid concentrations during fermentation, the pH of some fruit juices may remain unchanged due to their inherent buffering capacity as well to the malolactic fermentation (Filannino et al. 2014). Exposure to high levels of carbohydrates leads to inefficient metabolism and/or catabolic repression and bacteria need to equilibrate the extra- and intracellular concentrations. Indeed, depending on whether the environmental conditions were favorable (carrot and tomato juices) or unfavorable (cherry and pineapple juices), the trend of *Lb. plantarum* strains seemed to turn from pathways mainly devoted to growth (fermentation of carbohydrates) to routes that mainly allowed maintenance (e.g. malolactic fermentation) (Filannino et al. 2014). The catabolism of free amino acids is another mechanism for *Lb. plantarum* adaptation to plant juices. Branched chain amino acids (BCAA) decreased during fermentation of juices and seemed to be converted into their respective branched alcohols, whose levels concomitantly increased. Consumption of BCAA into their corresponding 2-ketoacids leads to gain of ATP and allows the regeneration of Glu from -ketoglutarate. A decreasing trend was also found for his the decarboxylation of His into histamine contributes to pH homeostasis and provides energy through the generation of proton motive force. The increase of Tyr, which was found during storage of cherry juice fermented with *Lb. plantarum*, may deserve specific consideration. Quinate is largely present in cherries and it act as a precursor of Tyr through a number of reactions. Tyr is a stimulatory amino acid for the growth of *Lb. plantarum* and, in general, its catabolism is involved in the mechanism of intracellular pH regulation (Filannino et al. 2014). Regarding the role of proteins on cell survival, there is very little information in literature. In a study by Nualkaekul et al. (2011) the protective effect of protein was confirmed by using vegetable peptone in model solutions, although it must be emphasized that each type of fruit juice will probably contain different types of proteins, peptides and free amino acids. Several volatile compounds identified during juice fermentation (e.g. alcohols, ketones, ketoacids and terpenes) are synthesized by lactic acid bacteria when subjected to environmental stresses. The synthesis of neutral diacetyl is induced at the transcriptional level by acidic conditions, which presumably contribute to intracellular pH regulation by decreasing the level of pyruvate. Activation of the acetate kinase route of the phosphogluconate pathway by *Lb. plantarum* during plant juice fermentation strictly relies on the availability of external acceptors of electrons, such as aldehydes, which were reduced to the corresponding alcohols (Filannino et al. 2014).

High levels of phenolic compounds are another factor likely to influence the growth and survival lactic acid bacteria in juices and smoothies. Phenolics play a double role—their anti-microbial features may influence the growth of lactic acid bacteria and other microorganisms, but also, they may serve as substrates being transformed into low-molecular weight structure by lactic acid bacteria metabolism. The capacity to tolerate and metabolize phenolic compounds is species- or strain-dependent. The conversion of phenolic compounds is based on: (i) glycosyl hydrolases, such as β-glucosidase and α-rhamnosidase, which convert flavonoid glycosides into corresponding aglycones; (ii) esterases degrading methyl gallate, tannins or phenolic acid esters; (iii) decarboxylases and reductases; and (iv) benzyl alcohol dehydrogenase, which catalyzes the reversible oxidation of some aromatic alcohols to aldehydes with the concomitant reduction of NAD^+ (Rodríguez et al. 2009). During cherry juice and broccoli puree fermentation, *Lactobacillus* spp. exhibited strain-specific metabolism of phenolic acids, including hydroxybenzoic acids (protocatechuic acid), hydroxycinnamic acids (caffeic and p-coumaric acids) and hydroxycinnamic acid derivatives (chlorogenic acid). As the reduced or decarboxylated products of phenolic acids have lower anti-microbial activity compared to their precursors, the metabolism of phenolic acids was identified as an effective mechanism of detoxification. Besides, it could be assumed that the reduction of phenolic acids is involved in the reoxidation of the co-factor NADH, which provides a metabolic advantage through the NAD^+ regeneration (Filannino et al. 2015).

Due to the importance of high viability of lactic acid bacteria in probiotic juices, several options are currently pursued for improving the bacterial survival. The exposure of bacteria to a sub-lethal stress could induce a kind of resistance and an adaptive stress response. Cultivation in an acidified laboratory medium (acid stress) or containing vanillic acid (phenol stress) or in a substrate added with variable amounts of fruit juice was proposed as a successful protocol to increase *Lb. reuteri* survival in fruit juices (Perricone et al. 2015).

References

Altay, F., F. Karbancıoglu-Güler, C. Daskaya-Dikmen, D. Heperkan Filiz Altay, F. Karbancıoglu-Güler and D. Daskaya-Dikmen Heperkan. 2013. A review on traditional Turkish fermented non-alcoholic beverages: Microbiota, fermentation process and quality characteristics. *International Journal of Food Microbiology* 167: 44-56.

Bernat, N., M. Cháfer, A. Chiralt, C. González-Martínez. 2014a. Vegetable milks and their fermented derivative products. *International Journal of Food Studies* 3: 93-124.

Bernat, N., M. Cháfer, A. Chiralt, C. González-Martínez. 2014b. Hazelnut milk fermentation using probiotic *Lactobacillus rhamnosus* GG and inulin. *International Journal of Food Science and Technology* 49(12): 2553-2562.

Bernat, N., M. Cháfer, A. Chiralt, J.M. Laparra, C. González-Martínez. 2015. Almond milk fermented with different potentially probiotic bacteria improves iron uptake by intestinal epithelial (Caco-2) cells. *International Journal of Food Studies* 4(1): 49-60.

Beuchat, L.R., B.J. Nail. 1978. Fermentation of peanut milk with. *Journal of Food Science* 43(4): 1109-1112.

Coda, R., A Lanera, A. Trani, M. Gobbetti, R. Di Cagno R. 2012. Yoghurt-like beverages made of a mixture of cereals, soy and grape must: Microbiology, texture, nutritional and sensory properties. *International Journal of Food Microbiology* 155(3): 120-127.

Dauchet, L., E. Kesse-Guyot, S. Czernichow, S. Bertrais, C. Estaquio, S. Peneau, A.C. Vergnaud, S. Chat-Yung, K. Castetbon, V. Deschamps, P. Brindel and S. Hercberg. 2007. Dietary patterns and blood pressure change over 5-y follow-up in the SU.VI.MAX cohort. *American Journal of Clinical Nutrition* 85: 1650-1656.

Di Cagno, R., F. Mazzacane, C.G. Rizzello, M. De Angelis, G.M. Giuliani, M. Meloni, B. De Servi and M. Gobbetti. 2010a. Synthesis of γ-aminobutyric acid (GABA) by *Lactobacillus plantarum* DSM19463: Functional grape must beverage and dermatological applications. *Applied Microbiology and Biotechnology* 86: 731-41.

Di Cagno, R., G. Cardinali, G. Minervini, L. Antonielli, C.G. Rizzello, P. Ricciuti and M. Gobbetti. 2010b. Taxonomic structure of the yeasts and lactic acid bacteria microbiota of pineapple (*Ananas comosus* L. Merr.) and use of autochthonous starters for minimally processing. *Food Microbiology* 27: 381-389.

Di Cagno, R., G. Minervini, C.G. Rizzello, M. De Angelis and M. Gobbetti. 2011b. Effect of lactic acid fermentation on antioxidant, texture, color and sensory properties of red and green smoothies. *Food Microbiology* 28: 1062-1071.

Di Cagno, R., G. Minervini, C.G. Rizzello, R. Lovino, M. Servili, A. Taticchi, S. Urbani, S., Gobbetti, M., 2011a. Exploitation of sweet cherry (*Prunus avium* L.) puree added of stem infusion through fermentation by selected autochthonous lactic acid bacteria. *Food Microbiology* 28: 900-909.

Di Cagno, R., R. Coda, M. De Angelis and M. Gobbetti. 2013. Exploitation of vegetables and fruits through lactic acid fermentation. *Food Microbiology* 33: 1-10.

Di Cagno, R., R.F. Surico, A. Paradiso, M. De Angelis, J.C. Salmon, S. Buchin, L. De Gara and M. Gobbetti. 2009. Effect of autochthonous lactic acid bacteria starters on health-promoting and sensory properties of tomato juices. *International Journal of Food Microbiology* 128: 473-483.

Di Cagno, R., R.F. Surico, S. Siragusa, M. De Angelis, A. Paradiso, F. Minervini, L. De Gara and M. Gobbetti. 2008. Selection and use of autochthonous mixed starter for lactic acid fermentation of carrots, French beans or marrows. *International Journal of Food Microbiology* 127: 220-222.

Erten, H., H. Tangüler and A. Canbas. 2008. A traditional Turkish lactic acid fermented beverage: Shalgam (salgam). *Food Reviews International* 24: 352-359.

Esteve, M.J., A. Frígola, C. Rodrigo and D. Rodrigo. 2005. Effect of storage period under variable conditions on the chemical and physical composition and color of Spanish refrigerated orange juices. *Food and Chemical Toxicology* 43: 1413-1422.

Filannino P., L. Azzi, I. Cavoski, O. Vincentini, C.G. Rizzello, M. Gobbetti and R. Di Cagno. 2013. Exploitation of the health-promoting and sensory properties of organic pomegranate (*Punica granatum* L.) juice through lactic acid fermentation. *International Journal of Food Microbiology* 163: 184-192.

Filannino, P., G. Cardinali, C.G. Rizzello, S. Buchin, M. De Angelis, M. Gobbetti, R. Di Cagno. 2014. Metabolic responses of *Lactobacillus plantarum* strains during fermentation and storage of vegetable and fruit juices. *Applied and Environmental Microbiology* 80(7): 2206-2215.

Filannino, P., Y. Bai, R. Di Cagno, M. Gobbetti, M.G. Gänzle. 2015. Metabolism of phenolic compounds by *Lactobacillus* spp. during fermentation of cherry juice and broccoli puree. *Food Microbiology* 46: 272-279.

Gobbetti, M., R. Di Cagno and M. De Angelis. 2010. Functional microorganisms for functional food quality. *Critical Reviews in Food Science and Nutrition* 508: 716-727.

Golomb, B.L., M.L. Marco. 2015. *Lactococcus lactis* metabolism and gene expression during growth on plant tissues. *Journal of Bacteriology* 197(2): 371-381.

Kammerer, D., R. Carle and A. Schieber. 2004. Quantification of anthocyanins in black carrot extracts (*Daucus carota* ssp. *sativus* var. *atrorubens* Alef.) and evaluation of their color properties. *European Food Research and Technology* 219: 479-486.

Kim, Y., S. Yoon, S.B. Lee, H.W. Han, H. Oh, W.J. Lee, S. M. Lee. 2014. Fermentation of soy milk via *Lactobacillus plantarum* improves dysregulated lipid metabolism in rats on a high cholesterol diet. *PloS one* 9(2): e88231.

Lal, G.G. 2007. Getting specific with functional beverages. Food Technology 61: 24-28.

Max-Rubner-Institut, 2008. *Bundesforschungsinstitut für Ernährung und Lebensmittel.* Haid-und-Neu-Str. 9, 76131 Karlsruhe, Germany, *Nationale Verzehrsstudie* (NVS), Teil 2.

McCorquodale, K., L. Damian, A. Richardson, D. Gee. 2006. Evaluation of sensory and objective changes to a fruit-based smoothie after the addition of two different quantities of tri-calcium citrate. *Journal of the American Dietetic Association* 106(8): A56.

Nualkaekul, S., I. Salmeron, D. Charalampopoulos. 2011. Investigation of the factors influencing the survival of *Bifidobacterium longum* in model acidic solutions and fruit juices. *Food Chemistry* 129(3): 1037–1044.

Perricone, M., A. Bevilacqua, C. Altieri, M. Sinigaglia, M.R. Corbo. 2015. Challenges for the production of probiotic fruit juices. *Beverages* 1(2): 95-103.

Qian, N. 2006. Fruit and vegetable smoothies, and its processing method. *Faming Zhuanli Shen.* Gong. Shuom. CN 1817192.

Rizzello, C. G., P. Filannino, R. Di Cagno, M. Calasso, M. Gobbetti. 2014. Quorum-sensing regulation of constitutive plantaricin by *Lactobacillus plantarum* strains under a model system for vegetables and fruits. *Applied and Environmental Microbiology* 80(2): 777-787.

Rizzello. C.G., R. Coda, D. Sánchez-Macías, D. Pinto, B. Marzani, P. Filanino, G.M. Giuliani, V.M. Paradiso, R. Di Cagno and M. Gobbetti. 2013. Lactic acid fermentation as a tool to enhance the functional features of *Echinacea* spp. *Microbial Cell Factories* 12: 44.

Robinson, D., A. Ogawa. 1999. Nutritious defenses to stress eating. *Journal of the American Dietetic Association* 99(9): A43.

Rodríguez, E., J.D. Martin and C.D. Romero. 2010. *Aloe vera* as a functional ingredient in foods. *Critical Reviews in Food Science and Nutrition* 50: 305-328.

Rodríguez, H., J.A. Curiel, J.M. Landete, B. de las Rivas, F.L. de Felipe, C. Gómez-Cordovés, C., J.M. Mancheño, R. Muñoz. 2009. Food phenolics and lactic acid bacteria. *International Journal of Food Microbiology* 132(2): 79-90.

Tangüler, H. and H. Erten. 2012. Occurrence and growth of lactic acid bacteria species during the fermentation of shalgam (salgam), a traditional Turkish fermented beverage. *LWT– Food Science and Technology* 46: 36-41.

Titus, D. 2008. *Smoothies! The Original Smoothies Book.* Juice Gallery, Chino Hills, CA, USA.

Wakil, S.M., O.T., Ayenuro, K.A. Oyinlola. 2014. Microbiological and nutritional assessment of starter-developed fermented tigernut milk. *Food and Nutrition Sciences* 5(6): 2014.

Watzl, B. 2008. *Smoothies e wellness aus der Flasche? Ernährungsumschau* 6: 352-353.

The Future of Lactic Acid Fermentation of Fruits and Vegetables

Spiros Paramithiotis[1,*] and Eleftherios H. Drosinos[1]

1. Introduction

Advances in methodological approach, equipment, as well as the social needs dictate, in most of the cases, the direction in which research is focused. The field of fermented fruits and vegetables has been no exception from the very first studies in the beginning of the 20th century that were conducted with minimal use of equipment and mostly based on observation and description, to the very latest use of high-throughput technologies that is far more equipment-dependent.

Nowadays, research on fermented fruits and vegetables seems to be tailored to consumer demands as well as to the research advances in the field of molecular biology. Consumer trends are the epicenter of intensive study since the last decade. Safety and health have been recognized among the food attributes that direct consumer behavior (Brewer and Rojas 2008, Feldmann and Hamm 2015). On the other hand, widespread use of modern molecular techniques has allowed in-depth characterization of a large number of traditional products. All these establish the fact that the fields in which intensive study is expected to take place in the following years are microecosystem characterization and nutritional value assessment of fermented fruits and vegetables.

2. Microecosystem Characterization

During the last decade a wide range of traditional products were studied. The motivating force behind such study was manifold: preservation of traditional knowledge, characterization of yet unexplored microecosystems and improvement in knowledge of the already studied ones. The

---

[1] Department of Food Science and Human Nutrition, Agricultural University of Athens, Athens, Greece.
* Corresponding author: E-mail: sdp@aua.gr

integration of this wealth of knowledge that was generated resulted in a thorough documentation of the dynamics of microbial population and in some cases, in the understanding of factors that determine them.

The new trend to occur recently and most likely to continue for the next few years, regarding the studies of microbial ecosystems in general and of lactic acid fermented fruits and vegetables in particular, calls for use of more advanced approaches to assess the microbial ecology and approaches that offer a more integrated perspective on the ecosystem under study. The latter was served mainly by metabolite profiling.

The metabolite kinetics of a wide range of traditional Romanian vegetable fermentations was studied by Wouters et al. (2013b). Sucrose was reported to be the sole carbon source in the beginning of fermentation with glucose, while fructose appeared on the second day of fermentation. Depending on the raw materials used, carbohydrates get depleted and the main metabolic products are lactic and acetic acids, ethanol, mannitol and glycerol. Metabolite profiling indicated the major contribution of both homo- and hetero-fermentative lactic acid bacteria and a more restricted to the first days of fermentation by yeast. Similar results were obtained during spontaneous leek fermentation (Wouters et al. 2013a). Glucose and fructose were reported as the main carbohydrates present with sucrose constituting only a minor proportion. The end products were lactic and acetic acids, ethanol and mannitol.

The metabolite changes during dongchimi fermentation were assessed by Jeong et al. (2013a). Glucose and fructose were reported as the only free sugars present during the process. Their concentration was rather low at the beginning of fermentation but increased rapidly after six days, despite the simultaneous production of lactic and acetic acids and mannitol. Then, after 30 days, a rapid decrease in their concentration was observed; at the same time, lactic and acetic acids as well as mannitol concentrations remained constant while glycerol and ethanol were produced due to yeast development. Regarding amino acid content, the concentration of most of them increased rapidly till the middle of fermentation but gradually decreased subsequently. These results were verified by Jeong et al. (2013b) who studied the metabolite changes during long-term storage of kimchi. Furthermore, the production of gamma-aminobutyric acid (GABA) was reported with *Lactobacillus sakei* identified as the producer. Slowing of the fermentation process was revealed by metabolite analysis of kimchi fermentation upon addition of red pepper powder. This was most probably due to effect on microbial succession and the production of metabolites, especially during the early fermentation period (Jeong et al. 2013c).

Paramithiotis et al. (2014a, b) studied two rather unusual fermentations, namely of green tomato and of asparagus. In the first case, *Leuconostoc mesenteroides* was reported to initiate fermentation despite the low initial pH value (3.8–5.4, depending on the ripening stage) and dominated the fermentations when initial pH ranged from 3.8 to 4.8 and co-dominated with *Ln. citreum* and *Lb. casei* when initial pH value was more than 4.8. These results offered a new insight into the role of *Ln. mesenteroides* in spontaneous vegetable fermentations since it is mostly associated with the initial stages of vegetable fermentations on account of the shorter generation time and ability to tolerate a wide range of salt concentrations. However, it is characterized by sensitivity to acidic conditions. On the other hand, *Lb. plantarum*, that is generally more acid-tolerant and distinguished by a large metabolic capacity that enables growth on a wide range of carbon sources and therefore rightfully associated with the final stages of fermentation (Daeschel et al. 1987), was totally absent in this type of fermentation. In both studies, the succession at species level was accompanied by one at subspecies level as well. Such a succession was also previously reported in must fermentations (Egli et al. 1998, Sabate et al. 1998) and should be taken into consideration when designing starter cultures.

Next generation sequencing was extensively applied for assessment of the microecosystem of lactic acid fermented fruits and vegetables (Jung et al. 2011, Park et al. 2012, Jeong et al. 2013a, b, c, Kyung et al. 2015) verifying in most cases the results obtained by classical microbiological as well as other culture-independent techniques. A new insight, towards understanding of the trophic relationships taking place during kimchi fermentation, was offered by Jung et al. (2013). In that study, the mRNA sequences obtained were assigned to metabolic genes verifying from that perspective the prevalence of carbohydrate metabolism and lactic acid fermentation. Increased metabolic activity of *Ln. mesenteroides* was detected during the onset of fermentation though *Lb. sakei* and *Weissella koreensis* were verified during the later stages. Moreover, many features of kimchi fermentation, such as species interaction, phage infection and biosynthesis of compounds important for organoleptic as well as nutritional quality of the product came to the fore.

3. Assessment of Nutritional Value

The nutritional value of lactic acid fermented fruits and vegetables was traditionally evaluated through the qualitative and quantitative assessment of nutrients and anti-nutritional factors. The effect of the whole production procedure was the epicenter of intensive study over the last decades and is thoroughly discussed in Chapter 3.

Advances in the field of molecular biology have allowed the field of nutritional genomics to develop. With this rapidly-evolving discipline, the interactions between nutrient intake and gene expression are studied using high-throughput genomic tools. Nutritional genomics is divided into two branches: nutrigenomics, which studies the effect of nutrients on gene expression and, nutrigenetics, which studies the effect of individual genotypes on nutrient intake and functionality.

The effect of several bioactive compounds in fruits and vegetables used regularly as a substrate in fermentation, such as lycopene, betanin, protocatechoric acid, sinapic acid, 3,3-diindolylmethane (DIM) and isothiocyanates on the progress of various degenerative diseases, were studied to some extent.

Carotenoids in general and lycopene in particular exert a series of health benefits by reducing the risk of several diseases. Modulation of gene expression has been recently added to the proposed mechanisms for this action. Regulation of the expression of several genes in breast cancer cell lines was exhibited by Chalabi et al. (2007a, b). Moreover, King-Batton et al. (2008) reported that treatment with lycopene at dietary relevant concentrations partially demethylated the promoter of GSTP1 tumor-suppressor gene, resulting in the restoration of its expression in MDA-MB-468 breast cancer cell lines. Regulation of gene expression was found to exert protective action against prostate cancer (Lee and Foo 2013).

Betanin, protocatechoric acid and sinapic acid are polyphenols present in beetroot, olives and mustard, respectively. Their effect on the methylation of histone H3 and the promoter regions of RASSF1A, GSTP1 and HIN-1 genes, the activity of DNA methyltransferases (DNMTs) and the expression of DNMT1 encoding DNA methyltransferase 1 and GSTP1 tumor-suppressor gene of human epithelial breast cancer MCF7 cell line were studied by Paluszczak et al. (2010). They reported that all phytochemicals inhibited DNMTs with betanin being the weakest, but none of them affected significantly the levels of DNMT1 and GSTP1 transcripts, the levels of DNMT1 and the methylation of H3 and the promoter regions of RASSF1A, GSTP1 and HIN-1 genes. *In vitro* inhibition of DNMTs was also reported for apigenin and luteolin, the flavanols that are found in parsley and celery (Fang et al. 2007). Another polyphenol, namely 6-methoxy-2E,9E-humuladien-8-one, present in the roots of *Zingiber zerumbet* (L.) was reported to exhibit growth-inhibitory activity on six human tumor cell lines and potent inhibitory activity against the MDA-MB-231 breast tumor cell line (Chung et al. 2008).

The anti-cancer activity of DIM, a compound present in *Brassica* vegetables has been extensively studied. Treatment with DIM was found to:

- inhibit AKT phosphorylation resulting in decreased activation of FLICE-like-inhibitory-protein and promoted Fas-mediated apoptosis of cholangiocarcinoma cells (Chen et al. 2006).

- activate caspases and concomitantly decrease growth and induce apoptosis of HCT116 and HT-29 human colon cancer cells, as well as downregulate survivin and enhance butyrate effects in APC$^{min/+}$ mice (Kim et al. 2007, Bhatnagar et al. 2009).

- inhibit pancreatic cancer cell invasion through upregulation of miR-146a and downregulate EGFR, MTA-2, IRAK-1 and NF-κB (Li et al. 2009, 2010).

- induce apoptosis in MDA-MB-231 breast cancer cells through survivin, Bcl-2 and cdc25A downregulation, p21[WAFI] upregulation and modulate aryl hydrocarbon receptor preventing thus epigenetic activation of COX-2 (Rahman et al. 2006, Gong et al. 2006, Degner et al. 2009).

- inhibit cell proliferation and induction of apoptosis of prostate cancer cells partly through the downregulation of AR, Akt and NF-κB signaling (Bhuiyan et al. 2006).

Isothiocyanates is a group of compounds that are found in *Brassica* vegetables and occur after hydrolysis of glucosinolates by murosinase. Sulforaphane, benzyl isothiocyanate (BITC), phenethyl isothiocyanate (PEITC), 3-phenylpropyl isothiocyanate (PPITC), 4-phenylbutyl isothiocyanate (PBITC), phenylhexyl isothiocyanate (PHI) and allyl isothiocyanate are among the most studied isothiocyanates. Apart from their antimicrobial action that has been well documented (Brabban and Edwards 1995, Delaquis and Mazza 1995, Brandi et al. 2006), their effect on DNA methylation and histone modification has also been extensively studied.

The inhibitory effect of BITC, PEITC, PPITC and PBITC on N-nitrosomethylbenzylamine-induced esophageal tumorigenesis in male Fischer rats was assessed by Wilkinson et al. (1995). PEITC and PPITC exhibited inhibition of tumor multiplicities at all dietary concentrations tested, whereas BITC and PBITC showed only marginal inhibitory effect. PEITC was reported to induce growth arrest and apoptosis in prostate cancer cells by acting on both DNA and chromatin (Chiao et al. 2004, Wang et al. 2007, 2008). Treatment with allyl isothiocyanate resulted in increased acetylation of histones occurring though stimulation of histone acetylation rather than inhibiting histone deacetylation (Lea et al. 2001). Exposure of prostate cancer cells to PHI resulted in diminished activity of histone deacetylases 1 and 2 and enhanced histone acetylation, resulting in increased accessibility of p21 promoter leading to cell cycle arrest and apoptosis (Beklemisheva et al. 2006).

Sulforaphane promotes growth arrest and apoptosis in human colon, prostate and breast cancer cell lines. The proposed mechanisms include inhibition of histone acetylase and decrease of estrogen receptor-A, epidermal growth factor receptor, as well as human epidermal growth factor receptor-2 synthesis (Parnaud et al. 2004, Myzak et al. 2004, Pledgie-Tracy et al. 2007).

Given that this field is rapidly developing and fermented fruits and vegetables are very important from a nutritional point of view, it is very likely that the nutritional genomics perspective will provide new insights into the nutritional value of these products.

4. Conclusions

Lactic acid fermented fruits and vegetables have been constantly studied for many decades. New insights are continuously offered, matching the scientific advances of each time-period. This exciting process results in an endless improvement and integration of existing knowledge regarding these products. The era that is currently underway is characterized by specific advances in the field of molecular biology that have already expanded our understanding of the trophic relationships taking place in the microcommunity and holds a promise of many more to be explored.

References

Beklemisheva, A.A., Y. Fang, J. Feng, X. Ma, W. Dai and J.W. Chiao. 2006. Epigenetic mechanism of growth inhibition induced by phenylhexyl isothiocyanate in prostate cancer cells. *Anti-cancer Research* 26: 1225-1230.

Bhatnagar, N., X. Li, Y. Chen, X. Zhou, S.H. Garrett and B. Guo 2009. 3,3'-diindolylmethane enhances the efficacy of butyrate in colon cancer prevention through down-regulation of surviving. *Cancer Prevention Research* 2: 581-589.

Bhuiyan, M.M.R., Y. Li, S. Banerjee, F. Ahmed, Z. Wang, S. Ali and F.H. Sarkar 2006. Down-regulation of androgen receptor by 3,3'-diindolylmethane contributes to inhibition of cell proliferation and induction of apoptosis in both hormone-sensitive LNCaP and insensitive C4-2B prostate cancer cells. *Cancer Research* 66: 10064-10072.

Brabban, A.D. and C. Edwards. 1995. The effects of glucosinolates and their hydrolysis products on microbial growth. *Journal of Applied Bacteriology* 79: 171-177.

Brandi G., G. Amagliani, G.F. Schiavano, M. De Santi and M. Sisti 2006. Activity of *Brassica oleracea* leaf juice on food-borne pathogenic bacteria. *Journal of Food Protection* 69: 2274-2279.

Brewer, M.S. and M. Rojas 2008. Consumer attitudes toward issues in food safety. *Journal of Food Safety* 28: 1-22.

Chalabi N., L. Delort, S. Satih, P. Dechelotte, Y.J. Bignon and D.J. Bernard-Gallon. 2007b. Immunohistochemical expression of RARa, RARb, and Cx43 in breast tumor cell lines after treatment with lycopene and correlation with RT-qPCR. *Journal of Histochemistry and Cytochemistry* 55: 877-883.

Chalabi, N., S. Satih, L. Delort, Y.J. Bignon and D.J. Bernard-Gallon 2007a. Expression profiling by whole-genome microarray hybridization reveals differential gene expression in breast cancer cell lines after lycopene exposure. *Biochimica et Biophysica Acta* 1769: 124-130.

Chen, Y., J. Xu, N. Jhala, P. Pawar, Z.B. Zhu, L. Ma, C.H. Byon and J.M. McDonald. 2006. Fas-mediated apoptosis in cholangiocarcinoma cells is enhanced by 3,3'-diindolylmethane through inhibition of AKT signaling and FLICE-like inhibitory protein. *American Journal of Pathology* 169: 1833-1842.

Chiao, J.W., H. Wu, G. Ramaswamy, C.C. Conaway, F.L. Chung, L. Wang and D. Liu. 2004. Ingestion of an isothiocyanate metabolite from cruciferous vegetables inhibits growth of human prostate cancer cell xenografts by apoptosis and cell cycle arrest. *Carcinogenesis* 25:1403-1408.

Chung, I.M., M.Y. Kim, W.H. Park and H.I. Moon. 2008. Histone deacetylase inhibitors from the rhizomes of *Zingiber zerumbet*. *Die Pharmazie* 63: 774-776.

Daeschel, M.A., R.E. Andersson and H.P. Fleming. 1987. Microbial ecology of fermenting plant materials. *FEMS Microbiology Reviews* 46: 357-367.

Degner, S.C., A.J. Papoutsis, O. Selmin and D.F. Romagnolo 2009. Targeting of aryl hydrocarbon receptor-mediated activation of cyclooxygenase-2 expression by the indole-3-carbinol metabolite 3,30-diindolylmethane in breast cancer cells. *Journal of Nutrition* 139: 26-32.

Delaquis, P.Q. and G. Mazza. 1995. Antimicrobial properties of isothiocyanates in food preservation. *Food Technology* 49: 73-78.

Egli, C.M., W.D. Edinger, C.M. Mitrakul and T. Henick-Kling. 1998. Dynamics of indigenous and inoculated yeast populations and their effects on the sensory character of Riesling and Chardonnay wines. *Journal of Applied Microbiology* 85: 779-789.

Fang, M., D. Chen and C.S. Yang. 2007. Dietary polyphenols may affect DNA methylation. *Journal of Nutrition* 137: 223S-228S.

Feldmann, C. and U. Hamm. 2015. Consumers' perceptions and preferences for local food: A review. *Food Quality and Preference* 40: 152-164.

Gong, Y., H. Sohn, L. Xue, G.L. Firestone and L.F. Bjeldanes. 2006. 3,3'-diindolylmethane is a novel mitochondrial H+-ATP synthase inhibitor that can induce p21Cip1/Waf1 expression by induction of oxidative stress in human breast cancer cells. *Cancer Research* 66: 4880-4887.

Jeong, S.H., H.J. Lee, J.Y. Jung, S.H. Lee, H.Y. Seo, W.S. Park and C.O. Jeon 2013c. Effects of red pepper powder on microbial communities and metabolites during kimchi fermentation. *International Journal of Food Microbiology* 160: 252-259.

Jeong, S.H., J.Y. Jung, S.H. Lee, H.M. Jin and C.O. Jeon. 2013a. Microbial succession and metabolite changes during fermentation of dongchimi, traditional Korean watery kimchi. *International Journal of Food Microbiology* 164: 46-53.

Jeong, S.H., S.H. Lee, J.Y. Jung, E.J. Choi and C.O. Jeon. 2013b. Microbial succession and metabolite changes during long-term storage of kimchi. *Journal of Food Science* 78: M763-M769.

Jung, J.Y., S.H. Lee, J.M. Kim, M.S. Park, J.W. Bae, Y. Hahn, E.L. Madsen and C.O. Jeon. 2011. Metagenomic analysis of kimchi, a traditional Korean fermented food. *Applied and Environmental Microbiology* 77: 2264-2274.

Jung, J.Y., S.H. Lee, H.M. Jin, Y. Hahn, E.L. Madsen and C.O. Jeon. 2013. Metatranscriptomic analysis of lactic acid bacterial gene expression during kimchi fermentation. *International Journal of Food Microbiology* 163: 171-179.

Kim, E.J., S.Y. Park, H.K. Shin, D.Y. Kwon, Y.J. Surh and J.H.Y. Park. 2007. Activation of caspase-8 contributes to 3,3'-diindolylmethane-induced apoptosis in colon cancer cells. *Journal of Nutrition* 137: 31-36.

King-Batoon, A., J.M. Leszczynska and C.B. Klein. 2008. Modulation of gene methylation by genistein or lycopene in breast cancer cells. *Environmental and Molecular Mutagenesis* 49:36-45.

Kyung, K.H., E.M. Pradas, S.G. Kim, Y.J. Lee, K.H. Kim, J.J. Choi, J.H. Cho, C.H. Chung, R. Barrangou and F. Breidt. 2015. Microbial ecology of watery kimchi. *Journal of Food Science* 80: M1031-M1038.

Lea, M.A., V.M. Randolph, J.E. Lee and C. Desbordes. 2001. Induction of histone acetylation in mouse erythroleukemia cells by some organosulfur compounds including allyl isothiocyanate. *International Journal of Cancer* 92: 784-789.

Lee, L.K. and K.Y. Foo. 2013. An appraisal of the therapeutic value of lycopene for the chemoprevention of prostate cancer: A nutrigenomic approach. *Food Research International* 54: 1217-1228.

Li, Y., T.G. Vanden Boom II, D. Kong, Z. Wang, S. Ali, P.A. Philip and F.H. Sarkar. 2009. Up-regulation of miR-200 and let-7 by natural agents leads to the reversal of epithelial-to-mesenchymal transition in gemcitabine-resistant pancreatic cancer cells. *Cancer Research* 69: 6704-6712.

Li, Y., T.G. Vanden Boom II, Z. Wang, D. Kong, S. Ali, P.A. Philip and F.H. Sarkar. 2010. miR-146a suppresses invasion of pancreatic cancer cells. *Cancer Research* 70: 1486-1495.

Myzak, M.C., P.A. Karplus, F.L. Chung and R.H. Dashwood. 2004. A novel mechanism of chemoprotection by sulforaphane: inhibition of histone deacetylase. *Cancer Research* 64: 5767-5774.

Paluszczak, J., V. Krajka-Kuzniak and W. Baer-Dubowska. 2010. The effect of dietary polyphenols on the epigenetic regulation of gene expression in MCF7 breast cancer cells. *Toxicology Letters* 192: 119-125.

Paramithiotis S., K. Kouretas and E.H. Drosinos. 2014b. Effect of ripening stage on the development of the microbial community during spontaneous fermentation of green tomatoes. *Journal of the Science of Food and Agriculture* 94: 1600-1606.

Paramithiotis, S., A.I. Doulgeraki, A. Karahasani and E.H. Drosinos. 2014a. Microbial population dynamics during spontaneous fermentation of *Asparagus officinalis* L. young sprouts. *European Food Research and Technology* 239: 297-304.

Park E.J., J.S. Chun, C.J. Cha, W.S. Park, C.O. Jeon and J.W. Bae. 2012. Bacterial community analysis during fermentation of ten representative kinds of kimchi with barcoded pyrosequencing. *Food Microbiology* 30: 197-204.

Parnaud, G., P. Li, G. Cassar, P. Rouimi, J. Tulliez, L. Combaret and L. Gamet-Payrastre 2004. Mechanism of sulforaphane-induced cell cycle arrest and apoptosis in human colon cancer cells. *Nutrition and Cancer* 48: 198-206.

Pledgie-Tracy, A., M.D. Sobolewski and N.E. Davidson. 2007. Sulforaphane induces cell type-specific apoptosis in human breast cancer cell lines. *Molecular Cancer Therapeutics* 6: 1013-1021.

Rahman, K.W., Y. Li, Z. Wang, S.H. Sarkar and F.H. Sarkar. 2006. Gene expression profiling revealed survivin as a target of 3,3'-diindolylmethane-induced cell growth inhibition and apoptosis in breast cancer cells. *Cancer Research* 66: 4952-4960.

Sabate, J., J. Cano, A. Querol and J.M. Guillamon. 1998. Diversity of *Saccharomyces cerevisiae* strains in wine fermentations: Analysis for two consecutive years. *Letters in Applied Microbiology* 26: 452-455.

Wang, L.G., A. Beklemisheva, X.M. Liu, A.C. Ferrari, J. Feng and J.W. Chiao. 2007. Dual action on promoter demethylation and chromatin by an isothiocyanate restored GSTP1 silenced in prostate cancer. *Molecular Carcinogenesis* 46: 24-31.

Wang, L.G., X.M. Liu, Y. Fang, W. Dai, F.B. Chiao, G.M. Puccio, J. Feng, D. Liu and J.W. Chiao 2008. De-repression of the p21 promoter in prostate cancer cells by an isothiocyanate via inhibition of HDACs and c-Myc. *International Journal of Oncology* 33: 375-380.

Wilkinson, J.T., M.A. Morse, L.A. Kresty and G.D. Stoner. 1995. Effect of alkyl chain length on inhibition of N-nitrosomethylbenzylamine-induced esophageal tumorigenesis and DNA methylation by isothiocyanates. *Carcinogenesis* 16: 1011-1015.

Wouters, D., N. Bernaert, W. Conjaerts, B. Van Droogenbroeck, M. De Loose and L. De Vuyst. 2013a. Species diversity, community dynamics and metabolite kinetics of spontaneous leek fermentations. *Food Microbiology* 33: 185-196.

Wouters, D., S. Grosu-Tudor, M. Zamfir and L. De Vuyst. 2013b. Bacterial community dynamics, lactic acid bacteria species diversity and metabolite kinetics of traditional Romanian vegetable fermentations. *Journal of the Science of Food and Agriculture* 93: 749-760.

Index

Printed and bound by CPI Group (UK) Ltd, Croydon, CR0 4YY

01/11/2024

01782622-0005